T0299117

Dynamic Multilevel Methods and the Numerical Simulation of Turbulence

This book describes the implementation of multilevel methods in a dynamical context, with application to the numerical simulation of turbulent flows. The general ideas for the algorithms presented stem from dynamical systems theory and are based on the decomposition of the unknown function into two or more arrays corresponding to different scales in the Fourier space.

Before describing in detail the numerical algorithm, survey chapters, on the mathematical theory of the Navier–Stokes equations and on the physics of the conventional theory of turbulence, are included. The multilevel methods are applied here to the simulation of homogeneous isotropic turbulent flows as well as turbulent channel flows. The implementation issues are discussed in detail, and numerical simulations of the flows cited above are presented and analyzed. The methods have been applied in the context of the direct numerical simulation and are therefore compared to such simulations.

This timely monograph should appeal to graduate students and researchers alike, providing a background for applied mathematicians as well as engineers.

Dynamic Multilevel Methods and the Numerical Simulation of Turbulence

THIERRY DUBOIS
Centre National de la Recherche Scientifique and Université Blaise Pascal

FRANÇOIS JAUBERTEAU
Université Blaise Pascal

ROGER TEMAM
Université Paris-Sud and Indiana University, Bloomington

CAMBRIDGE
UNIVERSITY PRESS

CAMBRIDGE UNIVERSITY PRESS
Cambridge, New York, Melbourne, Madrid, Cape Town,
Singapore, São Paulo, Delhi, Tokyo, Mexico City

Cambridge University Press
The Edinburgh Building, Cambridge CB2 8RU, UK

Published in the United States of America by
Cambridge University Press, New York

www.cambridge.org
Information on this title: www.cambridge.org/9780521621656

© Cambridge University Press 1999

This publication is in copyright. Subject to statutory exception
and to the provisions of relevant collective licensing agreements,
no reproduction of any part may take place without the written
permission of Cambridge University Press.

First published 1999

A catalogue record for this publication is available from the British Library

Library of Congress Cataloguing in Publication data
Dubois, Thierry.
Dynamic multilevel methods and the numerical simulation of
turbulence / Thierry Dubois, Francois Jauberteau, Roger Temam.
p. cm.
Includes bibliographical references.
ISBN 0-521-62165-8 (hb)
1. Turbulence. 2. Navier–Stokes equations–Numerical solutions.
3. Differentiable dynamical systems. I. Jauberteau, François,
1959– . II. Temam, Roger. III. Title.
QA913.D88 1999
532′.0527–dc21 98-36472
CIP

ISBN 9780521621656 Hardback

Cambridge University Press has no responsibility for the persistence or
accuracy of URLs for external or third–party internet websites referred to in
this publication, and does not guarantee that any content on such websites is,
or will remain, accurate or appropriate. Information regarding prices, travel
timetables, and other factual information given in this work is correct at
the time of first printing but Cambridge University Press does not guarantee
the accuracy of such information thereafter.

Contents

Preface

The purpose of these notes is to describe the implementation of multilevel methods for the numerical simulation of turbulent flows. Multilevel methods have proved to be a successful tool for the treatment on parallel computers of large problems involving numerous scales, problems that now become accessible: wavelets, multigrid methods, hierarchical bases in finite elements, and the numerical treatment of some heterogeneous media are examples of multiresolution treatments of such problems.

However, in all the examples above the treatment has been most often a "static" one devoted to stationary problems. The utilization of multilevel methods for time-dependent problems in the context of a complex dynamics is relatively unexplored, and this book is a small contribution to this vast and complex subject.

The general ideas for the algorithms presented here stem from the dynamical systems theory and are based on the decomposition of the unknown function into two or more arrays corresponding to different scales in the Fourier space. These subsets of unknowns are treated in differentiated ways adapted to the different scales. Although the concepts of exact and approximate inertial manifolds and the nonlinear Galerkin method underline the present study, we actually make little use of these concepts, but retain and further develop the idea of decomposing the unknown function into different arrays with different magnitudes and in treating the multilevel components differently and in an adaptive and dynamical way.

The authors realize all too well how much remains to be done for the numerical simulation of turbulent flows, from the laboratory prototype to the industrial models and to geophysical flows. Nevertheless we believe that multilevel methods are a necessary and useful tool for the treatment on parallel computers of the large problems involving numerous scales. From the mathematical and numerical analysis point of view, we believe also that new chapters of numerical analysis will have to be written in relation with the multilevel treatment of large evolutionary problems.

Although these notes do not pretend, in any way, to give a definitive answer to the hard problem of turbulence, the authors hope that they can help bring a different perspective in numerical simulations. To make the book accessible to a broader audience, we have included some survey chapters or sections, in particular on the mathematical theory of the Navier–Stokes equations and on the physics of the conventional theory of turbulence.

Bloomington, Clermont-Ferrand,
Paris, Stanford,
January 1998

Acknowledgments

This work was partially supported by the Department of Mathematical Sciences of the National Sciences Foundation, Grants NSF-DMS 9400615 and NSF-DMS 9705229 and by the Research Fund of Indiana University.

The research presented in this book was also partially supported by the Université de Paris-Sud and the Université Blaise Pascal (Clermont-Ferrand) and the Centre National de la Recherche Scientifique (CNRS) through the Laboratoire d'Analyse Numérique d'Orsay, and the Laboratoire de Mathématiques Appliquées (Clermont-Ferrand).

Part of the work was done while one of the authors (T. D.) was visiting the Institute for Computer Applications in Science and Engineering (ICASE, NASA Langley Research Center), and this book was completed, in the academic year 1997–98, while the same author benefited from the hospitality of the Center For Turbulence Research (CTR) at Stanford University and NASA Ames Research Center, as a one-year postdoctoral fellow on leave from the CNRS. Also, at the time where this book was completed, during the academic year 1997–98, the second author (F. J.) benefited from a visiting research position at the CNRS, on leave from the Université Blaise Pascal.

At various times, parts of the computations presented in this book were performed on the Cray Y-MP C90 (16 processors) of the NSF Pittsburgh Super-Computing Center (Pennsylvania, USA), on the Cray Y-MP C90 (8 processors) and the Fujitsu VPP300 of the Institut du Développement des Ressources en Informatique Scientifique (IDRIS, CNRS, France). The Cray YMP-EL (4 processors) of the Laboratoire de Mathématiques Appliquées (Université Blaise Pascal, Clermont-Ferrand, France) was also intensively used for long time integration of simulations at low resolutions (48^3, 64^3, and 96^3).

The authors would like to specially thank E. Gondet (IDRIS, CNRS, France) for his helpful comments on the vector and parallel optimization of the codes on the Cray computers.

The authors would like also to thank all those with whom they discussed or interacted about this work, and those who showed interest in it; in particular many in the "Inertial Manifolds" community, and, in the mathematical or fluid mechanics and engineering communities, Jerry Bona, Jean-Jacques Chattot, Haecheon Choi, Jean-Michel Ghidaglia, David Gottlieb, Mohamed Hafez, Thomas Hughes, Youssuf Hussaini, Maurice Ménéguzzi, Parviz Moin, Tinsley Oden, Jie Shen, and last but not least Alan Harvey from Cambridge University Press.

Introduction

In various engineering and environmental problems, there is a serious need to calculate turbulent flows. However, the Reynolds numbers for these applications are usually very high and the geometry is complicated. Generally, for problems in industry or meteorology, the Reynolds numbers are greater than several millions. High Reynolds numbers imply that a wide range of scales takes place between the small and large scales of the flow. Furthermore the characteristic times of the small scales are small by comparison with that of the large ones (see Chapter 2). Hence, in order to compute accurately all the scales of a turbulent flow, we must take a grid with very small mesh size, and a very small time step. The direct numerical simulation (DNS) of the turbulence is not possible at present time on problems of industrial interest, even on the most powerful computers currently available. The number of degrees of freedom needed for DNS can be estimated as $N^n \simeq \mathrm{Re}_L^{n^2/4}$, with $n = 2$ or 3 the space dimension, Re_L being the integral scale Reynolds number. As for the time step, we have $\Delta t \simeq \mathrm{Re}_L^{-n/4}$ (see Chapter 4).

Direct numerical simulations of homogeneous turbulent flows have been performed extensively to increase the understanding of the mechanisms involving the small-scale structures (intermittency). Results of numerical simulations provide many different ways for investigating turbulent flows. Indeed, important quantities, such as high order correlations, cannot be easily evaluated in the laboratory. By direct numerical simulation, various details of the small scale behavior can be obtained. Moreover, by comparing results of direct and modeled simulations, the validity and the limits of the closure models can be estimated. Hence, the modeling can be improved and new models can be developed.

In homogeneous turbulence, no boundary layers are present, nor complex geometry. Therefore reasonably high Reynolds numbers can be reached and the turbulence can be fully developed. Orszag and Patterson (1972) have conducted DNS of homogeneous turbulence at Reynolds number Re_λ, based on the Taylor microscale, smaller than 30 and with a resolution of 32^3. Siggia and Patterson (1978) have performed DNS at Reynolds number $\mathrm{Re}_\lambda \simeq 40$ with a resolution of 32^3. At this Reynolds number there is no separation between the energy-containing eddies and the small (dissipative) scales. During the subsequent decade, the increase of computer capacities allowed simulations at higher Reynolds numbers for which a small inertial subrange of the energy spectrum exists. Siggia (1981) studied the small-scale intermittency in three-dimensional turbulence. He collected data related to intermittency such as the flatness and, among many other quantities, those related to the first and second velocity derivatives. Kerr (1985) has presented various simulations corresponding to Reynolds numbers Re_λ from 18 to 83 with a number of modes varying from 32^3 to 128^3. For the 128^3 simulation, the inertial range extended over five modes. The author studied in detail the velocity derivative statistics and the statistics of a passive scalar. Correlations of fourth and higher orders were presented. She, Jackson, and Orszag (1988) studied the dependence of the skewness and flatness factors of the velocity derivatives upon the different scales of motion. The skewness factor is found to be a large-scale property, while the flatness factor depends mainly on scales lying in the dissipation range.

Vincent and Ménéguzzi (1991) obtained a simulation at Re_λ of the order of 150 which corresponds to 240^3 unknowns. This result was obtained on a Cray-2 using the four available processors. In this case, the inertial range extended over one decade. The authors confirm that the probability distribution of the velocity derivative is strongly non-Gaussian and that it is close to an exponential distribution. Finer resolutions, namely 512^3 modes, have been more recently reported by Chen et al. (1993) and Jiménez et al. (1993). These simulations were obtained on massively parallel computers. Even more recently, a 1024^3 simulation was done on a cluster of workstations (Woodward et al. (1995)). In Vincent and Ménéguzzi (1991), the authors reported several statistical quantities and studied the spatial structure of the flow. A similar but more detailed analysis was presented in Jiménez et al. (1993).

The choice of the number of unknowns that have to be retained in order to accurately describe the turbulence statistics, such as the energy spectrum function and the high-order moments of the velocity and its derivatives, is of great importance in DNS. In Kerr (1985), resolution refinement at fixed Reynolds number was used in order to estimate the suitable mesh size. In Vincent and Ménéguzzi (1991), the authors analyzed the transfer terms and the energy spectrum; in Jiménez et al. (1993) the effect of the resolution on the high order statistics was measured. It appears that the cutoff wavenumber must be chosen of the order of $2/\eta$, where η is the Kolmogorov length scale. This results in very strong computational restrictions on the size of the physical mesh and therefore on the time step. Due to the $k^{-5/3}$ decrease of the energy spectrum function in the inertial range, most of the energy is concentrated in a few low modes, so that most of the computational effort consists in computing very small scales. Indeed, less than 2% of the computations are required in order to compute the scales corresponding to wavenumbers $\eta k \leq \frac{1}{2}$. In DNS, all the scales (from the energy-containing to the dissipative ones) are computed with the same numerical scheme. This does not take into consideration the fact that the small and large scales have different physical characteristics. For instance, the small scales are known to reach a statistically steady state much faster than the large ones (see Batchelor (1971), Orszag (1973)).

Although much (indeed, most) computational effort is devoted to the small scales, the interesting structures are often the large scales of the flow, since they contain most of the kinetic energy and they control physical properties like turbulent diffusion of momentum or heat. Yet, for high Reynolds numbers, the energy containing eddies and the Kolmogorov scales are well separated. Furthermore, the small scales are more homogeneous and isotropic. In fact, various experiments and simulations have shown the universal character of the small scales. So one may want to model the small structures so as to properly describe their action on the large structures without fully computing them. Different types of modeling have been developed. They are based on some decomposition of the flow field

$$\mathbf{u} = \overline{\mathbf{u}} + \mathbf{u}' \tag{0.1}$$

where $\overline{\mathbf{u}}$ is the averaged velocity and $\mathbf{u}'$ is the fluctuating counterpart. The purpose is to estimate $\overline{\mathbf{u}}$ without fully computing $\mathbf{u}'$. This problem being not a closed one, we must use a model of closure for the terms depending on $\mathbf{u}'$. The models that are usually proposed depend on the definition of the averaging and on the choice of the closure hypothesis. Many models make an eddy-viscosity assumption (Boussinesq's hypothesis). If the average satisfies the Reynolds conditions, the term to model is the Reynolds stress tensor. Several models such as the zero-equation model (mixing length), the one and two-equation models ($K-\ell$ model, $K-\epsilon$ model) have been proposed to model the Reynolds stress tensor (see Chapter 4). Other models, like the two-point closure models, are based on modeling the two-point correlation tensor or its spectral representation (energy spectrum tensor), in order to provide more details than the Reynolds stress models. Totally different approaches to this modeling problem include the renormalization-group (RNG) methods and the PDF models (see Chapter 4).

In large eddy simulations (LESs), the approach is slightly different. Instead of computing the mean flow and related quantities as in Reynolds-stress closures, LES models compute the large scales of the flow while modeling the effect of the subgrid scales on the resolved (large) ones. Therefore, a low-dimensional dynamical system is resolved, while in the case of Reynolds-stress closures, a steady equation is most often solved. The LES approach then aims to reproduce the dynamic behavior of the flow, at least of some quantities; such information is necessary in some problems, such as acoustic ones. The decomposition $\mathbf{u} = \overline{\mathbf{u}} + \mathbf{u}'$ is chosen to separate the small and large scales contained in the velocity field. This separation is achieved by using a filter function (see Chapters 4 and 6). The fluctuating part $\mathbf{u}'$ is called the subgrid-scale (SGS) velocity, and the term to model in the filtered equations is the subgrid stress tensor. The first approach for this closure, introduced by Smagorinsky, involves an eddy-viscosity assumption. However, other models, such as the scale-similarity model and the linear combination model (LCM), have been proposed (see Chapter 4) to overcome some deficiencies of the Smagorinsky model. Generally, LES models consist in computing the large scales of motion, containing most of the energy (80% of the total energy). So LES models provide a more accurate description of the flow, since more scales are retained. However, if $\mathbf{k}_c$ is the cutoff wavenumber in the energy spectrum, a backscatter transfer, due to the nonlinear interaction with modes near $\mathbf{k}_c$, appears. We speak of an inverse error cascade, corresponding to a decorrelation between two different realizations of the flow that differ initially only on the small scales. The errors in the modeling of the small scales will gradually contaminate the larger scales through this error cascade.

The Kolmogorov theory of turbulence is based on phenomenological considerations and uses little information concerning the Navier–Stokes equations (see Chapter 2). On the contrary, the mathematical theory of the Navier–Stokes equations is aimed at studying mathematical properties of the solutions (see Chapter 1). In space dimension 2, the mathematical theory of the Navier–Stokes equations is quite complete in the sense that there is a satisfactory and coherent set of results on existence of solutions, uniqueness, regularity, continuous dependence on the data, and so on. No such thing is available for the actual three-dimensional (3D) flows, and there are still several gaps in the mathematical theory of the three-dimensional Navier–Stokes equations. The main issue for the mathematical theory in space dimension 3 is whether the enstrophy or, equivalently, the maximum of the magnitude of the velocity vector can become infinite at certain places and at certain times for a flow.

For the mathematical study of turbulence, the conventional approach is based on a statistical study of the flow (see Chapter 2), using ergodicity assumption.

Another approach for turbulence is the dynamical system approach, with the concept of attractor (see Chapter 5). The global attractor is a subset in the phase space that attracts all orbits as $t \to \infty$. It is the mathematical object describing the permanent regime, or large-time behavior. The study of the attractor and its approximation yields some information on the properties of the flow. The attractor has finite dimension. Hence, each orbit of the phase space converges to a finite-dimensional set, and the corresponding permanent regime can be described by a finite number of parameters, recovering in this way the fact predicted by Kolmogorov's theory of turbulence that turbulent flows depend on a finite number of degrees of freedom. The dimension of the attractor coincides with the estimates of the number of degrees of freedom of a turbulent flow. To approximate the attractor, several mathematical objects, such as inertial manifolds (IMs) and approximate inertial manifolds (AIMs), have been developed (see Chapter 5). These concepts are based on the following decomposition of the velocity field $\mathbf{u}$ into a large-scale and a small-scale component, or into a low-frequency and a high-frequency component, of the type

$$\mathbf{u} = \mathbf{y} + \mathbf{z}. \tag{0.2}$$

IM and AIM give an exact or approximate slaving law of $\mathbf{z}$ as a function of $\mathbf{y}$, namely

$$\mathbf{z}(t) = \Phi(\mathbf{y}(t)). \tag{0.3}$$

The utilization of such decompositions of the vector field has led to new multilevel schemes in numerical analysis (see for instance Foias, Manley, and Temam (1987, 1988), Jolly and Xiong (1995), Jones, Margolin, and Titi (1995)). Furthermore, for the practical utilization of these multilevel methods for the simulation of turbulence, the need occurred to implement the decomposition (0.2) of $\mathbf{u}$ in a dynamical adaptive way. The dynamical multilevel (DML) methodology (see Chapters 8, 9, and 10) stems from theoretical properties of the decomposition $\mathbf{u} = \mathbf{y} + \mathbf{z}$ of the solution of the Navier–Stokes equations, which is at the origin of the construction of the AIMs. It is also related to multigrid methods (V-cycles). Encompassing these mathematical and numerical notions, the DML methodology takes into account the fact that, in turbulent flows, the small scales reach a statistically steady equilibrium faster than the large ones.

In LES models, the velocity fluctuation $\mathbf{u}'$ is not resolved and the subgrid stress tensor, representing the interaction between the resolved and the subgrid scales, is modeled. In the DML methodology, the small scales $\mathbf{z}$ are computed with less accuracy and are updated less often than in a DNS simulation. Their approximate values are used to correct the dynamic behavior of the low-dimensional dynamical system corresponding to $\mathbf{y}$ (large scales). The DML methods have been implemented and used only in the context of DNS, that is, all physically relevant scales are computed. In such a case, the cutoff levels used in the DML methodology to separate the small and large scales varies in time so as to avoid energy accumulation near the cutoff levels, excessive dissipation, or backscatter transfer. Fundamental in the DML methods is the strategy for changing the cutoff level while time evolves. Indeed, it is important to control the errors in order to not disturb the large scales for which the statistically steady equilibrium is long to reach. The DML methods can be viewed as intermediate methods between LES and DNS.

Two different type of turbulent flows are considered in this work, namely two- and three-dimensional homogeneous turbulent flows (fully periodic flows) and three-dimensional nonhomogeneous turbulent flows, more specifically flows in an infinite channel. In the homogeneous case, the flow is forced in the large scales in order to sustain the turbulence and to avoid a decay of the kinetic energy. By integrating the discretized Navier–Stokes equations over a long time interval (30 eddy-turnover times τ_e), a statistically steady state can be reached. Therefore, the statistics are collected, and the steady states obtained with different algorithms can be compared. DML simulations with different parameters are compared with DNS ones at different resolutions, namely, $\eta k_N \in [0.8, 1.6]$. The effects of the DML method on the small scales are measured, and the efficiency of the DML methodology in this context of direct simulation is evaluated. The comparisons of the different simulations concerned global quantities such as the kinetic energy; the energy dissipation rate; and statistical properties of the flow such as the energy spectrum functions, the skewness and flatness factors, higher-order moments, and the probability distribution functions (see Chapter 10).

For the channel flow problem, the flow is sustained in the streamwise direction (x_1) by applying a constant pressure gradient in that direction. A simulation with the same configuration (i.e. the same channel lengths in the periodic directions and the same Reynolds number) as in Kim, Moin, and Moser (1987) is reported. The velocity–vorticity formulation of the Navier–Stokes equation, leading to a fourth-order equation for the velocity in the direction normal to the walls, is used. However, this formulation is discretized here in a different way than in Kim, Moin, and Moser (1987). Indeed, instead of a tau Chebyshev approximation in the direction normal to the walls, a Galerkin basis formed with Legendre polynomials has been implemented. Such basis is well suited for a scale decomposition of the velocity field in the normal direction. The study of multilevel schemes in the normal direction is under progress and will be reported in further works. Here, a DML algorithm in the periodic directions is proposed, and it has been tested. Again, the statistically steady states reached by the DNS and DML simulations are compared by analyzing different turbulent statistics such as the mean flow properties, the root mean square of the velocity, vorticity, and pressure fluctuations, the one-dimensional spectrum functions, the Reynolds shear stresses, and the high-order moments of the velocity fluctuations. The memory size required for DNS and DML spectral codes, as well as the vectorization and parallelization (multitasking) performance, obtained on a Cray YMP C90, are also compared. Furthermore, several quantities are computed with the results of the DNS simulations for comparisons and to validate the hypothesis of the DML methodology (see Chapters 8 and 9).

This book is organized as follows. In Chapter 1, the main mathematical results on the Navier–Stokes equations are recalled and stated without proofs. Chapter 2 concerns the theory of turbulent flows (statistical study) in the spirit of the conventional theory of turbulence. In Chapter 3, DNS algorithms (spectral methods) are described for homogeneous and nonhomogeneous turbulence. The practical limits of DNS for these problems are discussed in Chapter 4. Chapter 5 presents the theoretical results obtained by applying the dynamical system approach to the study of turbulence. The main results are recalled: attractor, inertial manifolds, approximate inertial manifolds. In Chapter 6, the problem of the scale separation is studied in homogeneous and nonhomogeneous directions. In Chapter 7, the numerical analysis of an algorithm with different treatments of $\mathbf{y}$ and $\mathbf{z}$ is conducted for a simple problem. In Chapters 8 and 9, the theory and algorithms of DML methodologies are presented. Furthermore several estimates are derived and used to motivate the multilevel strategy on which are based the DML algorithms presented here. Different DML algorithms are described, in the homogeneous and nonhomogeneous cases. Finally, numerical results obtained with the DML methods and the comparison with DNS results are presented in Chapter 10.

1

The Navier–Stokes Equations and Their Mathematical Background

This chapter contains a brief presentation of the Navier–Stokes equations and a brief survey of the present status of the mathematical theory as far as existence, uniqueness, and regularity of solutions are concerned. We will return to the mathematical theory of the Navier–Stokes equations in Chapter 5, where we will address questions related to the large-time behavior of their solutions in relation with the dynamical systems approach to turbulence: attractors, inertial manifolds, their dimension, and their approximation. These results are recalled for the sake of completeness, and we refer the interested reader to the mathematically oriented literature for more complete results: see for example Constantin and Foias (1989), Ladyzhenskaya (1969), Lions (1969), Temam (1984, 1995).

1.1. The Equations

Fluids obey the general laws of mechanics, namely conservation of momentum and conservation of mass. As for any other continuous medium, the conservation of momentum is expressed in terms of the Cauchy stress tensor $\boldsymbol{\sigma}$ with components σ_{ij}:

$$\rho\gamma_i = \sum_{j=1}^{3} \frac{\partial \sigma_{ij}}{\partial x_j} + f_i, \qquad i = 1, 2, 3. \tag{1.1}$$

Here ρ is the density of the fluid, $\boldsymbol{\gamma} = (\gamma_1, \gamma_2, \gamma_3)$ is the acceleration vector, $\mathbf{f} = (f_1, f_2, f_3)$ represents the volume forces applied to the fluid; $\boldsymbol{\sigma} = \boldsymbol{\sigma}(\mathbf{x}, t)$ is the Cauchy stress tensor at $\mathbf{x}$ at time t, $\mathbf{x} = (x_1, x_2, x_3)$.

In the Eulerian representation of the flow that we consider, $\rho = \rho(\mathbf{x}, t)$ is the density at $\mathbf{x}$ at time t, and $\mathbf{u} = \mathbf{u}(\mathbf{x}, t)$, $\mathbf{u} = (u_1, u_2, u_3)$ is the velocity of the particle of fluid at $\mathbf{x}$ at time t. Then the conservation of mass is expressed by the continuity equation

$$\frac{\partial \rho}{\partial t} + \operatorname{div}(\rho \mathbf{u}) = 0. \tag{1.2}$$

The acceleration vector $\gamma = \gamma(\mathbf{x}, t)$ of the particle at $\mathbf{x}$ at time t is expressed, by purely kinematic arguments, as

$$\begin{aligned} \gamma &= \frac{\partial \mathbf{u}}{\partial t} + (\mathbf{u} \cdot \boldsymbol{\nabla})\mathbf{u} \quad \text{or} \\ \gamma_i &= \frac{\partial u_i}{\partial t} + \sum_{j=1}^{3} u_j \frac{\partial u_i}{\partial x_j}, \qquad i = 1, 2, 3. \end{aligned} \tag{1.3}$$

Inserting this expression in the left-hand side of (1.1) we obtain the term $\rho(\mathbf{u} \cdot \boldsymbol{\nabla})\mathbf{u}$, which will be the only nonlinear term in the Navier–Stokes equations, also called the *inertial term.* The Navier–Stokes equations are among the very few equations of mathematical physics for which the nonlinearity does not occur from physical assumptions but just from mathematical (kinematical) arguments. Further transformations of the conservation-of-momentum equation necessitate further physical arguments and assumptions. Assuming that the fluid is Newtonian, the stress tensor is expressed in terms of the velocity vector and the pressure $p = p(\mathbf{x}, t)$ by the formula

$$\sigma_{ij} = \mu \left\{ \frac{\partial u_i}{\partial x_j} + \frac{\partial u_j}{\partial x_i} \right\} + \{\lambda \operatorname{div} \mathbf{u} - p\}\delta_{ij}, \tag{1.4}$$

where δ_{ij} is the Kronecker symbol and $\mu,\ \lambda$ are constants, $\mu > 0,\ 3\lambda + 2\mu \geq 0,\ \mu$ being the shear viscosity coefficient and $3\lambda + 2\mu$ the dilatation viscosity coefficient.

We assume also that the fluid is incompressible and homogeneous, so that $\rho(\mathbf{x}, t) = \rho_0$ is constant and the continuity equation reduces to

$$\operatorname{div} \mathbf{u} = 0. \tag{1.5}$$

Therefore the term in $\operatorname{div} \mathbf{u}$ disappears in (1.4) and, with (1.3) and (1.4), the conservation-of-momentum equation (1.1) becomes

$$\frac{\partial \mathbf{u}}{\partial t} + (\mathbf{u} \cdot \boldsymbol{\nabla})\mathbf{u} - \nu \Delta \mathbf{u} + \boldsymbol{\nabla} p = \mathbf{f}, \tag{1.6}$$

where $\Delta = \nabla^2$ is the Laplace operator. Equations (1.5) and (1.6) are the Navier–Stokes equations of viscous incompressible homogeneous fluids that will be considered from now on. In (1.6) we have set $\rho_0 = 1$ and $\nu = \mu$; alternatively for $\rho_0 \neq 1$, we obtain (1.6) by setting $\nu = \mu/\rho_0$ and replacing $\mathbf{f}/\rho_0$ by $\mathbf{f}$ (the mass density of body forces).

Alternative forms of the viscosity and inertial terms can be obtained by introducing the vorticity vector

$$\omega = \operatorname{curl} \mathbf{u} = \nabla \times \mathbf{u}, \tag{1.7}$$

so that

$$\Delta \mathbf{u} = -\nabla \times \omega, \tag{1.8}$$

and

$$(\mathbf{u} \cdot \nabla)\mathbf{u} = \nabla \cdot (\mathbf{u}\mathbf{u}) = \omega \times \mathbf{u} + \tfrac{1}{2}\nabla q^2, \tag{1.9}$$

where q is the modulus of $\mathbf{u}$; in this case $\frac{1}{2}q^2$ is added to the pressure term, $p + \frac{1}{2}q^2$ being a total pressure of Bernoulli type.

Implicit in (1.5) and (1.6) are the independent variables $\mathbf{x}$ and t; $\mathbf{x} = (x_1, x_2, x_3)$ belongs to Ω, which is the domain of $\mathbb{R}^3$ filled by the fluid, and t, the time, belongs to some interval of $\mathbb{R}$ during which the fluid is observed/studied, usually $t \geq 0$. Two-dimensional flows correspond to the case where the fluid fills a cylinder $\Omega \times \mathbb{R}$, $\Omega \subset \mathbb{R}^2$, and velocities are independent of x_3 and parallel to the Ω-plane; hence $\mathbf{u} = (u_1, u_2, 0)$ if Ω is included in the plane x_1, x_2. It is usually considered that two-dimensional flows effectively occur in the following situations:

- flow in a very thin layer of fluid,
- flow in a thick layer of fluid before instability occurs in the thick direction.

As we will see, it is sometime convenient for physical discussions to consider a nondimensional form of the conservation of momentum equation. For that purpose we introduce a reference length L_* and a reference time T_* for the flow (see numerous choices in Chapter 2), and set

$$\mathbf{x} = L_*\mathbf{x}', \qquad t = T_*t', \qquad p = P_*p'$$

$$\mathbf{u} = U_*\mathbf{u}', \qquad \mathbf{f} = \frac{L_*}{T_*^2}\mathbf{f}',$$

where $P_* = U_*^2$ and $U_* = L_*/T_*$ are a reference pressure and a reference velocity. By substitution into (1.6) we obtain exactly the same equation for the reduced quantities $\mathbf{u}'(\mathbf{x}', t')$, $p'(\mathbf{x}', t')$, $\mathbf{f}'(\mathbf{x}', t')$ but, in this case, ν is replaced by Re^{-1}, where Re is a nondimensional number, the Reynolds number:

$$\mathrm{Re} = \frac{L_* U_*}{\nu}. \tag{1.10}$$

As we will see in Chapter 2, various choices of L_*, U_* can be appropriate for a given flow, leading to various definitions of the Reynolds number. But, with all definitions, Re is large for turbulent flows (from 10^2 to 10^6 to 10^{12} in industrial flows, and even larger for geophysical flows).

When we set $\mathrm{Re} = +\infty$ (or $\nu = 0$), we obtain the case of inviscid flows. Here (1.5) is retained but (1.6) becomes the Euler equation of inviscid (perfect) flows:

$$\frac{\partial \mathbf{u}}{\partial t} + (\mathbf{u} \cdot \boldsymbol{\nabla})\mathbf{u} + \boldsymbol{\nabla} p = \mathbf{f}. \tag{1.11}$$

Energy and Enstrophy

Let Ω be the domain filled by the fluid in motion and let $\mathbf{u}$ be a given velocity field defined on Ω. Since the density $\rho = 1$, the corresponding kinetic energy of the fluid is

$$e(\mathbf{u}) = \frac{1}{2} \int_\Omega |\mathbf{u}(\mathbf{x})|^2 \, d\mathbf{x}. \tag{1.12}$$

Another quantity of interest is the enstrophy

$$E(\mathbf{u}) = \int_\Omega |\boldsymbol{\nabla} \mathbf{u}(\mathbf{x})|^2 \, d\mathbf{x} = \sum_{i,j=1}^{3} \int_\Omega \left| \frac{\partial u_i}{\partial x_j}(\mathbf{x}) \right|^2 \, d\mathbf{x}. \tag{1.13}$$

The importance of enstrophy lies in that it determines the rate of dissipation of kinetic energy. Indeed, let us assume for simplicity that $\Omega = \mathbb{R}^3$ and that $\mathbf{u}$ and p vanish and decay sufficiently rapidly at infinity. Then we take the scalar product of (1.6) with $\mathbf{u}$ integrate over Ω and integrate by parts certain terms. We obtain

$$\frac{d}{dt} e(\mathbf{u}) = -\nu E(\mathbf{u}) + \int_\Omega \mathbf{f} \cdot \mathbf{u} \, d\mathbf{x}. \tag{1.14}$$

When there are no volume forces (i.e. $\mathbf{f} = 0$), (1.14) gives the decay of kinetic energy due to viscosity effects:

$$\frac{d}{dt} e(\mathbf{u}) = -\nu E(\mathbf{u}). \tag{1.15}$$

It is noteworthy also that (1.14) and the computations leading to (1.14) play an essential role and are repeatedly used in the mathematical theory of the Navier–Stokes equations even when $\Omega \neq \mathbb{R}^3$.

1.2. Boundary Value Problems

Equations (1.5), (1.6) are supplemented with initial and boundary conditions depending on the physical problem under consideration. Throughout this book we will mainly consider the following boundary conditions: the no-slip boundary condition, the space-periodic boundary condition, and the channel flow.

No-Slip Boundary Condition

This corresponds to the case where the fluid fills a domain Ω whose boundary Γ is materialized, Ω being smooth and bounded. Assuming that the motion of the boundary Γ is prescribed (velocity $= \varphi$), the no-slip boundary condition reads

$$\mathbf{u} = \varphi \quad \text{on } \Gamma; \tag{1.16}$$

if the boundary is at rest, then

$$\mathbf{u} = 0 \quad \text{on } \Gamma. \tag{1.17}$$

The condition (1.16) or (1.17) is a Dirichlet boundary condition.

Space-Periodic Case

Clearly space-periodic boundary conditions are not accessible in realistic physical situations, but they arise in the study of homogeneous turbulence when we assume that the wall effects are not important. These boundary conditions do not contain the difficulties related to boundaries (e.g. the presence of boundary layers) but retain the complexities of the nonlinearity characterizing the Navier–Stokes equations.

For space-periodic flows we assume that the fluid fills the whole space $\mathbb{R}^3$ (or $\mathbb{R}^2$)

$$\begin{aligned}&\mathbf{u},\ \mathbf{f} \text{ and } p \text{ are periodic in each direction } x_i\\&\text{with period } L_i > 0.\end{aligned} \tag{1.18}$$

In this case we denote by Ω the period, $\Omega = (0, L_1) \times (0, L_2) \times (0, L_3)$ in space dimension 3, or $\Omega = (0, L_1) \times (0, L_2)$ in space dimension 2.

Channel Flow

We consider the flow in a channel parallel to the plane Ox_1x_2. Hence the channel is the region $\Omega = (0, L_1) \times (-h, +h) \times (0, L_3)$, the streamwise direction is the direction Ox_1, and the walls are the planes $x_2 = -h$ and $x_2 = +h$, where h is the channel half width. A pressure gradient K_p is applied in the streamwise direction Ox_1, and the flow is assumed to be periodic in the streamwise and spanwise directions $Ox_1,\ Ox_3$. Assuming that the walls $x_2 = -h$ and $+h$ are at rest, we obtain the boundary conditions

$$\mathbf{u}|_{x_2=-h} = \mathbf{u}|_{x_2=+h} = 0, \tag{1.19}$$

$$\mathbf{u}|_{x_1+L_1} = \mathbf{u}|_{x_1}, \qquad \mathbf{u}|_{x_3+L_3} = \mathbf{u}|_{x_3},$$

$$p|_{x_1+L_1} = p|_{x_1} + K_p L_1, \qquad p|_{x_3+L_3} = p|_{x_3}.$$

Then by the incompressibility condition, the integral

$$\int_{\Omega\cap\{x_1=a\}} u_1\, dx_2\, dx_3$$

is independent of a; it represents the flux in the channel. Nothing is changed for two-dimensional channels except that the direction x_3 is not taken into consideration.

Initial Conditions

When studying the evolution of a flow we must also prescribe the initial distribution of velocities

$$\mathbf{u}_0(\mathbf{x}, 0) = \mathbf{u}_0(\mathbf{x}), \qquad \mathbf{x} \in \Omega, \tag{1.20}$$

where $\mathbf{u}_0$ is given.

In order to obtain a well-posed initial boundary value problem for the Navier–Stokes equations, we do not prescribe the boundary value of the pressure or its initial value. As we will see below, the pressure is in fact fully and uniquely determined by the evolution of the velocity field $\mathbf{u}$ (uniquely up to an additive constant).

The results on existence and uniqueness that we state below are related to any of the following boundary and initial value problems:

($\mathcal{P}_{\text{ns}}$) consisting of (1.5), (1.6), (1.17), (1.20),
($\mathcal{P}_{\text{per}}$) consisting of (1.5), (1.6), (1.18), (1.20),
($\mathcal{P}_{\text{ch}}$) consisting of (1.5), (1.6), (1.19), (1.20).

These problems are well posed, subject however to some restrictions in space dimension 3.

A Boundary Value Problem for the Pressure

As we mentioned earlier, the pressure is fully determined in terms of the velocity field $\mathbf{u}$. More precisely at each instant of time the pressure can be expressed in terms of the velocity field at the same time via the solution of a suitable boundary value problems.[1] We briefly recall this important theoretical and computational fact and we show how it leads to writing the Navier–Stokes equations as an evolution equation involving $\mathbf{u}$ only.

We consider a solution $\mathbf{u}$ of the full Navier–Stokes equations, that is, of one of the problems $(\mathcal{P}_{\text{ns}})$, $(\mathcal{P}_{\text{per}})$, or $(\mathcal{P}_{\text{ch}})$, and we take the divergence of each side of equation (1.6). We obtain

$$\Delta p = \operatorname{div}\{\mathbf{f} - (\mathbf{u}\cdot\boldsymbol{\nabla})\mathbf{u}\} = \operatorname{div}\mathbf{f} - \sum_{i,j=1}^{3} \frac{\partial u_j}{\partial x_i}\frac{\partial u_i}{\partial x_j}. \tag{1.21}$$

In the space-periodic case (1.18), p is fully defined in terms of $\mathbf{u}$ and $\mathbf{f}$ by equation (1.21) and the periodicity of p. In the no-slip case or in the channel case we obtain a boundary condition for p on the rigid wall by a scalar multiplication of each side of (1.6) with $\boldsymbol{\nu}$ the unit outward normal on $\partial\Omega$; this yields

$$\frac{\partial p}{\partial \boldsymbol{\nu}} = \{\mathbf{f} + \nu\,\Delta\mathbf{u}\}\cdot\boldsymbol{\nu} \quad \text{on the wall.} \tag{1.22}$$

We conclude that p can be expressed in terms of $\mathbf{u}$ by solving the Neumann problem (1.21), (1.22) in the no-slip case; for a channel flow the Neumann problem consists of (1.21), (1.22) on the wall and space periodicity in directions x_1 and x_3. In all cases we obtain p as a quadratic function of $\mathbf{u}$:

$$p = \psi(\mathbf{u}). \tag{1.23}$$

Substituting this expression of p in (1.6) we obtain an evolution equation involving only $\mathbf{u}$. However, a slightly different and more interesting equation is obtained hereafter using the Helmholtz decomposition of $\mathbf{u}$.

[1] In particular, at time $t = 0$, by prescribing the initial spatial distribution of velocities $\mathbf{u}_0$ as in (1.20), we also prescribe, to within an additive constant, the initial pressure distribution $p(\mathbf{x}, 0)$.

1.3. The Functional Setting

We first loosely recall the definition of two function spaces H, V, which play an important role in the mathematical theory of Navier–Stokes equations: see Section 1.4 and the mathematically oriented literature for a more precise definition of these spaces.

The space H is the space of vector functions $\mathbf{u}$ defined on Ω such that

$$e(\mathbf{u}) < \infty, \qquad \operatorname{div} \mathbf{u} = 0,$$

and that satisfy some of the boundary conditions of the problem; namely, for the three cases under consideration,

$$\begin{aligned}
H = H_{\text{ns}} &= \{\mathbf{u},\ e(\mathbf{u}) < \infty,\ \operatorname{div}\mathbf{u} = 0,\ \mathbf{u}\cdot\nu = 0 \text{ on } \Gamma\},\\
H = H_{\text{per}} &= \Big\{\mathbf{u},\ e(\mathbf{u}) < \infty,\ \operatorname{div}\mathbf{u} = 0,\ u_i \text{ is periodic in}\\
&\qquad \text{direction } x_i,\ i = 1, 2, 3,\ \int_\Omega \mathbf{u}\, d\mathbf{x} = 0\Big\}, \qquad (1.24)\\
H = H_{\text{ch}} &= \{\mathbf{u},\ e(\mathbf{u}) < \infty,\ \operatorname{div}\mathbf{u} = 0,\\
&\qquad u_2 = 0 \text{ on the walls } (x_2 = \pm h),\\
&\qquad u_i \text{ is periodic in direction } x_i \text{ for } i = 1 \text{ and } 3\}.
\end{aligned}$$

The space V consists of vector fields $\mathbf{u}$ defined on Ω such that

$$E(\mathbf{u}) < \infty, \qquad \operatorname{div} \mathbf{u} = 0,$$

and which satisfy the boundary conditions of the problem; hence for the three cases under consideration

$$\begin{aligned}
V = V_{\text{ns}} &= \{\mathbf{u},\ E(\mathbf{u}) < \infty,\ \operatorname{div}\mathbf{u} = 0,\ \mathbf{u} = 0 \text{ on } \Gamma\},\\
V = V_{\text{per}} &= \Big\{\mathbf{u},\ E(\mathbf{u}) < \infty,\ \operatorname{div}\mathbf{u} = 0,\\
&\qquad \mathbf{u} \text{ is } \Omega-\text{periodic},\ \int_\Omega \mathbf{u}\, d\mathbf{x} = 0\Big\}, \qquad (1.25)\\
V = V_{\text{ch}} &= \{\mathbf{u},\ E(\mathbf{u}) < \infty,\ \operatorname{div}\mathbf{u} = 0,\ \mathbf{u} = 0 \text{ at } x_2 = \pm h,\\
&\qquad \mathbf{u} \text{ is periodic in directions } x_1 \text{ and } x_3\}.
\end{aligned}$$

We refer the reader to Section 1.2 for the conditions on the domain Ω in each of these cases.

We now describe the Helmholtz decomposition of a vector field.

Helmholtz Decomposition of a Vector Field

For a vector field defined on $\mathbb{R}^3$ (or $\mathbb{R}^2$) this decomposition resolves the vector into the sum of a gradient and a curl vector. There is an appropriate generalization of this concept that is valid for vector fields defined on a bounded set and that depends on the actual boundary conditions. More precisely, we decompose a vector $\mathbf{v}$ as

$$\mathbf{v} = \mathbf{grad}\, q + \mathbf{w}, \tag{1.26}$$

where $\mathbf{w}$ belongs to the corresponding space H. Hence $\operatorname{div} \mathbf{w} = 0$ (so that $\mathbf{w}$ is a curl vector, $\mathbf{w} = \operatorname{curl} \boldsymbol{\zeta}$), and this implies

$$\Delta q = \operatorname{div} \mathbf{v}. \tag{1.27}$$

Also, $\mathbf{w}$ satisfies the boundary conditions described in (1.24); therefore

- in the space-periodic case, $\mathbf{w}$ is space-periodic like $\mathbf{u}$; hence q is space-periodic too, and it is uniquely determined (up to an additive constant) by (1.27) and space periodicity;
- in the no-slip case, $\mathbf{w} \cdot \boldsymbol{\nu} = 0$ on $\partial\Omega$; hence

$$\frac{\partial q}{\partial \boldsymbol{\nu}} = \mathbf{v} \cdot \boldsymbol{\nu} \text{ on } \partial\Omega, \tag{1.28}$$

  and q is fully determined as the solution of the Neumann problem (1.27), (1.28);
- for the channel flow, we require space periodicity for $\mathbf{w}$ (and q) in directions x_1 and x_3 and we require $\mathbf{w} \cdot \boldsymbol{\nu} = 0$ on the wall, that is (1.28) holds on the wall. Here again these conditions define q and $\mathbf{w}$ uniquely.

Because of that, the mapping $\mathbf{v} \to \mathbf{w}$ is well defined; we call it P:

$$P : \mathbf{v} \to \mathbf{w}(\mathbf{v}). \tag{1.29}$$

By applying the operator P to both sides of (1.6), at each time t, we find

$$\frac{\partial \mathbf{u}}{\partial t} + \nu A\mathbf{u} + B(\mathbf{u}) = P\mathbf{f}, \tag{1.30}$$

where we have written

$$A\mathbf{u} = -P\,\Delta\mathbf{u}, \qquad B(\mathbf{u}) = B(\mathbf{u}, \mathbf{u}), \qquad B(\mathbf{u}, \mathbf{v}) = P((\mathbf{u} \cdot \nabla)\mathbf{v}). \tag{1.31}$$

The operator A is the Stokes operator; in the space-periodic case $P\,\Delta\mathbf{u} = \Delta\mathbf{u}$, but in the no-slip case, $P\,\Delta\mathbf{u} \neq \Delta\mathbf{u}$. The form (1.30) of the Navier–Stokes equation was first derived by Leray (1933), using a slightly different presentation based on the *weak formulation*, which we now recall.

Weak Formulation of the Navier–Stokes Equations

It is clear that at each instant of time the vector field

$$\mathbf{x} \in \Omega \to \mathbf{u}(\mathbf{x}, t),$$

also denoted by $\mathbf{u}(\cdot, t)$ or $\mathbf{u}(t)$, belongs to the function space V introduced in (1.25). Now let $\mathbf{v} = \mathbf{v}(\mathbf{x})$ be a test function belonging to V. To obtain the weak formulation of the Navier–Stokes equations we multiply equation (1.6) by $\mathbf{v}$, integrate over Ω, and integrate by parts:

$$\int_\Omega \frac{\partial \mathbf{u}}{\partial t}(\mathbf{x}, t)\cdot \mathbf{v}(\mathbf{x})\,d\mathbf{x} = \frac{\partial}{\partial t}\int_\Omega \mathbf{u}(\mathbf{x}, t)\cdot \mathbf{v}(\mathbf{x})\,d\mathbf{x},$$

$$\begin{aligned}
-\int_\Omega \Delta\mathbf{u}\,(\mathbf{x}, t)\cdot\mathbf{v}(\mathbf{x})\,d\mathbf{x} &= -\sum_{i=1}^{3}\int_{\partial\Omega}\frac{\partial u_i}{\partial \nu}(\mathbf{x}, t)v_i(\mathbf{x})\,d\Gamma \\
&\quad + \sum_{i,j=1}^{3}\int_\Omega \frac{\partial u_i}{\partial x_j}(\mathbf{x}, t)\frac{\partial v_i}{\partial x_j}(\mathbf{x})\,d\mathbf{x} \\
&= \sum_{i,j=1}^{3}\int_\Omega \frac{\partial u_i}{\partial x_j}(\mathbf{x}, t)\frac{\partial v_i}{\partial x_j}(\mathbf{x})\,d\mathbf{x}
\end{aligned}$$

(because of the boundary conditions),

$$\begin{aligned}
\int_\Omega \mathbf{grad}\; p(\mathbf{x}, t)\mathbf{v}(\mathbf{x})\,d\mathbf{x} &= \int_{\partial\Omega} p(x, t)\mathbf{v}(x)\cdot\nu(x)\,dx \\
&\quad - \int_\Omega p(\mathbf{x}, t)\,\operatorname{div}\mathbf{v}(\mathbf{x})\,d\mathbf{x} = 0
\end{aligned}$$

(using the boundary conditions and $\operatorname{div}\mathbf{v} = 0$).

Hence we find that

$$\begin{aligned}
&\frac{\partial}{\partial t}\int_\Omega \mathbf{u}(\mathbf{x}, t)\cdot\mathbf{v}(\mathbf{x})\,d\mathbf{x} + \nu\sum_{i,j=1}^{3}\int_\Omega \frac{\partial u_i}{\partial x_j}(\mathbf{x}, t)\frac{\partial v_i}{\partial x_j}(\mathbf{x})\,d\mathbf{x} \\
&\quad + \sum_{i,j=1}^{3}\int_\Omega u_i(\mathbf{x}, t)\frac{\partial u_j}{\partial x_i}(\mathbf{x}, t)v_j(\mathbf{x})\,d\mathbf{x} = \int_\Omega \mathbf{f}(\mathbf{x}, t)\cdot\mathbf{v}(\mathbf{x})\,d\mathbf{x}
\end{aligned} \tag{1.32}$$

for every test function in V. To simplify the notation, a more compact form may be used by dropping the dummy variables $\mathbf{x}$; we obtain

$$\frac{\partial}{\partial t}\int_\Omega \mathbf{u}(t)\cdot\mathbf{v}\,d\mathbf{x} + \nu\sum_{i,j=1}^{3}\int_\Omega \frac{\partial u_i}{\partial x_j}(t)\frac{\partial v_i}{\partial x_j}\,d\mathbf{x}$$

$$+\sum_{i,j=1}^{3}\int_\Omega u_i(t)\frac{\partial u_j}{\partial x_i}(t)v_j\,d\mathbf{x} = \int_\Omega \mathbf{f}(t)\cdot\mathbf{v}\,d\mathbf{x}.$$

The following notation will be also used:

$$(\boldsymbol{\varphi},\boldsymbol{\psi})_0 = \int_\Omega \boldsymbol{\varphi}(\mathbf{x})\cdot\boldsymbol{\psi}(\mathbf{x})\,d\mathbf{x}, \tag{1.33}$$

$$((\boldsymbol{\varphi},\boldsymbol{\psi})) = \sum_{i,j=1}^{3}\int_\Omega \frac{\partial\varphi_i}{\partial x_j}(\mathbf{x})\frac{\partial\psi_i}{\partial x_j}(\mathbf{x})\,d\mathbf{x} \tag{1.34}$$

for every pair of vector fields $\boldsymbol{\varphi}$, $\boldsymbol{\psi}$ defined on Ω, and if $\boldsymbol{\theta}$ is a third vector field,

$$b(\boldsymbol{\varphi},\boldsymbol{\psi},\boldsymbol{\theta}) = \sum_{i,j=1}^{3}\int_\Omega \varphi_i(\mathbf{x})\frac{\partial\varphi_j}{\partial x_i}(\mathbf{x})\theta_j(\mathbf{x})\,d\mathbf{x}. \tag{1.35}$$

Then (1.32) can be rewritten as follows:

The function $t \to \mathbf{u}(t)$ takes its values in V and satisfies

$$\frac{\partial}{\partial t}(\mathbf{u}(t),\mathbf{v})_0 + \nu((\mathbf{u}(t),\mathbf{v})) + b(\mathbf{u}(t),\mathbf{u}(t),\mathbf{v}) = (\mathbf{f}(t),\mathbf{v})_0 \tag{1.36}$$

for every $\mathbf{v} \in V$.

This is the weak formulation of the Navier–Stokes equations, which goes back to the pioneering work of Leray (1933, 1934a, 1934b). It plays an essential role in the mathematical theory of these equations as well as in numerical computations, at least when finite elements are used.

Finally, for two-dimensional flows, the weak formulation is the same as in the three-dimensional case except that the sums in (1.32), (1.34), (1.35) are for $i, j = 1, 2$ ($\Omega \subset \mathbb{R}^2$).

Energy Equality

An extension of the equation of energy conservation (energy equality) derived in Section 1.1 can be obtained for flows corresponding to the three types of boundary conditions (problems $(\mathcal{P}_{\text{ns}})$, $(\mathcal{P}_{\text{per}})$ or $(\mathcal{P}_{\text{ch}})$).

At each time t we replace in (1.32) (or (1.36)) $\mathbf{v} = \mathbf{v}(\mathbf{x})$ by $\mathbf{u}(t) = \mathbf{u}(\mathbf{x}, t)$. As for (1.14), we can show that

$$b(\mathbf{u}, \mathbf{u}, \mathbf{u}) = \int_\Omega [(\mathbf{u} \cdot \nabla)\mathbf{u}] \cdot \mathbf{u}\, d\mathbf{x} = 0,$$

leading to

$$\frac{1}{2}\frac{d}{dt}\int_\Omega |\mathbf{u}(\mathbf{x}, t)|^2\, d\mathbf{x} + \nu \sum_{i,j=1}^{3} \int_\Omega \left|\frac{\partial u_i}{\partial x_j}(\mathbf{x}, t)\right|^2 d\mathbf{x} \tag{1.37}$$
$$= \int_\Omega \mathbf{f}(\mathbf{x}, t) \cdot \mathbf{u}(\mathbf{x}, t)\, d\mathbf{x}.$$

Alternatively, setting, for every vector field φ,

$$|\varphi|_0 = \{(\varphi, \varphi)_0\}^{1/2}, \qquad \|\varphi\| = \{((\varphi, \varphi))\}^{1/2},$$

so that

$$e(\varphi) = \tfrac{1}{2}|\varphi|_0^2, \qquad E(\varphi) = \|\varphi\|^2,$$

we rewrite (1.37) as

$$\frac{1}{2}\frac{d}{dt}|\mathbf{u}(t)|_0^2 + \nu\|\mathbf{u}(t)\|^2 = (\mathbf{f}(t), \mathbf{u}(t))_0 \tag{1.38}$$

or

$$\frac{d}{dt}e(\mathbf{u}(t)) + \nu E(\mathbf{u}(t)) = (\mathbf{f}(t), \mathbf{u}(t))_0. \tag{1.39}$$

1.4. The Main Results on Existence and Uniqueness of Solutions

In this section we state without proof the main existence and uniqueness theorems for the solutions in space dimension 2 and 3. In the two-dimensional case the theory is fairly complete. Loosely speaking, weak solutions (i.e. non-smooth solutions) exist and are unique (see Theorem 1.1); strong solutions (i.e. smooth solutions) exist and are unique (see Theorem 1.2); furthermore, strong solutions can be as smooth as desired, up to $\mathcal{C}^\infty$ regularity, provided the data are sufficiently regular. By data we mean $\mathbf{f}$, $\mathbf{u}_0$, and ν as well as the shape of Ω in the no-slip case. As we will see below, the mathematical theory of the Navier–Stokes equations is not complete in dimension 3.

Function Spaces

We start by recalling the definition of the function spaces used in the mathematical theory. The definitions hereafter are valid in space dimension $n=2$ or 3.

We denote by $L^2(\Omega)^n$ the space of square-integrable vector functions from Ω into $\mathbb{R}^n$ ($n = 2$ or 3). As previously remarked, this is the space of finite kinetic energy vector fields on Ω. We endow the space with the usual scalar product

$$(\mathbf{u}, \mathbf{v})_0 = \sum_{i=1}^{n} \int_\Omega u_i(\mathbf{x}) v_i(\mathbf{x})\, d\mathbf{x}$$

and the associated norm

$$|\mathbf{u}|_0 = \{(\mathbf{u}, \mathbf{u})_0\}^{1/2} = \left\{\sum_{i=1}^{n} \int_\Omega |u_i(\mathbf{x})|^2\, d\mathbf{x}\right\}^{1/2}.$$

It is well known that $L^2(\Omega)^n$ endowed with such a scalar product and such a norm is a Hilbert space.

The space H defined in (1.24) is the space of vector functions in $L^2(\Omega)^n$ satisfying the boundary condition appearing in (1.24) (and which depend on the boundary value problem); H is a Hilbert subspace of $L^2(\Omega)^n$, and the projector P appearing in (1.29) is the orthogonal projector from $L^2(\Omega)^n$ onto H. The orthogonal complement of H in $L^2(\Omega)^n$ is a space G of gradient vectors (see e.g. Temam (1984, Ch. I, Sec. 1)).

The space V appearing in (1.25) is a subspace of the Sobolev space $H^1(\Omega)^n$. We denote by $H^1(\Omega)^n$ the Sobolev space which consists of square integrable vector functions whose gradients are square-integrable, that is,

$$H^1(\Omega)^n = \left\{\mathbf{v} \in L^2(\Omega)^n,\ \frac{\partial \mathbf{v}}{\partial x_i} \in L^2(\Omega)^n \quad \forall\, i\right\}.$$

The space $H^1(\Omega)^n$ is a Hilbert space for the scalar product

$$((\mathbf{u}, \mathbf{v}))_1 = \sum_{i,j=1}^{n} \int_\Omega \frac{\partial u_i}{\partial x_j} \frac{\partial v_i}{\partial x_j}\, d\mathbf{x} + \sum_{i=1}^{n} \int_\Omega u_i v_i\, d\mathbf{x},$$

and the norm

$$\|\mathbf{u}\|_1 = \{((\mathbf{u}, \mathbf{u}))_1\}^{1/2}.$$

The space V is then the Hilbert subspace of $H^1(\Omega)^n$ consisting of the vector functions satisfying the conditions in (1.25); these conditions depend on the boundary value problem under consideration. For all these spaces we set

$$((\mathbf{u}, \mathbf{v})) = \sum_{i,j=1}^{n} \int_\Omega \frac{\partial u_i}{\partial x_j} \frac{\partial v_i}{\partial x_j}\, d\mathbf{x}, \qquad \|\mathbf{u}\| = \{((\mathbf{u}, \mathbf{u}))\}^{1/2};$$

it can be proven that $\|\mathbf{u}\|$ is a norm on V equivalent to $\|\mathbf{u}\|_1$. Also, by mere integration by parts one can show that

$$\|\mathbf{u}\|^2 = E(\mathbf{u}) = \int_\Omega |\text{curl } \mathbf{u}(\mathbf{x})|^2\, d\mathbf{x}. \tag{1.40}$$

For more details on these spaces see for example Lions (1969), Temam (1984, 1995).

Existence and Uniqueness ($n = 2$)

We are interested in the solutions of the initial boundary value problem for the Navier–Stokes equations in their weak form (1.36) (or (1.30)), that is: To find the function $t \to \mathbf{u}(t)$ from $[0, T]$ into V satisfying

$$\frac{\partial}{\partial t}(\mathbf{u}(t), \mathbf{v})_0 + \nu((\mathbf{u}(t), \mathbf{v})) + b(\mathbf{u}(t), \mathbf{u}(t), \mathbf{v}) = (\mathbf{f}(t), \mathbf{v})_0 \tag{1.41}$$

for every $\mathbf{v} \in V$, and

$$\mathbf{u}(0) = \mathbf{u}_0. \tag{1.42}$$

We have

Theorem 1.1 (Weak Solutions). *We assume that* $\mathbf{u}_0$ *and* $\mathbf{f}$ *are given satisfying*

$$\mathbf{u}_0 \in H, \qquad \mathbf{f} \in L^2(\Omega \times (0, T))^2.$$

Then there exists a unique solution $\mathbf{u} = (u_1, u_2)$ *of* (1.41), (1.42) *such that*

$$u_i, \ \frac{\partial u_i}{\partial x_j} \in L^2(\Omega \times (0, T)), \qquad i, j = 1, 2, \tag{1.43}$$

and $\mathbf{u}$ *is a continuous function from* $[0, T]$ *into* $L^2(\Omega)^2$.

Theorem 1.2 (Strong (Smooth) Solutions). *We assume that* $\mathbf{u}_0$ *and* $\mathbf{f}$ *are given and satisfy*

$$\mathbf{u}_0 \in V, \qquad \mathbf{f} \in L^2(\Omega \times (0, T))^2.$$

Then there exists a unique solution $\mathbf{u} = (u_1, u_2)$ *of* (1.41), (1.42) *such that*

$$u_i, \frac{\partial u_i}{\partial t}, \frac{\partial u_i}{\partial x_j}, \frac{\partial^2 u_i}{\partial x_j \partial x_k} \in L^2(\Omega \times (0, T)), \qquad i, j, k = 1, 2, \tag{1.44}$$

and $\mathbf{u}$ *is a continuous function from* $[0, T]$ *into* V.

Remark 1.1. In space dimension 2, the solution is smoother if the data $\mathbf{u}_0$ and $\mathbf{f}$ are smoother (and Ω too for $(\mathcal{P}_{\rm ns})$); if the data are $\mathcal{C}^\infty$, then $\mathbf{u}$ and p are $\mathcal{C}^\infty$ too.

Also, the Navier–Stokes equations enjoy a smoothing property, that is, the solution can be smoother at $t > 0$ than the initial data $\mathbf{u}_0$, provided $\mathbf{f}$ is sufficiently smooth (and Ω too for $(\mathcal{P}_{\rm ns})$). □

Existence and Uniqueness $(n = 3)$

We now state the main theorems of existence and uniqueness of solution in space dimension 3; the function spaces are the same as for $n = 2$.

The mathematical theory of the Navier–Stokes equations is not yet complete in space dimension 3. A loose statement of the available results is as follows:

- There exists for all time a weak (nonsmooth) solution, but it may not be unique; see Theorem 1.3.
- A strong solution exists on a certain interval of time $(0, T_*)$, where T_* depends on the data; see Theorem 1.4. Furthermore, on $(0, T_*)$ the solution can be as smooth as the data permit; see e.g. the references previously mentioned.
- If a strong (smooth) solution exists for all time then it is unique; but we do not know if such a solution exists for all time; see Theorem 1.4.

Remark 1.2. A related problem in space dimension 3 is the possible appearance of singularities: *can the magnitude of the vorticity vector* curl $\mathbf{u}(\mathbf{x}, t)$ *become infinite at some points in space and time?* This question, not yet answered, was raised by J. Leray (see Leray (1933, 1934a, 1934b)). The possible occurrence of singularities motivated him to introduce and study weak solutions. Our present understanding of turbulence and chaos indicates that chaotic behavior may appear even in smooth systems. This however does not preclude the existence of chaos *and* singularities, but at this time this domain is completely unexplored. □

The weak formulation of the initial boundary value problem for the Navier–Stokes equations has the same form as (1.41), (1.42), and we have the following results.

Theorem 1.3 (Existence of Weak Solutions). *We assume that* $\mathbf{u}_0$ *and* $\mathbf{f}$ *are given satisfying*

$$\mathbf{u}_0 \in H, \qquad \mathbf{f} \in L^2(\Omega \times (0,T))^3.$$

Then there exists at least one solution $\mathbf{u} = (u_1, u_2, u_3)$ *of* (1.41), (1.42) *such that*

$$u_i, \frac{\partial u_i}{\partial x_j} \in L^2(\Omega \times (0,T)), \qquad i,j = 1,2,3,$$

and $\mathbf{u}$ *is weakly continuous from* $[0,T]$ *into* $L^2(\Omega)^3$, *that is, for every* $\varphi \in L^2(\Omega)^3$, *the function*

$$t \to (\mathbf{u}(t), \varphi)_0 = \int_\Omega \mathbf{u}(\mathbf{x},t) \cdot \varphi(\mathbf{x})\, d\mathbf{x}$$

is continuous.

Remark 1.3. As mentioned before, we do not know if the solutions of (1.41), (1.42) given by Theorem 1.3 are unique. □

For smooth solutions, as in Theorem 1.2, we have the following existence result:

Theorem 1.4 (Existence of Strong (Smooth) Solutions). *We assume that* $\mathbf{u}_0$ *and* $\mathbf{f}$ *are given satisfying*

$$\mathbf{u}_0 \in V, \qquad \mathbf{f} \in L^2(\Omega \times (0,T))^3.$$

Then there exists T_*, $0 < T_* \le T$, *depending on the data and, on* $(0, T_*)$, *there exists a unique solution* $\mathbf{u} = (u_1, u_2, u_3)$ *of* (1.41), (1.42) *such that*

$$u_i, \frac{\partial u_i}{\partial t}, \frac{\partial u_i}{\partial x_j}, \frac{\partial^2 u_i}{\partial x_j \partial x_k} \in L^2(\Omega \times (0,T_*)), \qquad i,j,k = 1,2,3,$$

and $\mathbf{u}$ *is a continuous function from* $[0, T_*)$ *into* V.

Remark 1.4. The solution $\mathbf{u}$ of the Navier–Stokes equations given by Theorem 1.4 on $[0, T_*)$ can be as smooth as the data permit. □

2

The Physics of Turbulent Flows

In this chapter, we recall important results of the conventional theory of turbulence. Two types of turbulent flows, namely homogeneous and nonhomogeneous ones, are considered; for the latter case, wall-bounded flows – more precisely, flows in a channel – are presented. Our discussion mainly concentrates on definitions and results that will be useful in the other parts of this volume. Therefore, this cannot be a complete review of the turbulence theory. Further developments can be found in the reference books of Batchelor (1971), Tennekes and Lumley (1972), Orszag (1973), and Monin and Yaglom (1975) among many others.

2.1. Some Probabilistic Tools

In this section, we first recall some basic definitions of the probability theory needed to define and derive properties of random variables and random functions. The theoretical framework presented hereafter will be useful in the next sections, where we will consider the solutions of the Navier–Stokes equations as random vector fields.

Probability Measure and Spaces

Let U be a space (a metric or functional space for instance). We denote by $\mathcal{B}$ a nonempty collection of subsets of U.

Definition 2.1. *$\mathcal{B}$ is called a σ-algebra if it enjoys the following properties:*

(i) If $A, B \in \mathcal{B}$ then $A \cup B \in \mathcal{B}$.
(ii) If $A \in \mathcal{B}$, then its complement belongs to $\mathcal{B}$, that is, $\complement A \in \mathcal{B}$.
(iii) For any countable sequence of sets $A_k \in \mathcal{B}$, $k \in \mathbb{N}$, we have $\bigcup_{k\in\mathbb{N}} A_k \in \mathcal{B}$.

From the above definition, we deduce several additional properties satisfied by a σ-algebra:

1. $U \in \mathcal{B}$; indeed $U = A \cup \mathcal{C}A \in \mathcal{B}$ if $A \in \mathcal{B}$.
2. As $U \in \mathcal{B}$, we have $\emptyset = \mathcal{C}U \in \mathcal{B}$.
3. From *(i)* and *(ii)*, we see that $A \cap B \in \mathcal{B}$ and $A \backslash B \in \mathcal{B}$ if A and B belong to $\mathcal{B}$.
4. The intersection of any countable sequence of subsets in $\mathcal{B}$ belongs to $\mathcal{B}$.

Note that in a metric space, the subset of the σ-algebra generated by the open sets are called the *Borel sets.*

We now consider set functions with real values, defined on a σ-algebra $\mathcal{B}$, that is,

$$\begin{aligned} \mu : \mathcal{B} &\longrightarrow \mathbb{R} \\ A &\longmapsto \mu(A). \end{aligned}$$

Definition 2.2. *A nonnegative set function μ defined on a σ-algebra $\mathcal{B}$ and satisfying*

(i) $\mu(\emptyset) = 0$,
(ii) for any countable sequence of subsets $A_k \in \mathcal{B}$, $k = 1, \ldots,$ that are pairwise disjoint, that is, $A_i \cap A_j = \emptyset$ for any i, j such that $i \neq j$, we have that $\mu(\bigcup_{k=1}^{+\infty} A_k) = \sum_{k=1}^{+\infty} \mu(A_k)$,

is called a measure.

A measure taking values in the range (0, 1) and such that $\mu(U) = 1$ is called a *probability measure*, and the triple $(U, \mathcal{B}, \mu)$ is called a *probability space.* In the following, we denote by P a probability measure.

Random Variables and Distribution Functions

Hereafter, we restrict ourselves to the case of a probability space $(U, \mathcal{B}, P)$.

Definition 2.3. *A real-valued function ζ defined on a subset $A \in \mathcal{B}$ is said to be $\mathcal{B}$-measurable if for any real x the set $\{y \in A; \zeta(y) \leq x\}$ belongs to $\mathcal{B}$. A finite $\mathcal{B}$-measurable function is called a* random variable.

We can associate with any random variable ζ, defined on a set $A \in \mathcal{B}$, a function F of real argument defined by

$$F(x) = P\{y \in A; \zeta(y) \leq x\} \quad \text{for any } x \in \mathbb{R}. \tag{2.1}$$

This function is called the *distribution function* of the random variable ζ. A distribution function F is nonnegative, is nondecreasing, and takes values in $[0, 1]$. Moreover it is continuous from the right, that is,

$$\lim_{\epsilon \to 0,\ \epsilon > 0} F(x+\epsilon) = F(x).$$

Definition 2.4. *The integral*

$$\langle \zeta \rangle = \int_U \zeta(y)\, P(dy), \tag{2.2}$$

which exists if $\langle |\zeta| \rangle = \int_U |\zeta(y)|\ P(dy) < +\infty$, *is called the* average *(or* mean*) of the random variable* ζ.

If the distribution function F of the random variable ζ is differentiable, then its derivative $f(x) = F'(x)$ is called the *probability density* of ζ and the integral above can be rewritten

$$\langle \zeta \rangle = \int_{\mathbf{R}} x f(x)\, dx.$$

Moreover, the probability density function satisfies $F(-\infty) = 0,\ F(+\infty) = 1$; hence,

$$F(x) = \int_{-\infty}^{x} f(y)\, dy \quad \text{and} \quad \int_{-\infty}^{+\infty} f(y)\, dy = 1. \tag{2.3}$$

We define the moment of order p of the random variable ζ by

$$\mu_p(\zeta) = \left\langle \zeta^p \right\rangle = \int_U \zeta(y)^p\ P(dy), \tag{2.4}$$

which exists whenever $\left\langle |\zeta|^p \right\rangle < +\infty$. The moment of order p of ζ can be rewritten as

$$\mu_p(\zeta) = \int_{\mathbf{R}} x^p f(x)\, dx,$$

when the probability density function exists.

Set of Random Variables and Joint Distribution Functions

Let us now consider a finite sequence of random variables $\zeta_j,\ j = 1, \dots, m$; then the function defined by

$$F_{\zeta_1 \dots \zeta_m}(x_1, \dots, x_m) = P\{y \in A;\ \ \zeta_1(y) \le x_1, \dots,\ \zeta_m(y) \le x_m\}, \tag{2.5}$$

is called the *joint distribution function* of the variables $\zeta_j,\ j = 1, \ldots, m$. The joint distribution function is nonnegative, is nondecreasing, and takes values in $[0, 1]$. Moreover it is continuous from the right with respect to each variable. It enjoys the obvious properties:

1. $F_{\zeta_1\ldots\zeta_m}(x_1, \ldots, x_k, -\infty, x_{k+2}, \ldots, x_m) = 0$,
2. $F_{\zeta_1\ldots\zeta_m}(x_1, \ldots, x_k, +\infty, \ldots, +\infty) = F_{\zeta_1\ldots\zeta_k}(x_1, \ldots, x_k)$,
3. $F_{\zeta_{j_1}\ldots\zeta_{j_m}}(x_{j_1}, \ldots, x_{j_m}) = F_{\zeta_1\ldots\zeta_m}(x_1, \ldots, x_m)$ for any permutation $j_1, \ldots, j_m$ of $1, \ldots, m$.

If the joint distribution function is sufficiently regular, that is, if its partial derivatives of order m exist and are continuous, then

$$f_{\zeta_1\ldots\zeta_m}(x_1, \ldots, x_m) = \frac{\partial^m F_{\zeta_1\ldots\zeta_m}}{\partial x_1 \cdots \partial x_m}(x_1, \ldots, x_m) \tag{2.6}$$

is called the *joint probability density* function, and it satisfies

$$F_{\zeta_1\ldots\zeta_m}(x_1, \ldots, x_m) = \int_{-\infty}^{x_1} \cdots \int_{-\infty}^{x_m} f_{\zeta_1\ldots\zeta_m}(y_1, \ldots, y_m)\, dy_1 \cdots dy_m, \tag{2.7}$$

$$\int_{\mathbf{R}^m} f_{\zeta_1\ldots\zeta_m}(y_1, \ldots, y_m)\, dy_1 \cdots dy_m = 1.$$

Furthermore, for any real-valued function φ defined on $\mathbb{R}^m$, we have

$$\langle \varphi(\zeta_1, \ldots, \zeta_m) \rangle = \int_{\mathbf{R}^m} \varphi(x_1, \ldots, x_m)\, f_{\zeta_1\ldots\zeta_m}(x_1, \ldots, x_m)\, dx_1 \cdots dx_m. \tag{2.8}$$

Definition 2.5. *Let $j_i > 0,\ i = 1, \ldots, m$, and consider a set of random variables $\{\zeta_i\}_{i=1,\ldots,m}$. If $\prod_{i=1}^m \zeta_i^{j_i}(y)$ is absolutely integrable, that is, $\langle \prod_{i=1}^m |\zeta_i|^{j_i} \rangle < +\infty$, then the* joint moment *$\mu_{j_1\ldots j_m}(\zeta_1, \ldots, \zeta_m)$ of $\{\zeta_i\}_{i=1,\ldots,m}$ is defined by*

$$\mu_{j_1\ldots j_m}(\zeta_1, \ldots, \zeta_m) = \left\langle \zeta_1^{j_1} \ldots \zeta_m^{j_m} \right\rangle. \tag{2.9}$$

The integer $q = j_1 + \cdots + j_m$ is called the order *of the joint moment.*

According to (2.8), the joint moment functions can be expressed in terms of the joint probability density function as follows:

$$\mu_{j_1\cdots j_m}(\zeta_1, \ldots, \zeta_m) = \int_{\mathbf{R}^m} x_1^{j_1} \cdots x_m^{j_m}\, f_{\zeta_1\ldots\zeta_m}(x_1, \ldots, x_m)\, dx_1 \cdots dx_m.$$

Random Functions and Random Vector Fields

Definition 2.6. *Let $(U, \mathcal{B}, P)$ be a probability space. A real-valued function $\zeta(x; u)$ defined on $X \times U$, where X is any space, is called a* random function *if, for every $x \in X$, $\zeta(x; \cdot)$ is $\mathcal{B}$-measurable.*

Definition 2.7. *Let $j_i > 0$, $i = 1, \dots, m$, and consider a sequence of points $x_i \in X$, $i = 1, \dots, m$. If $\prod_{i=1}^{m} \zeta(x_i)^{j_i}$ is absolutely integrable, then the moment function $m_{j_1 j_2 \dots j_m}(x_1, \dots, x_m)$ of the random function $\zeta(x; u)$ is defined by*

$$m_{j_1 j_2 \dots j_m}(x_1, x_2, \dots, x_m) = \left\langle \zeta(x_1)^{j_1} \zeta(x_2)^{j_2} \cdots \zeta(x_m)^{j_m} \right\rangle. \tag{2.10}$$

The integer $q = j_1 + \cdots + j_m$ is called the order *of the moment function.*

Note that a sufficient condition for the moment function of order q defined as above to exist is that, for every $x \in X$, $\langle |\zeta(x)|^q \rangle < +\infty$, that is, $\zeta(x; \cdot) \in L^q(U)$.

Definition 2.6 can be easily extended to the case of functions with values in a multidimensional space, for instance $\mathbb{R}^n$, $n>1$. In that case, a vector field $\boldsymbol{\zeta}(x; u)$ is called a random vector field if, for every $x \in X$ and $j = 1, \dots, n$, $\zeta_j(x; \cdot)$ is $\mathcal{B}$-measurable. By analogy with the case of a set of random variables previously discussed, we can define joint distribution functions and joint moment functions for a random vector $\boldsymbol{\zeta}(x; u)$. Let us consider a set of m points of X namely x_i, $i = 1, \dots, m$, and a sequence of integer indices ℓ_i, such that $1 \leq \ell_i \leq n$ for $i = 1, \dots, m$. We define the joint distribution function of the set of random variables $\{\zeta_{\ell_i}(x_i)\}_{i=1,\dots,m}$ by

$$F_{\zeta_{\ell_1}(x_1) \dots \zeta_{\ell_m}(x_m)}(y_1, \dots, y_m) = P\{u \in U;\ \zeta_{\ell_i}(x_i) \leq y_i,\ i = 1, \dots, m\}. \tag{2.11}$$

Assuming that the joint distribution function is smooth enough, we define as previously the joint probability function

$$f_{\zeta_{\ell_1}(x_1) \dots \zeta_{\ell_m}(x_m)}(y_1, \dots, y_m) = \frac{\partial^m F_{\zeta_{\ell_1}(x_1) \dots \zeta_{\ell_m}(x_m)}}{\partial y_1 \cdots \partial y_m}(y_1, \dots, y_m), \tag{2.12}$$

which satisfies

$$F_{\zeta_{\ell_1}(x_1) \dots \zeta_{\ell_m}(x_m)}(y_1, \dots, y_m) = \int_{-\infty}^{y_1} \cdots \int_{-\infty}^{y_m} f_{\zeta_{\ell_1}(x_1) \dots \zeta_{\ell_m}(x_m)}(z_1, \dots, z_m)\, dz_1 \cdots dz_m,$$

$$\int_{\mathbb{R}^m} f_{\zeta_{\ell_1}(x_1) \dots \zeta_{\ell_m}(x_m)}(z_1, \dots, z_m)\, dz_1 \cdots dz_m = 1.$$

Let $j_i > 0,\ i = 1, \ldots, m$, be such that $\sum_{j=1}^{m} j_i = q$; if $\prod_{i=1}^{m} \zeta_{\ell_i}(x_i; u)^{j_i}$ is absolutely integrable, then we define the joint moment function of $\{\zeta_{\ell_i}(x_i)\}_{i=1,\ldots,m}$ by setting

$$m_{j_1\ldots j_m}^{\ell_1\ldots\ell_m}(x_1, \ldots, x_m) = \left\langle \zeta_{\ell_1}(x_1)^{j_1} \cdots \zeta_{\ell_m}(x_m)^{j_m} \right\rangle \tag{2.13}$$

which can be reformulated as

$$m_{j_1\ldots j_m}^{\ell_1\ldots\ell_m}(x_1, \ldots, x_m)$$
$$= \int_{\mathbf{R}^m} z_1^{j_1} \ldots z_m^{j_m}\, f_{\zeta_{\ell_1}(x_1)\ldots\zeta_{\ell_m}(x_m)}(z_1, \ldots, z_m)\, dz_1 \cdots dz_m.$$

2.2. An Idealized Model of Turbulent Flows: Homogeneous (Isotropic) Turbulence

In the classical theory of turbulence, the vector field $\mathbf{u}(\mathbf{x}, t)$ of the flow is considered as a random vector field. This means that at a given point $(\mathbf{x}, t)$ the velocity field depends in a random way on the experience, that is, $\mathbf{u}(\mathbf{x}, t)$ takes values that cannot be determined from the data of the problem or of the experiment. As a consequence of this randomness, pointwise knowledge of $\mathbf{u}(\mathbf{x}, t)$ is not helpful (and not available). However, the study of average quantities related to $\mathbf{u}(\mathbf{x}, t)$ can lead to universal behaviors and properties of turbulent flows. We consider in this section flows in the whole space, that is, where the fluid fills the whole domain $\Omega = \mathbb{R}^n$. We assume that a probability space $(\Theta, \mathcal{B}, P)$ is known and that the velocity field (solution of the incompressible Navier–Stokes equations) depends on $\theta \in \Theta$, so that $\mathbf{u}(\mathbf{x}, t; \cdot)$ is a random vector as defined in Section 2.1. Even if we do not know how to formally construct the probability space $(\Theta, \mathcal{B}, P)$, we assume that the variable θ provides all the information (parameters) in order to uniquely determine an experiment. In an idealistic situation, θ describes the initial state of the flow, the viscosity, and the external volume force applied to sustain the flow.

2.2.1. Two-Point Correlation Tensors and Their Spectral Representation

The Two-Point Velocity Correlation Tensor

One of the most important mean quantities used by experimentalists and theoreticians to characterize a random function is the velocity correlation tensor for two separated points.

Definition and Properties

Definition 2.8. *The two-point velocity correlation tensor is defined by*

$$R_{ij}(\mathbf{x}_1, \mathbf{x}_2) = \langle u_i(\mathbf{x}_1) u_j(\mathbf{x}_2) \rangle, \qquad \mathbf{x}_1, \mathbf{x}_2 \in \mathbb{R}^n, \quad i, j = 1, \ldots, n. \quad (2.14)$$

We denote by $\mathbf{R}(\mathbf{x}_1, \mathbf{x}_2)$ *the matrix of coefficients* $R_{ij}(\mathbf{x}_1, \mathbf{x}_2)$.

In the above definition the dependence upon the time t of the velocity field is omitted for the sake of simplicity. From now on, as we will describe quantities characterizing only the spatial (or spectral) behavior of $\mathbf{u}(\mathbf{x}, t)$, the time dependence will not be explicitly specified.

The covariance tensor $\mathbf{R}(\mathbf{x}_1, \mathbf{x}_2)$ enjoys the following properties:

1. $\mathbf{R}(\mathbf{x}, \mathbf{x})$, for any $\mathbf{x} \in \mathbb{R}^n$, is a nonnegative matrix, that is,

$$\sum_{i,j=1}^{n} R_{ij}(\mathbf{x}, \mathbf{x}) \lambda_i \lambda_j \geq 0 \quad \text{for any } \boldsymbol{\lambda} \in \mathbb{R}^n.$$

2. $R_{ij}(\mathbf{x}_1, \mathbf{x}_2) = R_{ji}(\mathbf{x}_2, \mathbf{x}_1)$.
3. $|R_{ij}(\mathbf{x}_1, \mathbf{x}_2)| \leq |R_{ii}(\mathbf{x}_1, \mathbf{x}_1)|^{1/2} |R_{jj}(\mathbf{x}_2, \mathbf{x}_2)|^{1/2}$ for $i, j = 1, \ldots, n$.
4. Let m be any integer, and consider a sequence of points $\mathbf{x}_1, \mathbf{x}_2, \ldots, \mathbf{x}_m$ in $\mathbb{R}^n$; then

$$\sum_{i,j=1}^{m} \sum_{k=1}^{n} (\mathbf{R}(\mathbf{x}_i, \mathbf{x}_j) \boldsymbol{\lambda}_i)_k (\boldsymbol{\lambda}_j)_k \geq 0 \quad \text{for any } \boldsymbol{\lambda}_1, \ldots, \boldsymbol{\lambda}_m \in \mathbb{R}^n,$$

where $(\boldsymbol{\lambda}_j)_k$ denotes the kth component of the element $\boldsymbol{\lambda}_j \in \mathbb{R}^n$.

Properties 1, 2, and 4 follow immediately from the definition of the correlation tensor, while property 3 is a consequence of the Cauchy–Schwarz inequality. Note that we deduce from property 3 that $R_{ij}(\mathbf{x}_1, \mathbf{x}_2)$, $i, j = 1, \ldots, n$, as defined above, exists if $\mathbf{u}$ is square-integrable as a function of θ, $\mathbf{u}(\mathbf{x}; \theta) \in [L^2(\Theta)]^n$.

Homogeneous Turbulence

We now introduce a concept of spatial stationarity for the random vector field $\mathbf{u}(\mathbf{x}, t; \theta)$, namely, we introduce the concept of a spatially homogeneous random vector.

Definition 2.9. *A random field* $\mathbf{u}(\mathbf{x}; \theta)$, $\theta \in \Theta$, *is said to be* spatially homogeneous *when all its moments and joint moments are invariant under any space* *translation of the set of points* $\{\mathbf{x}_i\}_{i=1,\dots,m}$ *in* $\mathbb{R}^n$*, that is,*

$$m^{\ell_1\dots\ell_m}_{j_1\dots j_m}(\mathbf{x}_1+\mathbf{r},\dots,\mathbf{x}_m+\mathbf{r}) = m^{\ell_1\dots\ell_m}_{j_1\dots j_m}(\mathbf{x}_1,\dots,\mathbf{x}_m) \tag{2.15}$$

for any separation vector $\mathbf{r}\in\mathbb{R}^n$, $1\le \ell_i\le n,\ i=1,\dots,m.$

A first consequence of the homogeneity property is that the correlation tensor $\mathbf{R}(\mathbf{x}_1,\mathbf{x}_2)$ depends only on the separation vector $\mathbf{r}=\mathbf{x}_2-\mathbf{x}_1$, that is,

$$\mathbf{R}(\mathbf{x}_1,\mathbf{x}_2) = \mathbf{R}(\mathbf{x}_1,\mathbf{x}_1+\mathbf{r}) = \mathbf{R}(\mathbf{r}).$$

Moreover, the mean $\langle\mathbf{u}(\mathbf{x})\rangle$ of a homogeneous random field is independent of the location $\mathbf{x}$. Therefore, the study of such a field can be restricted to flow fields with zero mean. Another immediate property of a homogeneous random field is that

$$R_{ij}(\mathbf{r}) = R_{ji}(-\mathbf{r}). \tag{2.16}$$

Moreover, property 3 stated above can be rewritten as

$$|R_{ij}(\mathbf{r})| \le |R_{ii}(\mathbf{0})|^{1/2}\,|R_{jj}(\mathbf{0})|^{1/2},$$

with the particular case

$$|R_{ii}(\mathbf{r})| \le R_{ii}(\mathbf{0}) = \left\langle |u_i(\mathbf{x})|^2\right\rangle.$$

From the definition of the correlation tensor, we also deduce that

$$\mathrm{tr}(\mathbf{R}(\mathbf{0})) = \left\langle |\mathbf{u}(\mathbf{x})|^2\right\rangle,$$

where $\mathrm{tr}(\cdot)$ denotes the trace operator and $|\cdot|$ denotes the Euclidean norm in $\mathbb{R}^n$, given by $|\mathbf{u}(\mathbf{x})|^2=\sum_{j=1}^n u_j^2(\mathbf{x})$.

Let us now assume that the random field $\mathbf{u}(\mathbf{x};\theta)$ remains incompressible, that is, it satisfies

$$\mathrm{div}\,\mathbf{u} = \sum_{i=1}^{n}\frac{\partial u_i}{\partial x_i} = 0,$$

and that

$$\frac{\partial\mathbf{u}}{\partial x_j}(\mathbf{x};\theta)\in[L^2(\Theta)]^n, \qquad j=1,\dots,n.$$

Therefore, we have

$$\left\langle u_i(\mathbf{x}) \sum_{j=1}^{n} \frac{\partial u_j}{\partial r_j}(\mathbf{x}+\mathbf{r}) \right\rangle = 0$$

for any location $\mathbf{x} \in \mathbb{R}^n$ and separation vector $\mathbf{r} \in \mathbb{R}^n$. This clearly shows that

$$\sum_{j=1}^{n} \frac{\partial R_{ij}}{\partial r_j}(\mathbf{r}) = 0.$$

Moreover, it follows from (2.16) that

$$\sum_{j=1}^{n} \frac{\partial R_{ij}}{\partial r_j}(\mathbf{r}) = \sum_{j=1}^{n} \frac{\partial R_{ji}}{\partial r_j}(\mathbf{r}) = 0. \tag{2.17}$$

When the turbulence is spatially homogeneous, an ergodic hypothesis is often invoked. This suggests that statistical and time averages are equivalent:

$$\langle \varphi(\mathbf{x}, t) \rangle = \lim_{T \to +\infty} \frac{1}{T} \int_0^T \varphi(\mathbf{x}, s)\, ds, \tag{2.18}$$

for any random flow field $\varphi(\mathbf{x}, t)$. However, at this time, the existence of an ergodic theorem for the three-dimensional Navier–Stokes equations has not yet been proved, although it is usually assumed in the conventional theory of turbulence. Let us mention nevertheless a recent result obtained by Bercovici et al. (1995), who showed that in the two-dimensional case ($n = 2$) time and ensemble averages defined in a suitable sense lead to the same mean quantities. This result was obtained without any ergodic hypothesis, but its proof is based on the time analyticity of the solutions of the two-dimensional Navier–Stokes equations.

In the rest of Section 2.2, we will only consider homogeneous turbulent flows.

Spectral Representation of the Velocity Correlation Tensor

Let us first consider a homogeneous random variable $\zeta(x)$, $x \in \mathbb{R}$, with a correlation function $R(r)$, $r \in \mathbb{R}$. As a consequence of the Bochner–Khinchin theorem for continuous nonnegative functions (see Khinchin (1949), Lee (1960), Gikhman and Skorokhod (1969), and the references therein) we have the following result:

Theorem 2.1. *For a function $R(r)$, $r \in \mathbb{R}$, to be the covariance function of a homogeneous random variable $\zeta(x)$ that satisfies*

$$\lim_{r \to 0} \left\langle |\zeta(x+r) - \zeta(x)|^2 \right\rangle = 0, \tag{2.19}$$

it is necessary and sufficient that it have a representation of the form

$$R(r) = \int_{\mathbb{R}} e^{\mathbf{i}kr}\, d\phi(k),$$

where $\mathbf{i}$ *stands for* $\sqrt{-1}$.

Note that the continuity of $R(r)$ follows from the hypothesis (2.19). The function $\phi(k)$ in the representation above is called the spectral function of the random variable. If $\phi(k)$ is absolutely continuous, the function F defined by

$$\phi(k) = \int_{-\infty}^{k} F(r)\, dr$$

is called the spectral density of ζ. Moreover, it can be shown that if the covariance function is absolutely integrable, that is,

$$\int_{\mathbb{R}} |R(r)|\, dr < +\infty,$$

then there exists a spectral density given by

$$F(k) = \frac{1}{2\pi} \int_{\mathbb{R}} e^{-\mathbf{i}kr} R(r)\, dr.$$

A similar result holds in the case of homogeneous random vectors $\mathbf{u}(\mathbf{x})$, $\mathbf{x} \in \mathbb{R}^n$, except that the covariance function is now replaced by a correlation tensor $\mathbf{R}(\mathbf{r})$, $\mathbf{r} \in \mathbb{R}^n$; namely, we have:

Theorem 2.2. *For a matrix $\mathbf{R}(\mathbf{r}) = (R_{ij}(\mathbf{r}))_{i,j=1,\dots,n}$ to be the correlation tensor of a homogeneous random vector $\mathbf{u}(\mathbf{x})$ that satisfies*

$$\lim_{|\mathbf{r}| \to 0} \left\langle |\mathbf{u}(\mathbf{x}+\mathbf{r}) - \mathbf{u}(\mathbf{x})|^2 \right\rangle = 0, \tag{2.20}$$

it is necessary and sufficient that it have a representation of the form

$$\mathbf{R}(\mathbf{r}) = \int_{\mathbb{R}^n} e^{\mathbf{i}\mathbf{k}\cdot\mathbf{r}}\, d\boldsymbol{\Phi}(\mathbf{k}), \tag{2.21}$$

where $\boldsymbol{\Phi}(\mathbf{k}) = (\phi_{ij}(\mathbf{k}))_{i,j=1,\dots,n}$ satisfies

- *the matrix* $d\boldsymbol{\Phi}(\mathbf{k}) = (d\phi_{ij}(\mathbf{k}))_{i,j}$ *is nonnegative for any* $\mathbf{k} \in \mathbb{R}^n$,
- $\mathrm{tr}(\boldsymbol{\Phi}(+\infty)) - \mathrm{tr}(\boldsymbol{\Phi}(-\infty)) < +\infty$.

In (2.21), $\mathbf{k} \cdot \mathbf{r}$ stands for the standard Euclidean scalar product in $\mathbb{R}^n$. In the case of homogeneous turbulence, it is generally assumed that the spectral matrix $\boldsymbol{\Phi}(\mathbf{k})$ is differentiable, that is,

$$d\phi_{ij}(\mathbf{k}) = F_{ij}(\mathbf{k})\, d\mathbf{k} \quad \text{for } i, j = 1, \ldots, n,$$

where the $F_{ij}(\mathbf{k})$ are complex spectral densities. In such a case, the spectral representation given by Theorem 2.2 writes

$$R_{ij}(\mathbf{r}) = \int_{\mathbf{R}^n} e^{\mathbf{i}\mathbf{k}\cdot\mathbf{r}}\, F_{ij}(\mathbf{k})\, d\mathbf{k}. \tag{2.22}$$

If $R_{ij}(\mathbf{r})$, $i, j = 1, \ldots, n$, is absolutely integrable, the following reverse formula holds:

$$F_{ij}(\mathbf{k}) = \frac{1}{(2\pi)^n} \int_{\mathbf{R}^n} e^{-\mathbf{i}\mathbf{k}\cdot\mathbf{r}} R_{ij}(\mathbf{r})\, d\mathbf{r}. \tag{2.23}$$

Recalling (2.16), we obtain

$$F_{ij}(\mathbf{k}) = F_{ji}(-\mathbf{k}) = F_{ij}^*(\mathbf{k}), \tag{2.24}$$

where * denotes the complex conjugate. The last relation follows from the fact that $R_{ij}(\mathbf{r})$ is a real tensor.

In the particular case $\mathbf{r} = \mathbf{0}$, the spectral representation (2.22) of the correlation tensor shows that

$$R_{ij}(\mathbf{0}) = \langle u_i(\mathbf{x}) u_j(\mathbf{x}) \rangle = \int_{\mathbf{R}^n} F_{ij}(\mathbf{k})\, d\mathbf{k},$$

so that when $i = j$

$$R_{ii}(\mathbf{0}) = \langle |u_i(\mathbf{x})|^2 \rangle = \int_{\mathbf{R}^n} F_{ii}(\mathbf{k})\, d\mathbf{k}. \tag{2.25}$$

Therefore, we have

$$\mathrm{tr}(\mathbf{R}(\mathbf{0})) = \langle |\mathbf{u}(\mathbf{x})|^2 \rangle = \int_{\mathbf{R}^n} \mathrm{tr}(\mathbf{F}(\mathbf{k}))\, d\mathbf{k}.$$

Hence, the spectral density $F_{ii}(\mathbf{k})$ describes a distribution of the energy associated with a velocity component in the wavenumber space. Then $F_{ij}(\mathbf{k})$ is called the energy spectrum tensor. From (2.17) and the spectral representation of $R_{ij}(\mathbf{r})$, we deduce the following spectral condition:

$$\sum_{j=1}^{n} k_j F_{ij}(\mathbf{k}) = \sum_{j=1}^{n} k_j F_{ji}(\mathbf{k}) = 0. \tag{2.26}$$

Hence, the incompressibility constraint induces conditions on the correlation tensor as well as on the energy spectrum tensor; we will see later on that in some particular cases, these conditions imply specific forms of the correlation and energy spectrum tensors.

Example 2.1. A particular case of velocity field in $\mathbb{R}^n$ is the case of periodic flows introduced in Chapter 1. Let us assume that the periods in all spatial directions are the same, say L_1, so that the study of the flow can be restricted to the domain $\Omega = (0, L_1)^n$. We assume that the velocity field has a zero spatial average. Instead of the statistical average as it is used above, we consider here the spatial average,[1] that is, we define the operator $\langle\cdot\rangle_\Omega$ by

$$\langle\varphi(\mathbf{x})\rangle_\Omega = \frac{1}{|\Omega|}\int_\Omega \varphi(\mathbf{x})\, d\mathbf{x} \quad \text{for any flow variable } \varphi,$$

where $|\Omega| = L_1^n$ here. Therefore, the two-point velocity correlation tensor depends only on the separation vector and is periodic of period L_1. According to the Cauchy–Schwarz inequality

$$|R_{ij}(\mathbf{r})| \le \left\langle |u_i|^2\right\rangle_\Omega^{1/2} \left\langle |u_j|^2\right\rangle_\Omega^{1/2} \qquad \forall \mathbf{r}\in\Omega, \quad i, j = 1, \dots, n.$$

Hence, $R_{ij}(\mathbf{r})$ defined by (2.14) exists and belongs to $L^2(\Omega)$ if $\mathbf{u}(t) \in [L^2(\Omega)]^n$ for $t > 0$. Weak solutions of (1.5)–(1.6) enjoying this property exist[2] in the case $n = 3$ if (see Chapter 1, Theorem 1.3)

$$\begin{aligned} \mathbf{u}_0 &\in \left[L^2_{\text{per}}(\Omega)\right]^3 \quad \text{and} \quad \nabla\cdot\mathbf{u}_0 = 0,\\ \mathbf{f}(t) &\in \left[L^2_{\text{per}}(\Omega)\right]^3 \qquad \forall t\in[0,T],\ T>0. \end{aligned} \tag{2.27}$$

Moreover, the conditions (2.27) insure that $(\partial u_i/\partial x_j)(t) \in L^2_{\text{per}}(\Omega)$ for $t \in [0, T]$ and $i, j = 1, 2, 3$. Hence, $(\partial R_{ij}/\partial r_j)(\mathbf{r}) = \langle u_i(\mathbf{x})(\partial u_j/\partial r_j)(\mathbf{x}+\mathbf{r})\rangle_\Omega$, $i, j = 1, 2, 3$, and belongs to $L^2_{\text{per}}(\Omega)$.

[1] In the case of fully periodic flows, an ergodic hypothesis is often invoked in order to replace statistical averages by spatial ones. For our purpose, such statement is not necessary. Indeed, our aim here is only to show that in the particular case of periodic flow all the definitions and properties previously derived hold when a spatial average operator is considered.

[2] We restrict ourselves here to the description of the three-dimensional case. The two-dimensional case can be considered as well, and most of the results presented below hold.

Since $R_{ij}(\mathbf{r})$ is periodic, it can be expanded in Fourier series:

$$R_{ij}(\mathbf{r}) = \sum_{\mathbf{k}\in\mathbb{Z}^n} \hat{R}_{ij}(\mathbf{k})\, e^{i\mathbf{k}_{L_1}\cdot\mathbf{r}} \quad \text{for } \mathbf{r}\in\Omega, \tag{2.28}$$

where $\mathbf{k}_{L_1} = 2\pi\mathbf{k}/L_1$. By replacing u_i and u_j by their Fourier series in (2.14), it follows that

$$\hat{R}_{ij}(\mathbf{k}) = \hat{u}_i^*(\mathbf{k})\hat{u}_j(\mathbf{k}) \quad \text{for } \mathbf{k}\in\mathbb{Z}^n. \tag{2.29}$$

Hence, in the case of periodic solutions of (1.5)–(1.6), the two-point velocity correlation tensor has a discrete spectrum, while in the general case of random processes the spectrum of R_{ij}, whenever it exists, is continuous. If $\nabla u_i \in [L^2_{\text{per}}(\Omega)]^3$, that is, if (2.27) holds, then $|\hat{u}_i(\mathbf{k})| < K/\,|\mathbf{k}|$ where K is a constant, so that, from (2.29), $|\hat{R}_{ij}(\mathbf{k})| < K^2/|\mathbf{k}|^2$ insuring that

$$\begin{aligned} &\mathbf{r} \longrightarrow R_{ij}(\mathbf{r}) \text{ is a continuous function on } \Omega,\\ &\frac{\partial^2 R_{ij}}{\partial r_l \partial r_k} \in L^2_{\text{per}}(\Omega), \qquad i,j,k,l = 1,2,3. \end{aligned} \tag{2.30}$$

Moreover, (2.28) and (2.29) imply that

$$\operatorname{tr}(\mathbf{R}(\mathbf{r})) = \sum_{\mathbf{k}\in\mathbb{Z}^n} |\hat{\mathbf{u}}(\mathbf{k})|^2\, e^{i\mathbf{k}_{L_1}\cdot\mathbf{r}} \quad \text{for } \mathbf{r}\in\Omega. \tag{2.31}$$

Note that (2.31) coincides with the Parseval identity when $\mathbf{r} = 0$,

$$\operatorname{tr}(\mathbf{R}(\mathbf{0})) = \left\langle |\mathbf{u}|^2 \right\rangle_\Omega = \sum_{\mathbf{k}\in\mathbb{Z}^n} |\hat{\mathbf{u}}(\mathbf{k})|^2. \tag{2.32}$$

The Two-Point Vorticity Correlation Tensor

As in Chapter 1, we denote by $\boldsymbol{\omega}(\mathbf{x},t)$ the vorticity vector defined by $\boldsymbol{\omega}(\mathbf{x},t) = \nabla\times\mathbf{u}(\mathbf{x},t)$ in the three-dimensional case. In space dimension 2, the vorticity has only one nonzero component, and it is orthogonal to the velocity; that component is a scalar function that we denote by $\omega(\mathbf{x},t)$. In what follows, unless it is otherwise specified, we will mainly describe the three-dimensional case.

Definition 2.10. *The two-point vorticity correlation tensor* $\boldsymbol{\Omega}(\mathbf{r}) = (\Omega_{ij}(\mathbf{r}))_{i,j=1,2,3}$ *or the scalar function* $\Omega(\mathbf{r})$, *if* $n = 2$, *of a homogeneous random field* $\mathbf{u}(\mathbf{x})$ *is defined by*

$$\begin{aligned} \Omega_{ij}(\mathbf{r}) &= \left\langle \omega_i(\mathbf{x})\omega_j(\mathbf{x}+\mathbf{r}) \right\rangle, \qquad \mathbf{r}\in\mathbb{R}^3, \quad i,j = 1,2,3,\\ \Omega(\mathbf{r}) &= \langle \omega(\mathbf{x})\omega(\mathbf{x}+\mathbf{r}) \rangle, \qquad \mathbf{r}\in\mathbb{R}^2. \end{aligned} \tag{2.33}$$

As was done for the two-point velocity correlation tensor, we deduce from the Cauchy–Schwarz inequality that

$$|\Omega_{ij}(\mathbf{r})| \leq \left\langle |\omega_i(\mathbf{x})|^2 \right\rangle^{1/2} \left\langle |\omega_j(\mathbf{x}+\mathbf{r})|^2 \right\rangle^{1/2}$$

so that $\boldsymbol{\Omega}(\mathbf{r})$ defined by (2.33) exists if, for $i = 1, \ldots, n$, $\omega_i(\mathbf{x};\theta) \in L^2(\Theta)$, namely if $(\partial \mathbf{u}/\partial x_i)(\mathbf{x};\theta) \in [L^2(\Theta)]^n$, $i = 1, \ldots, n$. Note also that the latter assumption insures that $(\partial^2 \mathbf{R}/\partial r_l \partial r_k)(\mathbf{r})$ exists at least in the distribution sense. Therefore, recalling the definition of the two-point velocity correlation tensor as well as the homogeneity property, the vorticity correlation tensor can be easily related to the velocity correlation:

$$\begin{aligned} \Omega_{ij}(\mathbf{r}) &= -\delta_{ij}\ \Delta \mathrm{tr}(\mathbf{R}(\mathbf{r})) + \frac{\partial^2}{\partial r_i \partial r_j}\mathrm{tr}(\mathbf{R}(\mathbf{r})) + \Delta R_{ji}(\mathbf{r}), \qquad n = 3, \\ \Omega(\mathbf{r}) &= -\Delta \mathrm{tr}(\mathbf{R}(\mathbf{r})), \qquad n = 2, \end{aligned} \tag{2.34}$$

so that

$$\mathrm{tr}(\boldsymbol{\Omega}(\mathbf{r})) = -\Delta \mathrm{tr}(\mathbf{R}(\mathbf{r})), \qquad n = 3.$$

Assuming that

$$\lim_{|\mathbf{r}|\to 0} \left\langle |\boldsymbol{\omega}(\mathbf{x}+\mathbf{r}) - \boldsymbol{\omega}(\mathbf{x})|^2 \right\rangle = 0,$$

we can apply Theorem 2.2 to the two-point vorticity correlation tensor leading to the following spectral representation:

$$\Omega_{ij}(\mathbf{r}) = \int_{\mathbf{R}^n} \Lambda_{ij}(\mathbf{k})\, e^{\mathrm{i}\mathbf{k}\cdot\mathbf{r}}\, d\mathbf{k}. \tag{2.35}$$

Moreover, if $\boldsymbol{\Omega}(\mathbf{r})$ is absolutely integrable, the reverse formula holds,

$$\Lambda_{ij}(\mathbf{k}) = \frac{1}{(2\pi)^n} \int_{\mathbf{R}^n} \Omega_{ij}(\mathbf{r})\, e^{-\mathrm{i}\mathbf{k}\cdot\mathbf{r}}\, d\mathbf{r}. \tag{2.36}$$

Similarly, in the case $n=2$ we define a spectrum function $\Lambda(\mathbf{k})$ as the spectral representation of the vorticity correlation function. From (2.34), the vorticity spectral tensor (and function) can be related to the velocity one, namely

$$\begin{aligned} \Lambda_{ij}(\mathbf{k}) &= \left(\delta_{ij} - \frac{k_i k_j}{k^2}\right) k^2\, \mathrm{tr}(\mathbf{F}(\mathbf{k})) - k^2\, F_{ji}(\mathbf{k}), \qquad n = 3, \\ \Lambda(\mathbf{k}) &= k^2\, \mathrm{tr}(\mathbf{F}(\mathbf{k})), \qquad n = 2, \end{aligned} \tag{2.37}$$

where $k = |\mathbf{k}|$. From (2.37), we have

$$\mathrm{tr}(\mathbf{\Lambda}(\mathbf{k})) = k^2\,\mathrm{tr}(\mathbf{F}(\mathbf{k})), \qquad n = 3. \tag{2.38}$$

Therefore, we obtain from (2.33), (2.34), (2.35), and (2.38) the following expression for the enstrophy:

$$\begin{aligned} \left\langle|\omega(\mathbf{x})|^2\right\rangle &= \mathrm{tr}(\mathbf{\Omega}(\mathbf{0})) = \int_{\mathbf{R}^3} k^2\,\mathrm{tr}(\mathbf{F}(\mathbf{k}))\,d\mathbf{k} = -\Delta\,\mathrm{tr}(\mathbf{R}(\mathbf{0})),\\ \left\langle|\omega(\mathbf{x})|^2\right\rangle &= \Omega(\mathbf{0}) = \int_{\mathbf{R}^2} k^2\,\mathrm{tr}(\mathbf{F}(\mathbf{k}))\,d\mathbf{k} = -\Delta\,\mathrm{tr}(\mathbf{R}(\mathbf{0})). \end{aligned} \tag{2.39}$$

Example 2.2. In the case of periodic flows considered in Example 2.1, the vorticity correlation tensor is defined as above with $\langle\cdot\rangle$ replaced by $\langle\cdot\rangle_\Omega$. Under the hypothesis (2.27), Ω_{ij} is defined and belongs to $L^2_{\mathrm{per}}(\Omega)$ for $t \in [0, T]$. The coefficients of the Fourier expansion of $\Omega_{ij}(\mathbf{r})$ can be easily obtained and are given by

$$\hat{\Omega}_{ij}(\mathbf{k}) = \left(\delta_{ij} - \frac{k_i\,k_j}{k^2}\right)\frac{4\pi^2}{L_1^2}\,k^2|\hat{\mathbf{u}}(\mathbf{k})|^2 - \frac{4\pi^2}{L_1^2}\,k^2\,\hat{u}_j^*(\mathbf{k})\,\hat{u}_i(\mathbf{k});$$

hence

$$\mathrm{tr}(\mathbf{\Omega}(\mathbf{0})) = \left\langle|\omega|^2\right\rangle = \sum_{\mathbf{k}\in\mathbb{Z}^n} \frac{4\pi^2}{L_1^2} k^2|\hat{\mathbf{u}}(\mathbf{k})|^2. \tag{2.40}$$

Energy and Enstrophy Spectrum Functions

By averaging the energy spectrum tensor $F_{ij}(\mathbf{k})$, $i, j, = 1, \ldots, n$, over the spheres S_k of radii $k = |\mathbf{k}|$ in $\mathbb{R}^n$, we define a one-dimensional spectral tensor $\mathbf{\Phi}$ by setting

$$\phi_{ij}(k) = \int_{S_k} F_{ij}(\mathbf{k})\,d\mathbf{k}. \tag{2.41}$$

We associate with this tensor an energy spectrum function defined as follows:

Definition 2.11. *The energy spectrum function $E(k)$ of a homogeneous velocity field $\mathbf{u}(\mathbf{x})$ is defined by*

$$E(k) = \frac{1}{2}\,\mathrm{tr}(\mathbf{\Phi}(k)) = \frac{1}{2}\int_{S_k} \mathrm{tr}(\mathbf{F}(\mathbf{k}))\,d\mathbf{k} \quad \textit{for} \quad k \in (0, +\infty). \tag{2.42}$$

By integrating $E(k)$ over the range $(0, +\infty)$ and recalling the spectral representation (2.22) of the velocity correlation tensor, we deduce that

$$\int_0^{+\infty} E(k)\, dk = \frac{1}{2}\int_{\mathbf{R}^n} \mathrm{tr}(\mathbf{F}(\mathbf{k}))\, d\mathbf{k} = \frac{1}{2}\mathrm{tr}(\mathbf{R}(\mathbf{0})) = \frac{1}{2}\left\langle |\mathbf{u}(\mathbf{x})|^2\right\rangle .$$

Hence, the total kinetic energy is

$$E = \frac{1}{2}\left\langle |\mathbf{u}(\mathbf{x})|^2\right\rangle = \int_0^{+\infty} E(k)\, dk. \tag{2.43}$$

The energy spectrum function then represents a density of the contributions to the total kinetic energy concentrated on wavenumbers $\mathbf{k}$ of modulus k. In the case of incompressible flows (see for instance Batchelor (1971)), the energy spectrum function $E(k)$ can be shown to behave like k^4 near the origin $k = 0$.

Similarly, we can define a one-dimensional spectrum function for the vorticity, namely

Definition 2.12. *The enstrophy spectrum function $\epsilon(k)$ of a homogeneous velocity field $\mathbf{u}(\mathbf{x})$ is defined by*

$$\epsilon(k) = \int_{S_k} \mathrm{tr}(\mathbf{\Lambda}(\mathbf{k}))\, d\mathbf{k} \qquad \textit{for } k \in (0, +\infty). \tag{2.44}$$

By integrating $\epsilon(k)$ over the range $(0, +\infty)$ and recalling (2.38) and (2.39) we find

$$\int_0^{+\infty} \epsilon(k)\, dk = \int_{\mathbf{R}^n} k^2\, \mathrm{tr}(\mathbf{F}(\mathbf{k}))\, d\mathbf{k} = \left\langle |\boldsymbol{\omega}(\mathbf{x})|^2\right\rangle .$$

Recalling (2.42), we also have

$$\int_{\mathbf{R}^n} k^2\, \mathrm{tr}(\mathbf{F}(\mathbf{k}))\, d\mathbf{k} = 2\int_0^{+\infty} k^2 E(k)\, dk,$$

so that the energy dissipation rate $\epsilon = \nu\langle|\boldsymbol{\omega}(\mathbf{x})|^2\rangle$ can be rewritten as

$$\epsilon = 2\nu \int_0^{+\infty} k^2\, E(k)\, d\mathbf{k}. \tag{2.45}$$

Example 2.3. In the case of periodic flows, as was previously mentioned, the correlation tensors have discrete spectrum so that the definitions given above have to be modified. We define the energy spectrum function by

$$E(k) = \frac{1}{2}\sum_{\mathbf{k}\in S_{k,1/2}} \mathrm{tr}(\hat{\mathbf{R}}(\mathbf{k})) = \frac{1}{2}\sum_{\mathbf{k}\in S_{k,1/2}} |\hat{\mathbf{u}}(\mathbf{k})|^2 \quad \text{for } k \in \mathbb{N}, \tag{2.46}$$

where $S_{k,1/2} = \{\mathbf{k} \in \mathbb{Z}^n / |\mathbf{k}| \in [k - \frac{1}{2}, k + \frac{1}{2})\}$, with the convention that $S_{0,1/2} = \{\mathbf{k} \in \mathbb{Z}^n / |\mathbf{k}| \in [0, \frac{1}{2})\}$.

From (2.46) and (2.32) we define the kinetic energy by

$$E = \sum_{k=0}^{+\infty} E(k) = \frac{1}{2} \sum_{\mathbf{k}\in\mathbb{Z}^n} |\hat{\mathbf{u}}(\mathbf{k})|^2 = \frac{1}{2} \left\langle |\mathbf{u}|^2 \right\rangle . \tag{2.47}$$

Similarly, we define the enstrophy spectrum function by

$$\epsilon(k) = \sum_{\mathbf{k}\in S_{k,1/2}} \operatorname{tr}(\hat{\mathbf{\Omega}}(\mathbf{k})) = \sum_{\mathbf{k}\in S_{k,1/2}} |\mathbf{k}_{L_1}|^2 |\hat{\mathbf{u}}(\mathbf{k})|^2 \quad \text{for } k \in \mathbb{N}. \tag{2.48}$$

Note that $\epsilon(k) \sim (8\pi^2/L_1^2)k^2\, E(k)$ for large values of k.

From (2.48) and (2.40) we deduce that

$$\sum_{k=0}^{+\infty} \epsilon(k) = \sum_{\mathbf{k}\in\mathbb{Z}^n} \frac{4\pi^2}{L_1^2} k^2 |\hat{\mathbf{u}}(\mathbf{k})|^2 = \left\langle |\omega|^2 \right\rangle . \tag{2.49}$$

Hence, the energy dissipation rate ϵ can be rewritten as

$$\epsilon = \nu \left\langle |\omega|^2 \right\rangle = \nu \sum_{\mathbf{k}\in\mathbb{Z}^n} \frac{4\pi^2}{L_1^2} k^2 |\hat{\mathbf{u}}(\mathbf{k})|^2,$$

and is related to the energy spectrum as follows:

$$2\nu \sum_{k=0}^{+\infty} \frac{4\pi^2}{L_1^2} \left(k - \frac{1}{2}\right)^2 E(k) \le \epsilon < 2\nu \sum_{k=0}^{+\infty} \frac{4\pi^2}{L_1^2} \left(k + \frac{1}{2}\right)^2 E(k).$$

2.2.2. Homogeneous Isotropic Turbulence

We now introduce a particular class of homogeneous random fields, namely the isotropic ones. Following Monin and Yaglom (1975), we define isotropy as follows:

Definition 2.13. *The vector field* $\mathbf{u}(\mathbf{x})$ *is said to be* isotropic *if all mean quantities are unaffected by any simultaneous rotation and reflection, as well as any translation of the coordinate system.*

An immediate consequence of this property is that $\langle \mathbf{u}(\mathbf{x}) \rangle = \mathbf{0}$; indeed, as the turbulence is assumed here to be homogeneous, the mean value $\langle \mathbf{u}(\mathbf{x}) \rangle$ is a constant vector, and as it must be invariant under rotations, it must be a zero vector.

Reduced Forms of the Correlation and Spectrum Tensors

In the case under consideration of homogeneous isotropic flows, the two-point velocity correlation tensor and so the corresponding spectrum tensor take a very simple form. In order to investigate this property further, it is convenient to define a special coordinate system so that the origin is translated with respect to the location vector $\mathbf{x}$ and the first axis lies along the separation vector $\mathbf{r}$; namely, we introduce the following reference frame:

$$\tilde{\mathbf{e}}_1 = \frac{\mathbf{r}}{r}, \qquad \tilde{\mathbf{e}}_2 = \frac{\mathbf{r} \times \mathbf{e}}{|\mathbf{r} \times \mathbf{e}|}, \qquad \tilde{\mathbf{e}}_3 = \frac{\mathbf{r} \times \tilde{\mathbf{e}}_2}{|\mathbf{r} \times \tilde{\mathbf{e}}_2|},$$

where r is the magnitude of the vector $\mathbf{r}$ and $\mathbf{e}$ is any unit vector in $\mathbb{R}^3$ orthogonal to $\mathbf{r}$. We then denote by $\tilde{R}_{ij}(\mathbf{r})$ the two-point velocity correlation tensor in the new coordinate system. It can be easily shown that in this new set of coordinates (see Monin and Yaglom (1975)) the tensor $\tilde{R}_{ij}(\mathbf{r})$ is symmetric with only two different and nonzero components, namely,

$$\tilde{R}_{11}(\mathbf{r}) = \langle [\mathbf{u}(\mathbf{x}) \cdot \tilde{\mathbf{e}}_1][\mathbf{u}(\mathbf{x}+\mathbf{r}) \cdot \tilde{\mathbf{e}}_1] \rangle,$$
$$\tilde{R}_{22}(\mathbf{r}) = \tilde{R}_{33}(\mathbf{r}) = \langle [\mathbf{u}(\mathbf{x}) \cdot \tilde{\mathbf{e}}_2][\mathbf{u}(\mathbf{x}+\mathbf{r}) \cdot \tilde{\mathbf{e}}_2] \rangle,$$

and all the nondiagonal elements $\tilde{R}_{ij}(\mathbf{r})$ with $i \neq j$ are equal to zero. Moreover, as $\tilde{R}_{ij}(\mathbf{r})$ must be invariant under any rotation of the coordinate system, it depends only on the magnitude of the separation vector:

$$\tilde{R}_{ij}(\mathbf{r}) = \tilde{R}_{ij}(r) \quad \text{for any } \mathbf{r} \text{ such that } |\mathbf{r}| = r.$$

We now introduce $u_L(\mathbf{x}) = \mathbf{u}(\mathbf{x}) \cdot \tilde{\mathbf{e}}_1$, the component of $\mathbf{u}(\mathbf{x})$ on the first axis, which then corresponds to a projection of $\mathbf{u}(\mathbf{x})$ onto $\mathbf{r}$, and $u_N(\mathbf{x}) = \mathbf{u}(\mathbf{x}) \cdot \tilde{\mathbf{e}}_2$. These components are respectively called the longitudinal and lateral velocity components. Hence, the correlation tensor $\tilde{R}_{ij}(r)$ takes a very simple form and depends only on two scalar functions

$$R_{LL}(r) = \langle u_L(\mathbf{x}) u_L(\mathbf{x}+\mathbf{r}) \rangle, \tag{2.50}$$

called the longitudinal correlation function, and

$$R_{NN}(r) = \langle u_N(\mathbf{x}) u_N(\mathbf{x}+\mathbf{r}) \rangle, \tag{2.51}$$

called the lateral correlation function.

By transforming $\tilde{R}_{ij}(r)$ back to the initial (Euclidean) coordinate system we obtain the expression for $R_{ij}(\mathbf{r})$ in terms of $R_{LL}(r)$ and $R_{NN}(r)$; indeed,

$$R_{ij}(\mathbf{r}) = [R_{LL}(r) - R_{NN}(r)]\frac{r_i r_j}{r^2} + \delta_{ij} R_{NN}(r). \tag{2.52}$$

Again, this particular form for the correlation tensor is a consequence of the homogeneity and isotropy properties of the random vector field $\mathbf{u}(\mathbf{x}, t)$.

By analogy, a similar procedure can be applied to the energy spectrum tensor showing that this tensor depends only on two scalar functions of $k = |\mathbf{k}|$, and that it can be written as

$$F_{ij}(\mathbf{k}) = [F_{LL}(k) - F_{NN}(k)]\frac{k_i k_j}{k^2} + \delta_{ij} F_{NN}(k), \tag{2.53}$$

where $F_{LL}(k)$ ($F_{NN}(k)$) is the longitudinal (lateral) spectrum function.

Let us now describe the effect of the incompressibility condition reflected by (2.17) and (2.26) on the correlation tensor and respectively on its spectral representation. From (2.52), we deduce that

$$\sum_{i=1}^{n} \frac{\partial R_{ij}(\mathbf{r})}{\partial r_i} = \frac{r_j}{r}\left[\frac{\partial R_{LL}}{\partial r}(r) + \frac{2}{r}[R_{LL}(r) - R_{NN}(r)]\right],$$

so that (2.17) implies

$$R_{LL}(r) - R_{NN}(r) + \frac{r}{2}\frac{\partial R_{LL}}{\partial r}(r) = 0. \tag{2.54}$$

Equations (2.52) and (2.54) show that the knowledge of the correlation tensor $R_{ij}(\mathbf{r})$ is reduced to the knowledge of the scalar function $R_{LL}(r)$. Going now to the energy spectrum tensor, we can show that (2.26) and (2.53) imply that

$$F_{LL}(k) = 0.$$

The spectral tensor then reads

$$F_{ij}(\mathbf{k}) = F_{NN}(k)\left(\delta_{ij} - \frac{k_i k_j}{k^2}\right) \tag{2.55}$$

so that it is fully defined up to a scalar function of k. Now, from Definition 2.11 and the relation (2.55) above, we deduce that

$$E(k) = \int_{S_k} F_{NN}(k)\, d\mathbf{k} = 4\pi k^2 F_{NN}(k), \tag{2.56}$$

which together with (2.55) shows that the energy spectrum tensor is fully determined by the energy spectrum function, namely

$$F_{ij}(\mathbf{k}) = \left(\delta_{ij} - \frac{k_i k_j}{k^2}\right)\frac{E(k)}{4\pi k^2}. \tag{2.57}$$

Remark 2.1. In the case of fully periodic flows considered in the previous examples, the same results hold for the correlation tensor. □

Relations Between the Correlation and Spectrum Functions

We now derive relations between the energy spectrum function and the one-dimensional functions $R_{LL}(r)$ and $R_{NN}(r)$. From (2.52), we deduce that

$$\operatorname{tr}(\mathbf{R}(\mathbf{r})) = 2R_{NN}(r) + R_{LL}(r), \tag{2.58}$$

which in view of (2.54) can be expressed only in terms of the lateral correlation function, that is,

$$\operatorname{tr}(\mathbf{R}(\mathbf{r})) = 3R_{LL}(r) + r\frac{\partial R_{LL}}{\partial r}(r). \tag{2.59}$$

From (2.57) and (2.23), we have

$$\frac{2E(k)}{4\pi k^2} = \operatorname{tr}(\mathbf{F}(\mathbf{k})) = \frac{1}{(2\pi)^n}\int_{\mathbf{R}^n} \operatorname{tr}(\mathbf{R}(\mathbf{r}))\, e^{-i\mathbf{k}\cdot\mathbf{r}}\, d\mathbf{r}, \tag{2.60}$$

so that, with the help of (2.59), this leads to

$$E(k) = \frac{k^2}{4\pi^2}\int_{\mathbf{R}^n} \left(3R_{LL}(r) + r\frac{\partial R_{LL}}{\partial r}(r)\right) e^{-i\mathbf{k}\cdot\mathbf{r}}\, d\mathbf{r}.$$

Under the natural assumption that $r^2 R_{LL}(r)$ tends to zero at infinity (i.e. $r \to +\infty$) and after a transformation of the elements of integration into spherical coordinates, we easily show that in the case $n = 3$,

$$E(k) = \frac{1}{\pi}\int_0^{+\infty} k^2 r^2 R_{LL}(r)\left(\frac{\sin kr}{kr} - \cos kr\right) dr. \tag{2.61}$$

Using (2.60) and recalling the spectral representation of the velocity correlation tensor, we can express the trace of $\mathbf{R}(\mathbf{r})$ in terms of $E(k)$, namely

$$\operatorname{tr}(\mathbf{R}(\mathbf{r})) = \int_{\mathbf{R}^3} \operatorname{tr}(\mathbf{F}(\mathbf{k}))\, e^{i\mathbf{k}\cdot\mathbf{r}}\, d\mathbf{k} = \int_{\mathbf{R}^3} \frac{E(k)}{2\pi k^2}\, e^{i\mathbf{k}\cdot\mathbf{r}}\, d\mathbf{k}.$$

Transforming the plane coordinates $\mathbf{k} = (k_1, k_2, k_3)$ into spherical coordinates, the last integral can be shown to be equal to

$$\operatorname{tr}(\mathbf{R}(\mathbf{r})) = 2\int_0^{+\infty} E(k)\,\frac{\sin kr}{kr}\, dk. \tag{2.62}$$

Similarly, by replacing (2.60) into (2.22), we can relate the one-dimensional correlation functions to the energy spectrum function

$$R_{LL}(r) = 2\int_0^{+\infty} \frac{E(k)}{kr}\left(\frac{\sin kr}{kr} - \cos kr\right) dk, \tag{2.63}$$

and

$$R_{NN}(r) = \int_0^{+\infty} \frac{E(k)}{kr}\left(\sin kr + \frac{\cos kr}{kr} - \frac{\sin kr}{(kr)^3}\right) dk. \tag{2.64}$$

At this point, we can easily evaluate the first moment of the longitudinal and lateral one-dimensional correlation functions. Indeed, from the relations (2.54), (2.59), and (2.62), it can be found that the first integral moments of $R_{LL}(r)$ and $R_{NN}(r)$ are equal to

$$\begin{aligned} \int_0^{+\infty} R_{NN}(r)\, dr &= \frac{1}{2}\int_0^{+\infty} R_{LL}(r)\, dr, \\ \int_0^{+\infty} R_{LL}(r)\, dr &= \frac{\pi}{2}\int_0^{+\infty} k^{-1} E(k)\, dk. \end{aligned} \tag{2.65}$$

Now, by multiplying (2.63) by r and integrating over the range $(0, +\infty)$, we obtain

$$\int_0^{+\infty} r R_{LL}(r)\, dr = 2\int_0^{+\infty} k^{-2} E(k)\, dk, \tag{2.66}$$

and we deduce from (2.54) that

$$\int_0^{+\infty} r R_{NN}(r)\, dr = 0. \tag{2.67}$$

Finally, by multiplying (2.61) by πk^{-3} and integrating over the range $(0, +\infty)$, we find

$$\int_0^{+\infty} r^2 R_{LL}(r)\, dr = \pi\int_0^{+\infty} k^{-3} E(k)\, dk. \tag{2.68}$$

Remark 2.2. Note that for the periodic flows previously considered in Examples 2.1–2.3, the energy spectrum function has a discrete spectrum, so that most of the results derived above do not hold. Indeed, integration rules such as integration by parts and change of coordinate were used to derive the relations (2.61)–(2.68). They cannot be directly applied to discrete sums. □

Behavior of the Velocity Correlation Functions Near $r = 0$: The Taylor Microscale

We now describe the behavior of the correlation function $R_{LL}(r)$ in the neighborhood of the origin $r = 0$, which leads to the important concept of Taylor microscale. Note that, by virtue of the Cauchy–Schwarz inequality, the origin is the point where $R_{LL}(r)$ reaches its maximum value. By replacing the trigonometric functions in (2.63) by their polynomial expansion, we obtain

$$R_{LL}(r) = \frac{2}{3}\int_0^{+\infty} E(k)\,dk - \frac{r^2}{15}\int_0^{+\infty} k^2 E(k)\,dk + O(r^4), \tag{2.69}$$

so that by assuming enough regularity for the longitudinal correlation function, we deduce by analogy with the Taylor–Young expansion of $R_{LL}(r)$ that

$$\begin{aligned} R_{LL}(0) &= \frac{2}{3}\int_0^{+\infty} E(k)\,dk = \frac{1}{3}\left\langle |\mathbf{u}(\mathbf{x})|^2 \right\rangle, \\ \frac{\partial R_{LL}}{\partial r}(0) &= 0, \\ \frac{\partial^2 R_{LL}}{\partial r^2}(0) &= -\frac{2}{15}\int_0^{+\infty} k^2 E(k)\,dk. \end{aligned} \tag{2.70}$$

The first length scale used to characterize a turbulent field is the Taylor microscale λ and it is defined by

$$\lambda = \left(-\frac{R_{LL}(0)}{\frac{\partial^2 R_{LL}}{\partial r^2}(0)} \right)^{1/2} = \sqrt{5}\left(\frac{\int_0^{+\infty} E(k)\,dk}{\int_0^{+\infty} k^2 E(k)\,dk} \right)^{1/2}. \tag{2.71}$$

The right-hand side of (2.71) can be related to the vorticity spectrum tensor. Indeed, from (2.60) and (2.39) we have

$$\left\langle |\boldsymbol{\omega}(\mathbf{x})|^2 \right\rangle = 2\int_0^{+\infty} k^2 E(k)\,dk, \tag{2.72}$$

so that from (2.43), we finally deduce

$$\lambda^2 = 5\,\frac{\left\langle |\mathbf{u}(\mathbf{x})|^2 \right\rangle}{\left\langle |\boldsymbol{\omega}(\mathbf{x})|^2 \right\rangle}, \tag{2.73}$$

which, according to (2.70), is equivalent to

$$\left\langle |\boldsymbol{\omega}(\mathbf{x})|^2 \right\rangle = 5\,\frac{\left\langle |\mathbf{u}(\mathbf{x})|^2 \right\rangle}{\lambda^2} = 15\,\frac{R_{LL}(0)}{\lambda^2}.$$

Therefore, the Taylor expansion of $R_{LL}(r)$ can be rewritten as

$$R_{LL}(r) = R_{LL}(0)\left(1 - \frac{r^2}{2\lambda^2}\right) + O(r^4). \tag{2.74}$$

Using (2.54), we deduce the Taylor expansion of the lateral one-dimensional correlation function

$$R_{NN}(r) = R_{NN}(0)\left(1 - \frac{r^2}{\lambda^2}\right) + O(r^4).$$

We now introduce a characteristic velocity u defined by

$$u^2 = \tfrac{1}{3}\langle|\mathbf{u}(\mathbf{x})|^2\rangle = \tfrac{2}{3}E, \tag{2.75}$$

which is independent of the location $\mathbf{x}$ due to the homogeneity property of the flow. Together with the characteristic length scale λ introduced above, this allows us to define a Reynolds number by setting

$$\mathrm{Re}_\lambda = \frac{u\lambda}{\nu}. \tag{2.76}$$

This Reynolds number, called the Taylor microscale Reynolds number, is often used in the literature to characterize a homogeneous turbulent flow.

Example 2.4. As was mentioned in Remark 2.2, the results (2.61)–(2.68) cannot be used in this form in the case of fully periodic flows. However, the correlation functions can be expanded as above near the origin $r = 0$. We have previously seen that under the hypothesis (2.27), the velocity correlation tensor is continuous with respect to $\mathbf{r}$. Assuming more regularity on the data insures that R_{ij} belongs to $C^2(\Omega)$. More precisely, according to Theorem 1.4, if

$$\begin{aligned} &\mathbf{u}_0 \in \left(L^2_{\mathrm{per}}(\Omega)\right)^3 \quad \text{and} \quad \nabla\cdot\mathbf{u}_0 = 0,\\ &\frac{\partial u_i}{\partial x_j}(t=0) \in L^2_{\mathrm{per}}(\Omega) \quad \text{for} \quad i, j = 1, 2, 3,\\ &\mathbf{f}(t) \in \left(L^2_{\mathrm{per}}(\Omega)\right)^3 \qquad \forall t \in [0, T],\ T > 0, \end{aligned} \tag{2.77}$$

then there exists $T_* < T$ and on $(0, T_*)$ there exists a weak solution $\mathbf{u}$ of (1.5)–(1.6) such that

$$u_i, \frac{\partial u_i}{\partial t}, \frac{\partial u_i}{\partial x_j} \quad \text{and} \quad \frac{\partial^2 u_i}{\partial x_j \partial x_k} \in L^2_{\mathrm{per}}(\Omega), \qquad i, j, k = 1, 2, 3.$$

In such a case, it can be shown that $|u_i(\mathbf{k})| < K/|\mathbf{k}|^2$, so that $|\hat{R}_{ij}(\mathbf{k})| < K^2/|\mathbf{k}|^4$, insuring that $R_{ij} \in C^2(\mathbb{R}^n)$ for $i, j = 1, 2, 3$. From (2.52), we have

$$R_{LL}(r) = R_{11}(r, 0, 0) \quad \text{and} \quad R_{NN}(r) = R_{22}(r, 0, 0),$$

so that the two scalar functions R_{LL} and R_{NN} belong to $C^2([0, L_1/2])$. We can then use Taylor–Young expansions of these functions in order to study their behavior near the origin $r = 0$. First, we note that since R_{LL} is symmetric and it reaches its maximum value at the origin, then $(\partial R_{LL}/\partial r)(0) = 0$. Hence, the Taylor expansions for R_{LL} and $\partial R_{LL}/\partial r$ read

$$\begin{aligned} R_{LL}(r) &= R_{LL}(0) + \frac{r^2}{2}\frac{\partial^2 R_{LL}}{\partial r^2}(0) + o(r^2), \\ \frac{\partial R_{LL}}{\partial r}(r) &= r\frac{\partial^2 R_{LL}}{\partial r^2}(0) + o(r), \\ R_{NN}(r) &= R_{LL}(0) + r^2\frac{\partial^2 R_{LL}}{\partial r^2}(0) + o(r^2). \end{aligned} \tag{2.78}$$

Therefore, from (2.52), we can derive a polynomial expansion of $\mathrm{tr}(\mathbf{R}(\mathbf{r}))$ near the origin, namely

$$\mathrm{tr}(\mathbf{R}(\mathbf{r})) = 3R_{LL}(0) + \frac{5r^2}{2}\frac{\partial^2 R_{LL}}{\partial r^2}(0) + o(r^2). \tag{2.79}$$

By analogy, we deduce that

$$\begin{aligned} \mathrm{tr}(\mathbf{R}(\mathbf{0})) &= 3R_{LL}(0) = 3R_{NN}(0), \\ \mathrm{tr}\left(\frac{\partial^2 \mathbf{R}}{\partial r^2}(\mathbf{0})\right) &= 5\frac{\partial^2 R_{LL}}{\partial r^2}(0). \end{aligned} \tag{2.80}$$

Recalling that $\mathrm{tr}(\mathbf{R})$ depends only on r, we have

$$\Delta \mathrm{tr}(\mathbf{R}(\mathbf{r})) = \mathrm{tr}\left(\frac{\partial^2 \mathbf{R}}{\partial r^2}(r)\right) + \frac{2}{r}\mathrm{tr}\left(\frac{\partial \mathbf{R}}{\partial r}(r)\right),$$

which is equivalent, due to (2.52) and (2.54), to

$$\Delta \mathrm{tr}(\mathbf{R}(\mathbf{r})) = \mathrm{tr}\left(\frac{\partial^2 \mathbf{R}}{\partial r^2}(r)\right) + \frac{8}{r}\frac{\partial R_{LL}}{\partial r}(r) + 2\frac{\partial^2 R_{LL}}{\partial r^2}(r).$$

As R_{LL} belongs to $C^2([0, L_1/2])$, we have

$$\lim_{r\to 0}\frac{1}{r}\frac{\partial R_{LL}}{\partial r} = \frac{\partial^2 R_{LL}}{\partial r^2}(0).$$

Hence, with the help of (2.80) we obtain

$$\Delta \mathrm{tr}(\mathbf{R}(\mathbf{0})) = 15\, \frac{\partial^2 R_{LL}}{\partial r^2}(0).$$

Recalling (2.31), (2.32), (2.40), (2.80) and defining u and λ as before, the expansions (2.78) become

$$\begin{aligned} R_{LL}(r) &= u^2\left(1 - \frac{r^2}{2\lambda^2}\right) + o(r^2), \\ R_{NN}(r) &= u^2\left(1 - \frac{r^2}{\lambda^2}\right) + o(r^2). \end{aligned} \tag{2.81}$$

Recalling (2.47) and (2.49), the Taylor microscale can be related to the energy and enstrophy spectrum functions as follows:

$$\lambda^2 = 10\nu\, \frac{E}{\epsilon}.$$

Note that the hypothesis (2.77) was necessary to derive (2.81), leading to the definition of λ, while only the hypothesis (2.27) is necessary for λ as defined by (2.73) to exist; λ can be defined even for weak solutions.

One-Dimensional Longitudinal and Lateral Spectrum Functions

Finally, we conclude this section by defining one-dimensional spectrum functions which are commonly used by experimentalists. In experiments, the velocity components are often measured along an axis, say $x_2 = x_3 = 0$. The correlation of the longitudinal (lateral) component u_1 (u_2) on this straight line is denoted $B_{LL}(r)$ ($B_{NN}(r)$). One-dimensional spectral transforms of $B_{LL}(r)$ and $B_{NN}(r)$ then lead to the definition of the longitudinal one-dimensional spectrum

$$\begin{aligned} F_1(k_1) &= \frac{1}{2\pi}\int_{-\infty}^{+\infty} e^{-ik_1 r} B_{LL}(r)\, dr \\ &= \frac{1}{\pi}\int_0^{+\infty} \cos(k_1 r)\, B_{LL}(r)\, dr, \end{aligned} \tag{2.82}$$

and of the lateral one-dimensional spectrum

$$\begin{aligned} F_2(k_1) &= \frac{1}{2\pi}\int_{-\infty}^{+\infty} e^{-ik_1 r} B_{NN}(r)\, dr \\ &= \frac{1}{\pi}\int_0^{+\infty} \sin(k_1 r)\, B_{NN}(r)\, dr. \end{aligned} \tag{2.83}$$

From the definition of the energy spectrum tensor, we also have

$$F_1(k_1) = \int_{-\infty}^{+\infty} \int_{-\infty}^{+\infty} F_{11}(k_1, k_2, k_3)\, dk_2\, dk_3 \tag{2.84}$$

and

$$F_2(k_1) = \int_{-\infty}^{+\infty} \int_{-\infty}^{+\infty} F_{22}(k_1, k_2, k_3)\, dk_2\, dk_3. \tag{2.85}$$

By transforming the integral in the right-hand side of (2.84) into polar coordinates in the (k_2, k_3) plan, we find that

$$F_1(k_1) = 2\pi \int_{k_1}^{+\infty} \left(1 - \frac{k_1^2}{k^2}\right) k F_{NN}(k)\, dk. \tag{2.86}$$

Similarly, the lateral one-dimensional spectrum function can be related to the lateral spectrum function with the integral equation

$$F_2(k_1) = \pi \int_{k_1}^{+\infty} \left(1 + \frac{k_1^2}{k^2}\right) k F_{NN}(k)\, dk. \tag{2.87}$$

Note that the one-dimensional spectrum functions satisfy

$$F_2(k_1) = \frac{1}{2}\left(F_1(k_1) - k_1 \frac{dF_1(k_1)}{dk_1}\right). \tag{2.88}$$

Jiménez et al. (1993) used this last relation to verify the isotropy of their direct simulations. Indeed, they computed the ratio

$$\frac{F_1(k_1) - k_1 \frac{dF_1(k_1)}{dk_1}}{2F_2(k_1)},$$

which should be equal or close to one for all k_1 when the flow is isotropic. They found this ratio close to one at high wavenumbers for a sufficiently high Reynolds number, in which case the energy-containing eddies and the Kolmogorov scales are well separated.

2.2.3. Dynamical Equations for the Correlation Tensor and the Energy Spectrum Function

The Two-Point Velocity Correlation Tensor

Let us consider the equation of motion (1.6) of the velocity component u_i (u_j) at the point $\mathbf{x}$ ($\mathbf{x}' = \mathbf{x} + \mathbf{r}$), namely

$$\frac{\partial u_i}{\partial t} - \nu\,\Delta u_i + \sum_{k=1}^{n} u_k \frac{\partial u_i}{\partial x_k} + \frac{\partial p}{\partial x_i} = 0,$$
$$\frac{\partial u'_j}{\partial t} - \nu\,\Delta u'_j + \sum_{k=1}^{n} u'_k \frac{\partial u'_j}{\partial x'_k} + \frac{\partial p'}{\partial x'_j} = 0,$$

where, for the sake of simplicity, we have denoted $u_k = u_k(\mathbf{x})$ and $u'_k = u_k(\mathbf{x}')$ for any $k = 1, \dots, n$. We then multiply the first equation by u'_j and the second one by u_i, and we apply the statistical average $\langle\cdot\rangle$ to the resulting equation. Using the homogeneity property and the incompressibility of the flow, we derive the following evolution equation for the velocity correlation tensor:

$$\frac{\partial R_{ij}}{\partial t}(\mathbf{r}) - 2\nu\,\Delta R_{ij}(\mathbf{r}) \;+\; T_{ij}(\mathbf{r}) \;+\; P_{ij}(\mathbf{r}) = 0; \tag{2.89}$$

here we have set

$$T_{ij}(\mathbf{r}) = \sum_{k=1}^{n} \frac{\partial}{\partial r_k} \left\langle u_i u'_k u'_j - u_i u_k u'_j \right\rangle,$$
$$P_{ij}(\mathbf{r}) = \frac{\partial}{\partial r_j} \langle p' u_i \rangle - \frac{\partial}{\partial r_i} \left\langle p u'_j \right\rangle.$$

Let us assume that both $\mathbf{T}(\mathbf{r})$ and $\mathbf{P}(\mathbf{r})$ possess a spectral representation as given by Theorem 2.2, namely

$$T_{ij}(\mathbf{r}) = \int_{\mathbf{R}^n} \Gamma_{ij}(\mathbf{k})\, e^{\mathbf{ik}\cdot\mathbf{r}}\, d\mathbf{k},$$
$$P_{ij}(\mathbf{r}) = \int_{\mathbf{R}^n} \Pi_{ij}(\mathbf{k})\, e^{\mathbf{ik}\cdot\mathbf{r}}\, d\mathbf{k}. \tag{2.90}$$

From (2.89) and the spectral representations above, we deduce an evolution equation for the spectral tensor:

$$\frac{\partial F_{ij}}{\partial t}(\mathbf{k}) + 2\nu k^2\, F_{ij}(\mathbf{k}) + \Gamma_{ij}(\mathbf{k}) + \Pi_{ij}(\mathbf{k}) = 0, \tag{2.91}$$

for all $\mathbf{k} \in \mathbb{R}^n$.

The Energy Equation

It can be easily shown that the tensors $\mathbf{T}(\mathbf{r})$ and $\mathbf{P}(\mathbf{r})$ satisfy the following properties

$$\begin{aligned} \mathrm{tr}(\mathbf{P}(\mathbf{r})) &= 0, \\ \mathrm{tr}(\mathbf{\Pi}(\mathbf{k})) &= 0, \\ T_{ij}(\mathbf{0}) &= \int_{\mathbf{R}^n} \Gamma_{ij}(\mathbf{k})\, d\mathbf{k} = 0, \qquad i, j = 1, \ldots, n. \end{aligned}$$

Therefore, applying the trace operator to (2.89) and (2.91) leads to

$$\begin{aligned} &\frac{\partial}{\partial t}\mathrm{tr}(\mathbf{R}(\mathbf{r})) - 2\nu\, \Delta \mathrm{tr}(\mathbf{R}(\mathbf{r})) + \mathrm{tr}(\mathbf{T}(\mathbf{r})) = 0, \\ &\frac{\partial}{\partial t}\mathrm{tr}(\mathbf{F}(\mathbf{k})) + 2\nu k^2\, \mathrm{tr}(\mathbf{F}(\mathbf{k})) + \mathrm{tr}(\mathbf{\Gamma}(\mathbf{k})) = 0. \end{aligned} \tag{2.92}$$

Recalling Definition 2.11 for the energy spectrum function $E(k)$, we derive from (2.92) the following evolution equation

$$\frac{\partial E(k)}{\partial t} + 2\nu k^2 E(k) + T(k) = 0, \tag{2.93}$$

where $T(k) = \int_{S_k} \mathrm{tr}(\mathbf{\Gamma}(\mathbf{k}))\, d\mathbf{k}$. Equation (2.93) shows that the small-scale components lose more energy by viscous dissipation than the large-scale ones. The effect of the inertial forces is to spread energy from the energy containing eddies through the small scales of turbulence, where the energy is lost by viscous dissipation. The process of energy transfer between scales takes an important role in turbulence theories. Many models are based on some assumptions about this process. The properties and behaviors of the transfer terms, as for instance the degrees of locality in these terms and their dependence on disparity of the interacting scales, are subject of recent studies (see Zhou (1993a, 1993b), Zhou, Yeung, and Brasseur (1994), Dubois, Jauberteau, and Zhou (1997), and the references therein). Hence, the nonlinear term together with the pressure tends to conserve the kinetic energy while the viscous term tends to dissipate it.

Recalling the properties of the tensor $\mathbf{T}(\mathbf{r})$ listed above, we find

$$\int_0^{+\infty} T(k)\, dk = \int_{\mathbf{R}^n} \mathrm{tr}(\mathbf{\Gamma}(\mathbf{k}))\, d\mathbf{k} = \mathrm{tr}(\mathbf{T}(\mathbf{0})) = 0.$$

Therefore, by integrating (2.93) over the interval $(0, +\infty)$ and recalling (2.43) we obtain

$$\frac{d}{dt}\left(\int_0^{+\infty} E(k)\, dk\right) + 2\nu \int_0^{+\infty} k^2 E(k)\, dk = 0, \tag{2.94}$$

which can be equivalently rewritten as

$$\frac{1}{2}\frac{d}{dt}\left\langle|\mathbf{u}(\mathbf{x})|^2\right\rangle+\nu\left\langle|\boldsymbol{\omega}(\mathbf{x})|^2\right\rangle=0. \tag{2.95}$$

Note that in the case of forced turbulence, a term should be added to the right-hand side of (2.95), namely the rate of injection of energy $\langle\mathbf{f}\cdot\mathbf{u}\rangle$ where $\mathbf{f}$ denotes an external force.

As we have previously seen, the shape of the one-dimensional correlation functions near the origin is determined by the Taylor microscale. Moreover, λ can be related to the rate of decrease of the total kinetic energy. Indeed, recalling (2.73) and (2.75), we derive from (2.95)

$$\frac{3}{2}\frac{du^2}{dt}=-15\nu\frac{u^2}{\lambda^2}.$$

Hence, the rate of decrease of the kinetic energy is

$$\frac{1}{u^2}\frac{du^2}{dt}=-\frac{10\nu}{\lambda^2}. \tag{2.96}$$

As a consequence of the above relation, the length scale λ is also called the dissipation length parameter.

The Enstrophy Equation

As a consequence of (2.95), when no injection of energy is supplied by a mean flow or an external force, homogeneous turbulence decays in finite time. However, in the three-dimensional case, during the first period of the decay when the inertial terms are significant (i.e. before the system becomes purely viscous), enstrophy can be enhanced by nonlinear effects. This can be shown by deriving the evolution equation for the enstrophy. For this purpose, we recall the vorticity equation

$$\frac{\partial\boldsymbol{\omega}}{\partial t}+(\mathbf{u}\cdot\boldsymbol{\nabla})\boldsymbol{\omega}=\nu\,\Delta\boldsymbol{\omega}+(\boldsymbol{\omega}\cdot\boldsymbol{\nabla})\mathbf{u}. \tag{2.97}$$

By multiplying (2.97) by $\boldsymbol{\omega}$ and averaging, a dynamical equation for the mean-square vorticity is obtained:

$$\frac{1}{2}\frac{d}{dt}\left\langle|\boldsymbol{\omega}|^2\right\rangle+\langle[(\mathbf{u}\cdot\boldsymbol{\nabla})\boldsymbol{\omega}]\cdot\boldsymbol{\omega}\rangle=\nu\,\langle\Delta\boldsymbol{\omega}\cdot\boldsymbol{\omega}\rangle+\langle[(\boldsymbol{\omega}\cdot\boldsymbol{\nabla})\mathbf{u}]\cdot\boldsymbol{\omega}\rangle. \tag{2.98}$$

By using the homogeneity and incompressibility properties of the flow, (2.98) reduces to

$$\frac{1}{2}\frac{d}{dt}\left\langle|\boldsymbol{\omega}|^2\right\rangle = \sum_{i,j=1}^{n}\left\langle \omega_i\omega_j\frac{\partial u_i}{\partial x_j}\right\rangle - \nu\sum_{j=1}^{n}\left\langle|\nabla\omega_j|^2\right\rangle. \tag{2.99}$$

The second term in the RHS of (2.99) is a dissipative term so only the first one can contribute to production of enstrophy. The latter term describes the mean stretching of vortex lines (see Orszag (1973)).

The situation is different in the two-dimensional case. From the equation of motion, the vorticity equation can be derived, namely

$$\frac{\partial\omega}{\partial t} - \nu\,\Delta\omega + \mathbf{u}\cdot\nabla\omega = 0, \tag{2.100}$$

when no external force is used to sustain the flow. In the three-dimensional vorticity equation (2.97) an additional nonlinear term $(\boldsymbol{\omega}\cdot\nabla)\mathbf{u}$, responsible of the stretching of the vorticity, is present. The absence of vortex stretching in the two-dimensional case is the major difference between the two- and three-dimensional turbulences. From (2.100), it can be shown that the vorticity satisfies a law (conservative for $\nu = 0$) similar to (2.95); indeed,

$$\frac{1}{2}\frac{d}{dt}\left\langle|\omega(\mathbf{x})|^2\right\rangle = \nu\left\langle\omega(\mathbf{x})\Delta\omega(\mathbf{x})\right\rangle = -\nu\left\langle|\nabla\omega(\mathbf{x})|^2\right\rangle. \tag{2.101}$$

Using the homogeneity property of the flow, the RHS of (2.101) can be rewritten as

$$\left\langle|\nabla\omega(\mathbf{x})|^2\right\rangle = -\Delta^2\mathrm{tr}(\mathbf{R}(\mathbf{0})).$$

Then, recalling (2.22) and (2.42) we finally obtain

$$\left\langle|\nabla\omega(\mathbf{x})|^2\right\rangle = \int_{\mathbf{R}^n} k^4\,\mathrm{tr}(\mathbf{F}(\mathbf{k}))\,d\mathbf{k} = 2\int_0^{+\infty} k^4 E(k)\,dk.$$

Hence, the spectral form of (2.101) reads

$$\frac{d}{dt}\left(\int_0^{+\infty} k^2 E(k,t)\,dk\right) = -2\nu\int_0^{+\infty} k^4 E(k,t)\,dk. \tag{2.102}$$

As for the kinetic energy, it then appears that the enstrophy is also an invariant quantity when $\nu = 0$.

2.2.4. The Universal Theory of Equilibrium

We now want to define a characteristic velocity and length scale for the range of wavenumbers containing most of the energy. For the velocity, we can use the characteristic velocity previously introduced, namely

$$u = \frac{1}{\sqrt{3}} \left\langle |\mathbf{u}(\mathbf{x})|^2 \right\rangle^{1/2}.$$

An important scale used for measurements in experiments is the longitudinal integral scale defined by

$$L = \frac{\int_0^{+\infty} R_{11}(r, 0, 0)\, dr}{R_{11}(0, 0, 0)},$$

which, in view of (2.65), (2.70), and (2.52), is related to the energy spectrum function in the following way:

$$L = \frac{3\pi}{4} \frac{\int_0^{+\infty} k^{-1} E(k)\, dk}{\int_0^{+\infty} E(k)\, dk}. \tag{2.103}$$

This length L can be used as a characteristic length of the energy-containing eddies. In order to characterize a turbulent flow, an integral scale Reynolds number is defined and related to L by

$$\mathrm{Re}_L = \frac{uL}{\nu}. \tag{2.104}$$

Moreover, together with the length L and the velocity u, a characteristic time τ_e of the energy-containing eddies is introduced,

$$\tau_e = \frac{L}{u}; \tag{2.105}$$

it is called the eddy-turnover time.

As was done in the previous sections, we consider the case of a decaying turbulence, that is, the system has a given level of energy at an initial state and it decreases to zero as time evolves. During the first period of the decay, when the inertial forces are not negligible in comparison with the dissipative ones, it was found experimentally (see e.g. Batchelor (1971, Fig. 6.1, p. 106)) that the rate of decay of the kinetic energy ϵ is independent of the Reynolds number and that

$$\frac{3}{2} \frac{du^2}{dt} = -c_1 \frac{u^3}{L}, \qquad \text{so that} \quad \epsilon = c_1 \frac{u^3}{L}, \tag{2.106}$$

where c_1 is a dimensionless constant of the order of unity. This (experimental) relation is one of the most important of homogeneous turbulence theory. Note that the experiments cited in Batchelor (1971) correspond to three-dimensional flows; this relation is not valid in the two-dimensional case.

Combining (2.106), (2.73), and (2.75), we can relate the Reynolds numbers Re_L and Re_λ by the algebraic relation

$$\mathrm{Re}_L = \frac{c_1}{15}\mathrm{Re}_\lambda^2. \tag{2.107}$$

An immediate and important consequence of (2.106) is that the time scale of the decay is of the order of τ_e since (2.106) implies

$$\frac{1}{u^2}\frac{du^2}{dt} = -\frac{2}{3}\, c_1 \tau_e^{-1}. \tag{2.108}$$

The final period of the decay, when the energy is wholly damped by the viscosity, is reached after a time of the order of the characteristic time of the energy-containing eddies.

Remark 2.3. Jiménez et al. (1993) considered a forced turbulence instead of a decaying one; the force only acts on the large scales of their simulations. In such a case, they have observed that, in the range of Reynolds numbers accessible by direct simulations (i.e. $\mathrm{Re}_\lambda < 200$), the ratio $\epsilon L/u^3$ tends to a constant of the order of 0.7 when Re_λ is increased. They have also noted that in order to reach a statistically steady state for the energy-containing eddies, the system has to be integrated over several eddy-turnover times τ_e. □

We now introduce a local characteristic time $\tau(k)$ corresponding to the wavenumber k. For instance, in the Oboukhov's theory (see Lesieur (1990)), such a time scale is defined as $[k^3E(k)]^{-1/2}$. It is generally assumed (see Batchelor (1971, p. 104)) that $\tau(k)$ decreases as the size of the corresponding eddies decreases, so that

$$\tau(k) \ll \tau_e \quad \text{for } k > k_L,$$

where $k_L = 1/L$ corresponds to the wavenumber where most of the energy is concentrated. Hence, for wavenumbers sufficiently large compared with the energy-containing ones, the characteristic time of the eddies is much smaller than the period of the decay or the period of the convergence toward a statistically steady state. According to this remark, one can reasonably assume that there exists a range of the spectrum associated with degrees of freedom which reach a statistical equilibrium after a short transient period; this fundamental hypothesis is one of the bases of the universal equilibrium theory first developed by Kolmogorov (1941a, 1941b).

Let us now introduce the wavenumber k_d that corresponds to the wavenumber where the viscous forces are concentrated, and let us assume that the Reynolds number is high enough so that $k_d \gg k_L$. In fact another assumption strongest than the previous one is made: *The small scales lying in the range of high wavenumbers responsible for most of the energy dissipation* (i.e. $k \sim k_d$) *are in approximate equilibrium and are statistically independent of the large ones.*

A consequence of these hypotheses is that the behavior of these scales, where the dissipation effects becomes significant compared with the inertial forces, depends only on two external parameters:

1. the viscosity ν, which governs the distribution of thc local dissipation through the wavenumbers,
2. the insertion of energy, which proceeds at rate ϵ.

In order to characterize these (small) scales we introduce a dissipation length $\eta = 1/k_d$ and a corresponding velocity v; note that η is currently referred to as the Kolmogorov length scale. From the previous discussion, η is a function of ν and ϵ only. Hence, by seeking for a form $\eta = \nu^\alpha \epsilon^\beta$ and by invoking a dimensional argument, we deduce that $\alpha = \frac{3}{4}$ and $\beta = -\frac{1}{4}$, so that

$$\eta = \frac{1}{k_d} = \left(\frac{\nu^3}{\epsilon}\right)^{1/4}. \tag{2.109}$$

A similar argument allows us to define v as $v = (\nu\epsilon)^{1/4}$. Note that the Reynolds number associated with η is $v\eta/\nu$ and, from the definitions of v and η, is equal to unity. From (2.106) and (2.109), we deduce that the ratio between the dissipation and the integral scales is of the order of $\mathrm{Re}_L^{-3/4}$, namely,

$$\frac{\eta}{L} = c_1^{-1/4}\left(\frac{\nu}{uL}\right)^{3/4}. \tag{2.110}$$

When the Reynolds number is large enough so that the lengths η and L are widely separated (i.e. the ratio $L/\eta \gg 1$), the range of wavenumbers that are in equilibrium is large enough so that the wavenumbers $\eta k \ll 1$ become independent of the viscous effects. The corresponding scales forming an inertial subrange are determined by the transfer term $T(k)$ and depend only on the rate of transfer of the energy from the larger scales to the smaller ones. The energy spectrum $E(k, t)$ corresponding to this range then depends only on ϵ and k, so that we may expect that

$$E(k,t) = C_K \epsilon^{\alpha} k^{\beta},$$

where C_K is a dimensionless constant named the Kolmogorov constant. As previously, a dimensional analysis shows that $\alpha = \frac{2}{3}$ and $\beta = -\frac{5}{3}$. Hence, the energy spectrum function takes the form

$$E(k,t) = C_K \epsilon^{2/3} k^{-5/3} \tag{2.111}$$

in the inertial range, that is, for $k_L \ll k \ll k_d$. From (2.111) a characteristic time for the eddies corresponding to the wavenumber k in the inertial range can be easily obtained:

$$\tau_i(k) = [k^3 E(k)]^{-1/2} = C_K^{-1/2} \epsilon^{-1/3} k^{-2/3}. \tag{2.112}$$

Note that this definition of $\tau_i(k)$ matches with the dissipation time scale $\tau_d(k) = (\nu k^2)^{-1}$ for wavenumbers k near k_d. Indeed, by recalling (2.109) we see that $\tau_i(k_d) = C_K^{-1/2} \tau_d(k_d)$. By recalling (2.105) and (2.106), we deduce that

$$\tau_i(k) < \tau_e \quad \text{for } k > \left(c_1 C_K^{3/2}\right)^{-1/2} k_L. \tag{2.113}$$

The expression (2.111) is one of the major results of the Kolmogorov theory, and it has been well confirmed by experiments and numerical simulations, up to possible small corrections. However, as will be seen later, the hypotheses stated above have been criticized because they do not allow the theory to reflect the intermittent properties of (some) turbulent flows. In this context a theory, among others, referred to as the Kolmogorov–Oboukhov–Yaglom theory, has been developed; the reader is referred to Monin and Yaglom (1975) and the references therein for detailed developments on this theory. The main idea of this refined theory is to consider the local dissipation function

$$\epsilon(\mathbf{x},t) = \frac{\nu}{2} \sum_{i,j} \left(\frac{\partial u_i}{\partial x_j} + \frac{\partial u_j}{\partial x_i} \right)^2,$$

as a random function with large fluctuations with respect to space and time instead of considering only the global energy dissipation rate ϵ. By making some assumptions on the statistical properties of $\epsilon(\mathbf{x},t)$, the probability distributions for the small-scale turbulent components can be corrected from its previous behavior. However, it can be noted that the Kolmogorov self-similarity hypothesis provides a correct first-order description of the energy spectrum in the inertial range, which has been well verified by experiments and simulations; the value of C_K in (2.111) has been found to be of the order of 1.5.

The behavior of the high wavenumbers lying in the far dissipation range ($\eta k \gg 1$) has been extensively studied (see Batchelor (1971) and Monin and Yaglom (1975)). In a recent article, Smith and Reynolds (1991) assumed an asymptotic form of the energy spectrum function, namely

$$E(k) = C_K \epsilon^{2/3} k^{-5/3} e^{-\alpha_m (k\eta)^m}, \tag{2.114}$$

where α_m is an absolute constant. In Smith and Reynolds (1991), the authors have shown that the skewness factor, which will be defined in Section 2.2.5, of the derivative velocity depends on m, so that a comparison with experimental data allow them to derive the proper value for m. More recently, Manley (1992) improved the above analysis and showed that $m = 1$ seems to provide a better match than $m = 2$ for large Reynolds numbers. In this case, the value of α_m is given by $\alpha_1 = [2C_K\Gamma(\frac{4}{3})]^{3/4} \simeq 1.545 C_K^{3/4}$.

Some Remarks on Two-Dimensional Homogeneous Turbulence

Some of the previous developments can be applied to the two-dimensional isotropic turbulence. As was previously shown in Section 2.2.3, the energy and enstrophy are invariant for inviscid two-dimensional flows. The first noninvariant quadratic form is the mean squared vorticity gradient, which is proportional to the enstrophy dissipation rate $\gamma(t)$ defined by

$$\gamma(t) = \nu \left\langle |\nabla\omega(\mathbf{x})|^2 \right\rangle = 2\nu \int_0^{+\infty} k^4 E(k,t)\, dk. \tag{2.115}$$

An important consequence of (2.102) is that the enstrophy remains bounded by its initial value so that $\epsilon(t) = (\nu/2)\langle|\omega(\mathbf{x},t)|^2\rangle$ will tend toward zero when ν goes to zero. This situation is completely different from the three-dimensional case where $\epsilon(t)$ is largely independent of the viscosity.

As noted in Lesieur (1990, p. 267), a consequence of the equations (2.94) and (2.102) is that the energy tends to cascade from large wavenumbers to small wavenumbers, while the enstrophy tends to cascade towards large wavenumbers. Hence, it is generally assumed that the role of the small scales is to increase $\langle|\omega(\mathbf{x},t)|^2\rangle$ and then to maintain the dissipation of the kinetic energy. Kraichnan (1967), Leith (1968), and Batchelor (1969) proposed a theory, similar to the Kolmogorov one, which is based on the existence of a range of wavenumbers that are in approximate quasiequilibrium and are dynamically independent of the large scales at large Reynolds number.

As was noted above, one of the fundamental (noninvariant) quantities of the two-dimensional turbulence is the rate of dissipation of enstrophy. Let us now introduce the wavenumber k_d that corresponds to the modes containing most of the mean squared vorticity gradient. We expect that above this wavenumber, the enstrophy is damped by the viscous effects. Following the argument of the Kolmogorov theory, the scale η corresponding to $k_d = 1/\eta$ can depend only on γ and on the viscosity ν, so that a dimensional analysis leads to

$$\eta = \left(\frac{\nu^3}{\gamma}\right)^{1/6} . \tag{2.116}$$

By recalling that, in the three-dimensional case, the energy dissipation rate $\epsilon(t)$ depends only on a characteristic velocity u and on the integral scale L, we find

$$L \sim \frac{u^3}{\epsilon} = c_1 \frac{\left(\frac{2}{3}E\right)^{3/2}}{\epsilon} .$$

By analogy, Lilly (1971) proposed to define a macroscale by setting

$$L = \frac{E^{1/2}}{\gamma^{1/3}} = \frac{\left(\int_0^{+\infty} E(k,t)\, dk\right)^{1/2}}{\left(2\nu \int_0^{+\infty} k^4 E(k,t)\, dk\right)^{1/3}} . \tag{2.117}$$

This length scale can be interpreted as an integral scale characteristic of the eddies which contain most of the enstrophy. We denote by $k_L = 1/L$ the associated wavenumber and by Re_L the Reynolds number

$$\mathrm{Re}_L = \frac{uL}{\nu}, \qquad \text{where} \quad u = \frac{1}{\sqrt{2}} \left\langle |\mathbf{u}(\mathbf{x})|^2 \right\rangle^{1/2} = E^{1/2} . \tag{2.118}$$

From the above definition of L, we have

$$\mathrm{Re}_L = \frac{E}{\nu \gamma^{1/3}} ,$$

and with the help of (2.116)

$$\frac{L}{\eta} = \mathrm{Re}_L^{1/2} . \tag{2.119}$$

When the Reynolds number Re_L is large enough, one can assume that the scales L and η are well separated so that $\eta \ll L$. In this case, the existence of an inertial subrange, corresponding to wavenumbers $k_L \ll k \ll k_d$ and independent of the viscosity, can be assumed. In this range, the energy spectrum function $E(k, t)$

depends on one parameter, namely the enstrophy dissipation rate $\gamma(t)$, and on the wavenumber k, that is,

$$E(k,t) = F(k,\gamma,t),$$

where F is a scalar function. A dimensional analysis shows that F takes the following form:

$$E(k,t) = C_K \gamma^{2/3} k^{-3}, \tag{2.120}$$

where C_K is a nondimensional constant. The prediction (2.120) was first proposed by Kraichnan (1967), so that the phenomenological theory of two-dimensional turbulence is often referred to as the Kraichnan theory. Two-dimensional isotropic turbulence has been investigated by many authors with the help of direct numerical simulations (DNS) of the Navier–Stokes equations. The k^{-3} rule has been partially verified. Indeed, it seems that at low resolutions (i.e. at low Reynolds numbers) the energy spectrum function obeys a decay law closer to k^{-4} than to k^{-3}. When the Reynolds number is increased, after a transition period during which $E(k,t)$ behaves like k^{-4}, the exponent tends to a value close to -3; this result has been observed by Brachet et al. (1988).

Note that, in contrast with the three-dimensional case, a consequence of (2.120) is that the energy- and enstrophy-containing eddies are not separated in the two-dimensional case. As in the three-dimensional case, a characteristic time scale $\tau_i(k)$ for k lying in the inertial subrange can be defined:

$$\tau_i(k) = [k^3 E(k)]^{-1/2} = C_K^{-1/2} \gamma^{-1/3}. \tag{2.121}$$

Hence, the eddy-turnover time is independent of k in the inertial range. In two-dimensional turbulence, the small scales have a different behavior than in the three-dimensional case; they do not reach an equilibrium state in a time much shorter than the large-eddy turnover time.

Finally, we introduce a microscale intermediate between the integral scale and the enstrophy dissipation scale and similar to the Taylor microscale. Let us recall that the Taylor microscale is proportional to the ratio of the root mean square (rms) of the velocity and the rms of the vorticity. By analogy, Herring et al. (1974) introduced the microscale

$$\ell = \left(\frac{\epsilon}{\gamma}\right)^{1/2} = \left(\frac{\langle \omega^2(\mathbf{x})\rangle}{\langle |\nabla\omega(\mathbf{x})|^2\rangle}\right)^{1/2} = \left(\frac{\int_0^{+\infty} k^2 E(k,t)\,dk}{\int_0^{+\infty} k^4 E(k,t)\,dk}\right)^{1/2}, \tag{2.122}$$

which characterizes the mean spatial extension of the vorticity gradients. An associated Reynolds number can then be defined by

$$\mathrm{Re}_\ell = \frac{u\ell}{\nu}. \tag{2.123}$$

2.2.5. The Statistical Properties of Homogeneous Turbulent Flows

In the previous sections, the discussion centered around one-point moments, such as the kinetic energy or the energy dissipation rate, and two-point quantities, such as the correlation tensor or the energy spectrum function. The study of these moments led to the universal theory of Kolmogorov. However, other quantities determined from the probability and joint probability distributions of the velocity and its derivatives have well-known properties and provide information on the mechanical processes of turbulent motion and energy transfers. The purpose of this section is to define these quantities and briefly recall their main properties.

The mth-order centered moment of the velocity component $u_i(\mathbf{x}, t; \theta)$ is defined by

$$\mu_m(\mathbf{x}, t) = \langle [u_i(\mathbf{x}, t) - \langle u_i(\mathbf{x}, t)\rangle]^m \rangle.$$

The second centered moment (or variance) represents the mean-square departure from the mean value; hence it provides a measure of the width of the probability density function of u_i. The third moment provides information on the symmetry of the density function. Indeed, if the probability density function is symmetric with respect to $\langle u_i \rangle$ then all odd moments $\mu_{2m+1}(\mathbf{x}, t)$ are equal to zero.

Nondimensional moments are usually defined by

$$F_m(u_i(\mathbf{x}, t)) = \frac{\mu_m}{\mu_2^{m/2}}.$$

For the sake of simplicity we omit the dependence on the space and time variables in what follows. The third nondimensionalized moment $F_3(u_i)$ is the skewness factor which provides a measure of the asymmetry of the probability density function. The fourth nondimensionalized moment $F_4(u_i)$ is called kurtosis or flatness factor. Density functions with large kurtosis are characterized by the fact that the associated random functions frequently take on both very large and very small values.

From experimental measurements, the distributions of the velocity components u_i in three-dimensional turbulence are known to be approximately Gaussian (see Batchelor (1971)). Direct numerical simulations confirmed this result (see Vincent and Ménéguzzi (1991) and Jiménez et al. (1993)). A normal (or Gauss–Laplace) distribution of mean value $\langle u_i \rangle$ has a density function given by

$$f_i(u) = \frac{1}{\sigma\sqrt{2\pi}} \exp\left\{ -\frac{(u - \langle u_i \rangle)^2}{2\sigma^2} \right\}, \tag{2.124}$$

where $\sigma = \mu_2^{1/2}$ is the standard deviation of $u_i(\mathbf{x}, t)$. Due to the symmetry of the Gauss–Laplace distribution, the odd moments μ_{2m+1} equal zero. Moreover, the even moments can easily be computed from their definition and (2.124):

$$\mu_{2m} = \frac{2^m \sigma^{2m}}{\sqrt{\pi}} \Gamma\left(k + \frac{1}{2}\right) = \frac{(2m)!}{2^m m!} \mu_2^m,$$

so that

$$F_{2m}(u_i(\mathbf{x}, t)) = \frac{(2m)!}{2^m m!}.$$

Hence, $F_4(u_i) = 3$.

We now consider two points separated by a vector $\mathbf{r} \in \mathbb{R}^n$. According to Section 2.1, we can define a joint distribution function by setting

$$F_{ij,r}(u_1, u_2) = P[u_i(\mathbf{x}, t) < u_1, \quad u_j(\mathbf{x} + \mathbf{r}, t) < u_2] \quad \text{where} \quad u_1, u_2 \in \mathbb{R}.$$

The joint probability density function f_{ij} is defined by

$$F_{ij,r}(u_1, u_2) = \int_{-\infty}^{u_1} \int_{-\infty}^{u_2} f_{ij}(z_1, z_2)\, dz_1\, dz_2. \tag{2.125}$$

The computation of two-point joint probability densities requires too many points and variables, so that it is never studied in either experiments or direct simulations. A related quantity is more generally studied, namely the distribution of velocity differences (see Batchelor (1971) and the references therein, and Vincent and Ménéguzzi (1991)):

$$\delta u_1(\mathbf{r}) = u_1(\mathbf{x} + \mathbf{r}) - u_1(\mathbf{x}).$$

The distribution of $\delta u_1(\mathbf{r})$ depends only on $r = |\mathbf{r}|$ when the turbulence is assumed homogeneous and isotropic.

The flatness of $\delta u_1(r)$ obtained from experimental measurements has been analyzed in Batchelor (1971). At large separation distances, $F_4(\delta u_1(r))$ is close to 3.0; when r goes to zero, $F_4(\delta u_1(r))$ increases. In Vincent and Ménéguzzi (1991), the authors have shown that the density of $\delta u_1(r)$ obtained from DNS (at Re $\simeq$ 150) is approximately Gaussian for large values of r and it is similar to the distribution of $\partial u_1/\partial x_1$ for values of r close to zero. It is known that the flatness of the longitudinal derivative $\partial u_1/\partial x_1$ is function of the Reynolds number of the form $c\ \mathrm{Re}_\lambda^{0.18}$ (see Kerr (1985), Vincent and Ménéguzzi (1991), and Jiménez et al. (1993), for instance) and is larger than 3.0 (in Vincent and Ménéguzzi (1991) a value of 5.9 was found). Higher-order even moments of the longitudinal and lateral derivatives of u_1 have been reported in Jiménez et al. (1993), showing a departure from Gaussian and confirming the previous results.

The skewness factor $F_3(\partial u_1/\partial x_1)$ seems to be independent of the Reynolds number and of the order of -0.5. Note that in Jiménez et al. (1993) a small increase of $F_3(\partial u_1/\partial x_1)$ with Re_λ has been reported. When the turbulence is isotropic and incompressible, the skewness and flatness factors can be related by the Betchov inequality (see Monin and Yaglom (1975)):

$$\left| F_3\left(\frac{\partial u_1}{\partial x_1}\right)\right| \leq \frac{2}{\sqrt{21}} F_4\left(\frac{\partial u_1}{\partial x_1}\right)^{1/2}. \tag{2.126}$$

Moreover, in isotropic turbulence, Batchelor and Townsend (1947) have shown that the skewness factor and the stretching term (see (2.99)) satisfy the following relation:

$$\sum_{i,j=1}^{n} \left\langle \omega_i \omega_j \frac{\partial u_i}{\partial x_j}\right\rangle = -\frac{35}{2} F_3\left(\frac{\partial u_1}{\partial x_1}\right)\left(\frac{\epsilon}{15\nu}\right)^{3/2}. \tag{2.127}$$

By recalling (2.72) and (2.94), we deduce that

$$\sum_{i,j=1}^{n} \left\langle \omega_i \omega_j \frac{\partial u_i}{\partial x_j}\right\rangle = \int_0^{+\infty} k^2 T(k)\, dk.$$

The relation (2.127) can then be rewritten as

$$F_3\left(\frac{\partial u_1}{\partial x_1}\right) = -\left(\frac{135}{98}\right)^{1/2} \frac{\int_0^{+\infty} k^2 T(k)\, dk}{\left(\int_0^{+\infty} k^2 E(k)\, dk\right)^{3/2}}. \tag{2.128}$$

Finally, we want to mention that the aliasing errors (see Canuto et al. (1988)) when computing the nonlinear terms by a pseudospectral method with a low resolution may deteriorate the qualitative computation of the high-order moments of the velocity derivative. Indeed, these moments are highly dependent on the smallest scales, which should then be computed with accuracy. Jiménez (1994) noted that a minimum resolution of $k_N \geq 3k_d/2$ is necessary in order to accurately compute the first six moments of the velocity derivative.

2.3. Nonhomogeneous Turbulence: The Channel Flow Problem

We now devote our attention to nonhomogeneous turbulence and consider flows in a channel. In this section, we describe the main properties of flows in an infinite channel with parallel walls, $\Omega = \mathbb{R} \times (-h, +h) \times \mathbb{R}$, where h is the half width of the channel. As in Section 2.2, we introduce and define characteristic velocities and length scales for such flows; again, most developments refer to three-dimensional flows. For more details on the turbulence theory related to this problem, the reader is referred for instance to the book of Tennekes and Lumley (1972).

Homogeneity in the Infinite Directions

First of all, we recall that the velocity field $\mathbf{u}(\mathbf{x}, t)$ satisfies the equations (1.5)–(1.6) where the external force $\mathbf{f}$ is set equal to zero. Moreover the following boundary conditions are imposed at the walls of the channel:

$$\mathbf{u}(x_1, \pm h, x_3, t) = \mathbf{0}, \tag{2.129}$$

so that the walls are at rest with respect to the coordinate system. We proceed as in Section 2.2, that is, we consider the flow as a random variable depending on some experimental parameters, and we introduce an averaging operator denoted by $\langle \cdot \rangle$. Moreover, we assume that the flow is homogeneous in the infinite directions x_1 and x_3, so that all mean quantities, with respect to the averaging operator, are independent of the location $(x_1, x_3) \in \mathbb{R}^2$. In what follows, we will extensively use this property, which considerably simplifies the analysis of the problem.

A first consequence of the homogeneity property, together with the incompressibility (1.5), is that

$$\langle u_2(\mathbf{x}, t) \rangle = 0 \quad \text{for all} \quad (\mathbf{x}, t) \in \Omega \times (0, +\infty). \tag{2.130}$$

Indeed, by averaging the continuity equation (1.5) we obtain

$$\frac{\partial}{\partial x_2} \langle u_2(\mathbf{x}, t) \rangle = 0, \quad \text{so that} \ \langle u_2(\mathbf{x}, t) \rangle = \langle u_2(t) \rangle .$$

Then (2.130) follows from the boundary condition (2.129).

Remark 2.4. Periodic boundary conditions are generally applied in the streamwise and spanwise directions, x_1 and x_3. These boundary conditions are considered to be valid if the streamwise and spanwise periods are long enough so that the values of the velocity inside the box become decorrelated with the boundary values (see for instance Kim, Moin, and Moser (1987)). These boundary conditions are not the only possible choice, but they are commonly used in the mathematical study and the numerical approximation of the problem (see Chapter 1). □

Mean Pressure Gradient and Flow Rate

If no further conditions are imposed, any initial perturbation with finite amplitude will decay in time (see (1.15)). In order to sustain the flow, a mean pressure gradient $K_p(\mathbf{x}, t) = \langle(\partial p/\partial x_1)(\mathbf{x}, t)\rangle$ can be applied in the streamwise direction. However, in order to avoid acceleration of the flow in the streamwise direction and to remain consistent with the homogeneity assumption, $K_p(\mathbf{x}, t)$ is assumed to be independent of x_1 and x_3. Moreover $K_p(\mathbf{x}, t)$ is independent of the distance to the wall. Indeed, by averaging the momentum equation for the normal velocity u_2, we find

$$\frac{\partial}{\partial x_2}\left(\left\langle|u_2(\mathbf{x}, t)|^2\right\rangle + \langle p(\mathbf{x}, t)\rangle\right) = 0.$$

After integration up to the wall, we obtain a relation between the mean pressure and the rms of the normal velocity, namely

$$\langle p(\mathbf{x}, t)\rangle + \left\langle|u_2(\mathbf{x}, t)|^2\right\rangle = \langle p(x_1, -h, x_3, t)\rangle . \tag{2.131}$$

By assumption, $\langle|u_2(\mathbf{x}, t)|^2\rangle$ is independent of x_1 and x_3, so that by differentiating (2.131) with respect to the streamwise direction, we have

$$\left\langle \frac{\partial p}{\partial x_1}(\mathbf{x}, t)\right\rangle = \left\langle \frac{\partial p}{\partial x_1}(x_1, -h, x_3, t)\right\rangle,$$

showing that $K_p(\mathbf{x}, t) = K_p(t)$. Equations (1.6) can then be reformulated as

$$\frac{\partial \mathbf{u}}{\partial t} - \nu\Delta\mathbf{u} + \sum_{i=1}^{n} u_i \frac{\partial \mathbf{u}}{\partial x_i} + \boldsymbol{\nabla} p + K_p\, \mathbf{e}_1 = \mathbf{0}, \tag{2.132}$$

where $p(\mathbf{x}, t)$ is a pressure with no mean gradient: $\langle(\partial p/\partial x_1)(\mathbf{x}, t)\rangle = 0$.

Definition 2.14. *The flow rate $Q(t)$ is the quantity*

$$Q(t) = \int_{-h}^{+h} \langle u_1(\mathbf{x}, t)\rangle \; dx_2. \tag{2.133}$$

By definition, $Q(t)$ is independent of x_2. Moreover, due to the homogeneity property, $\langle u_1(\mathbf{x}, t)\rangle$ is independent of x_1 and x_3 so that the flow rate depends only on the time variable. The mean pressure gradient and the flow rate are related by a dynamical equation which can be easily derived from the momentum equation. Let us average the equation of the streamwise component of the velocity u_1 in (2.132); we obtain

$$\frac{\partial \langle u_1\rangle}{\partial t} - \nu \frac{\partial^2 \langle u_1\rangle}{\partial x_2^2} + \left\langle \frac{\partial}{\partial x_2}(u_1 u_2) \right\rangle + K_p = 0,$$

where all partial derivatives with respect to x_1 and x_3 vanish on averaging, due to the homogeneity assumption. By integrating over the channel width and recalling the definition of the flow rate (2.133), we obtain

$$\begin{aligned} \frac{dQ}{dt} &- \nu \left(\frac{\partial \langle u_1\rangle}{\partial x_2}(h) - \frac{\partial \langle u_1\rangle}{\partial x_2}(-h) \right) \\ &+ \langle u_1 u_2\rangle (h) - \langle u_1 u_2\rangle (-h) + 2h\, K_p = 0. \end{aligned} \tag{2.134}$$

According to the boundary conditions, the cross terms $\langle u_1 u_2\rangle$ vanish at the wall. Invoking the symmetry of the geometry, we can assume that the mean flow is symmetric at the wall, so that its derivative with respect to x_2 is antisymmetric; hence (2.134) takes the following simple form:

$$\frac{dQ}{dt} + 2\nu \frac{\partial \langle u_1\rangle}{\partial x_2}(-h) + 2hK_p = 0. \tag{2.135}$$

Two formulations of the problem can then be considered: the first one assumes that the flow rate remains constant (i.e. $Q(t) = Q$), and the second one assumes that the mean pressure gradient is constant (i.e. $K_p(t) = K_p$). These conditions then lead to two types of flows, which are characterized by different Reynolds numbers. We denote by Re_Q the Reynolds number corresponding to the first condition,

$$\mathrm{Re}_Q = \frac{3Q}{4\nu},$$

and by Re_p the Reynolds number corresponding to the second one,

$$\mathrm{Re}_p = \frac{|K_p| h^3}{2\nu^2}.$$

Note that in the case of the Poiseuille flow, that is, $\mathbf{u}(\mathbf{x}, t) = (h - x_2^2, 0, 0)$, both Reynolds numbers are equal. This is not the case for other stationary or nonstationary flows.

Let us now investigate further the latter situation:

- *The case of constant flux*: From (2.135), the mean pressure gradient satisfies

$$K_p(t) = -\frac{\nu}{h}\frac{\partial \langle u_1 \rangle}{\partial x_2}(-h, t), \tag{2.136}$$

and it is then time-dependent as long as the flow is not stationary. Recalling the definition of the vorticity and the homogeneity assumption (2.136) becomes

$$K_p(t) = \begin{cases} -\frac{\nu}{h}\langle \omega(-h, t)\rangle, & n = 2, \\ \frac{\nu}{h}\langle \omega_3(-h, t)\rangle, & n = 3. \end{cases}$$

- *The case of constant mean pressure gradient*: By integrating (2.135) over any time interval $(0, t)$ we derive an expression for the flow-rate change

$$\begin{aligned} \Delta Q(t) &= Q(t) - Q_0 \\ &= -2K_p ht - 2\nu \int_0^t \frac{d\langle u_1 \rangle}{dx_2}(-h, s)\, ds. \end{aligned} \tag{2.137}$$

Thus $Q(t)$ will tend to a stationary value only if the RHS of (2.137) converges when t tends to infinity.

Remark 2.5. From the previous relations we see that, for a nonstationary flow, keeping the flow rate Q (respectively K_p) constant implies that the mean pressure gradient K_p (respectively Q) becomes time-dependent. In Rozhdestvensky and Simakin (1984), the authors note that flows with both constant flow rate and constant mean pressure gradient exist and in such a case $\mathrm{Re}_p > \mathrm{Re}_Q$. From (2.136), we can exhibit a relation between the two Reynolds numbers in this particular case:

$$\mathrm{Re}_p = \frac{2h^2}{3Q}\, |\langle \omega(-h) \rangle|\, \mathrm{Re}_Q.$$

Such flows are called two-dimensional secondary flows by Rozhdestvensky and Simakin (1984). □

Statistically Steady Flows and the Law of the Wall

We now assume that the flow reaches a statistically steady state so that the time average $(1/T)\int_0^T \langle u_1(x_2, t)\rangle\, dt$ remains bounded and converges when T tends to infinity; we denote by U its limit, namely

$$U(x_2) = \lim_{T\to+\infty} \frac{1}{T} \int_0^T \langle u_1(\mathbf{x}, t)\rangle\, dt. \tag{2.138}$$

For the sake of simplicity, we denote by $\langle\cdot\rangle$ the operator $\langle\cdot\rangle = \lim_{T\to+\infty}(1/T)\int_0^T \langle\cdot\rangle\, dt$.

By averaging the equation satisfied by u_1 in (2.132), we find

$$\frac{d}{dx_2}\left(\nu\frac{dU}{dx_2}(x_2) - \langle u_1 u_2\rangle\right) - K_p = 0,$$

and by integrating over the interval $(-h, x_2)$, we obtain the expression of the total shear stress:

$$\tau(x_2) = \nu\frac{dU}{dx_2}(x_2) - \langle u_1 u_2\rangle\,(x_2) = K_p(x_2 + h) + u_*^2, \tag{2.139}$$

where u_*^2 denotes the shear stress at the wall, namely $u_*^2 = \tau(-h)$. Due to the boundary conditions, the Reynolds stress $\langle u_1 u_2\rangle$ vanishes at the wall so that

$$u_*^2 = \nu\frac{dU}{dx_2}(-h) = \nu\omega_{\mathrm{w}} \qquad \text{with} \tag{2.140}$$

$$\omega_{\mathrm{w}} = \begin{cases} \langle\omega(-h)\rangle, & n = 2, \\ -\langle\omega_3(-h)\rangle, & n = 3. \end{cases}$$

It is generally assumed that both the Reynolds stress $\langle u_1 u_2\rangle$ and the viscous stress dU/dx_2 are antisymmetric and vanish at the center of the channel ($x_2 = 0$). In such a case, the friction velocity u_* and the mean pressure gradient are related by

$$u_*^2 = -K_p h,$$

so that

$$K_p = -\frac{\nu}{h}\frac{dU}{dx_2}(-h)$$

follows from (2.140). Then, the total stress satisfies

$$\nu\frac{dU}{dx_2}(x_2) - \langle u_1 u_2\rangle\,(x_2) = -u_*^2\frac{x_2}{h}. \tag{2.141}$$

Flows in a channel are often characterized by a Reynolds number based on the friction velocity,

$$\mathrm{Re}_* = \frac{u_* h}{\nu}, \tag{2.142}$$

and a Reynolds number based on the mean velocity $U_0 = U(0)$ at the center of the channel,

$$\mathrm{Re}_0 = \frac{U_0 h}{\nu}. \tag{2.143}$$

A wall length scale δ_τ and a nondimensional distance x_2^+ in the direction normal to the wall are usually defined by

$$\delta_\tau = \frac{\nu}{u_*} \quad \text{and} \quad x_2^+ = \frac{x_2 + h}{\delta_\tau} = \frac{u_*}{\nu}(x_2 + h).$$

Hence, when data analysis is performed, all mean quantities are normalized with respect to the friction velocity and all distances in the normal direction are expressed in terms of wall units, leading to universal forms.

Equation (2.141) can be rewritten as

$$-\frac{\langle u_1 u_2 \rangle (x_2)}{u_*^2} + \mathrm{Re}_*^{-1} \frac{d}{d\left(\frac{x_2}{h}\right)} \left(\frac{U}{u_*}\right) = -\frac{x_2}{h}. \tag{2.144}$$

Equation (2.144) shows that, for a sufficiently large Reynolds number, the viscous term can become negligibly small, so that we obtain a linear profile for the Reynolds stress, namely

$$\langle u_1 u_2 \rangle \simeq \frac{x_2\, u_*^2}{h}.$$

However, since the Reynolds stress vanishes at the wall, the above asymptotic behavior (when $\mathrm{Re}_* \to +\infty$) is only valid sufficiently far from the wall; this region is called the *core region.* Close to the wall (i.e. for x_2^+ of the order of unity), we reformulate (2.141) as

$$-\frac{u_1 u_2}{u_*^2} + \frac{d}{dx_2^+}\left(\frac{U}{u_*}\right) = 1 - \frac{x_2^+}{\mathrm{Re}_*}. \tag{2.145}$$

Assuming that the Reynolds stress is small enough to be neglected in the vicinity of the wall, we deduce that, when $\mathrm{Re}_* \to +\infty$, the mean velocity profile is linear, that is,

$$\frac{U(x_2)}{u_*} \simeq x_2^+.$$

This profile in a region $x_2^+ \in (0, 5)$ has been experimentally and numerically obtained (see for instance Tennekes and Lumley (1972)). This region is called the *viscous sublayer.*

Equation (2.141) does not provide information on the mean velocity profile in the core region. However, it has been experimentally found that the mean velocity satisfies the following logarithmic law:

$$\frac{U(x_2)}{u_*} = 2.5 \ln x_2^+ + 5.0 \qquad \text{for } x_2^+ > 30. \tag{2.146}$$

In Tennekes and Lumley (1972), the authors have shown that (2.146) can be derived by making several physical assumptions on the mean velocity in the center part of the channel and by matching equations (2.144) and (2.145) in the region $x_2^+ \in (5, 30)$.

Remark 2.6. Since the Kolmogorov theory (see Section 2.2.4) is very general, we may expect that it applies to any kind of flows. In Section 2.2.4, we have shown that based on a Fourier analysis of the velocity field and all related quantities (such as the correlation and energy spectrum tensors), the scale similarity hypothesis of Kolmogorov can be applied and it leads to the derivation of the $k^{-5/3}$ rule. However, due to the boundaries in the channel flow problem, Fourier theory cannot be used. Moser (1994) has shown that another basis, the Karhunen–Loeve (KL) eigenfunctions, which are eigenfunctions of the two-point velocity correlation tensor, can be used to define an energy spectrum and an inertial range for nonhomogeneous flows. Using the Kolmogorov similarity hypothesis, the $k^{-5/3}$ law was recovered. The author suggested that the KL eigenfunctions are probably the most appropriate basis for analyzing nonhomogeneous turbulent flows. □

Some Remarks on Two-Dimensional Channel Flows

The statistics of two-dimensional channel flows have been compared with real (three-dimensional) turbulence in Rozhdestvensky and Simakin (1984) and in Jiménez (1990); see also the references therein. In both papers, the authors study the evolution of finite amplitude perturbations of Poiseuille flows by means of DNS. Note that in Jiménez (1990) the constant flow-rate formulation was used, whereas in Rozhdestvensky and Simakin (1984) the constant mean pressure gradient formulation was used. Nevertheless, the conclusions in the two papers concerning the behavior of the statistics were similar, namely:

- In the three-dimensional case, the Reynolds and diffusive stresses match at the center of the channel, where the turbulence is more developed. In the two-dimensional case, it appears that the diffusive stress is stronger than the Reynolds (convection) stress, not only in the viscous layer (i.e. close to the boundary, $x_2^+ = 10\text{–}20$) but over the whole width of the channel.
- The fluctuation profiles $\langle \mathbf{u}' \rangle$ (where $\mathbf{u}' = \mathbf{u} - \langle \mathbf{u} \rangle$ is the velocity fluctuation) are different from the three-dimensional ones but not weaker; u_2' appears to be stronger in the two-dimensional case.
- Finally, the mean velocity profile $U(x_2)$ seems to be closer to a parabolic profile with an excess of velocity at the channel centerline. We recall that in space dimension $n = 3$ the mean velocity profile outside the viscous layer follows the well-known velocity-defect law and is much flatter than the parabolic profile.

Finally, Jiménez (1990) noted that the energy spectrum decays at the same rate as in homogeneous two-dimensional turbulence and obeys a k^{-4} law (see Section 2.2.4).

Even though two-dimensional channel flows differ in many respects from three-dimensional flows, for large enough Reynolds numbers (Re_Q larger than 7,000 in long boxes, $L_1 \geq 8\pi$), flows with chaotic time behavior and fine spatial structures can be obtained. Moreover, the shear stress evolves from a diffusive toward a more turbulent behavior.

3

Computational Methods for the Direct Simulation of Turbulence

In this chapter, we describe computational methods commonly used for the direct numerical simulation (DNS) of turbulent flows. In Section 3.1, we consider the case of fully periodic flows with no mean, that is, homogeneous isotropic turbulent flows; in Section 3.2, the case of wall-bounded flows is discussed. Finally, in Section 3.3, other type of flows are briefly described. In all cases, we restrict ourselves to the presentation of the spectral methods; more complete descriptions of these methods can be found in the reference books of Gottlieb and Orszag (1977) and Canuto et al. (1988).

3.1. The Fully Periodic Case: Homogeneous Turbulence

As a model for homogeneous isotropic or nearly isotropic turbulent flows we consider fully periodic flows in $\mathbb{R}^n$, $n = 2$ or 3. Recalling that, due to isotropy, all the statistical quantities are independent of the spatial direction (see Chapter 2), we assume that both the velocity field $\mathbf{u}(\mathbf{x}, t)$ and the pressure $p(\mathbf{x}, t)$ have the same period L_1 in each direction of space. Indeed, there is no reason here for choosing different length periods as was done in Chapter 1. Note that, in practice, as the integral scale provides the size of the largest scales in the flow, the period should be larger than the integral scale defined in Chapter 2.

As a homogeneous isotropic flow has no mean, we impose

$$\int_\Omega \mathbf{u}(\mathbf{x}, t)\, d\mathbf{x} = \mathbf{0}, \tag{3.1}$$

where $\Omega = \prod_{i=1}^{n}(0, L_1)$ is the domain filled by the flow. For convenience, we recall here the incompressible Navier–Stokes equations introduced in

Chapter 1:

$$\frac{\partial \mathbf{u}}{\partial t} - \nu\, \Delta \mathbf{u} + \omega \times \mathbf{u} + \nabla \left(p + \frac{1}{2}|\mathbf{u}|^2 \right) = \mathbf{f},$$

$$\nabla \cdot \mathbf{u} = 0, \quad \int_\Omega \mathbf{u}(\mathbf{x}, t)\, d\mathbf{x} = \mathbf{0}, \tag{3.2}$$

$$\mathbf{u}(\mathbf{x}, t = 0) = \mathbf{u}_0(\mathbf{x}),$$

where $\omega = \nabla \times \mathbf{u}$ is the vorticity, $\mathbf{f}(\mathbf{x}, t)$ is the external volume force, and ν is the kinematic viscosity. We set $\mathbf{x} = (x_1, x_2)$ in the case $n = 2$, and $\mathbf{x} = (x_1, x_2, x_3)$ in the case $n = 3$. Moreover $|\cdot|$ denotes the $\mathbb{R}^n$ Euclidean norm.

Note that in (3.2) the nonlinear term is written in its rotational form; this form has been used in all our numerical simulations of homogeneous turbulent flows. It enjoys some stability properties (see Canuto et al. (1988)), and its numerical computation requires less operations than the standard form $(\mathbf{u} \cdot \nabla)\mathbf{u}$.

Spectral Representation

Due to their periodicity, $\mathbf{u}$, p, and $\mathbf{f}$ can be expanded in Fourier series:

$$\mathbf{u}(\mathbf{x}, t) = \sum_{\mathbf{k} \in \mathbb{Z}^n} \hat{\mathbf{u}}(\mathbf{k}, t)\, e^{\mathrm{i}(2\pi/L_1)\mathbf{k}\cdot\mathbf{x}}, \qquad p(\mathbf{x}, t) = \sum_{\mathbf{k} \in \mathbb{Z}^n} \hat{p}(\mathbf{k}, t)\, e^{\mathrm{i}(2\pi/L_1)\mathbf{k}\cdot\mathbf{x}},$$

$$\mathbf{f}(\mathbf{x}, t) = \sum_{\mathbf{k} \in \mathbb{Z}^n} \hat{\mathbf{f}}(\mathbf{k}, t)\, e^{\mathrm{i}(2\pi/L_1)\mathbf{k}\cdot\mathbf{x}},$$

where $\mathbf{k} = (k_1, k_2)$ if $n = 2$ (or $\mathbf{k} = (k_1, k_2, k_3)$ if $n = 3$). Then (3.1) becomes

$$\hat{\mathbf{u}}(\mathbf{k} = \mathbf{0}, t) = \frac{1}{|\Omega|} \int_\Omega \mathbf{u}(\mathbf{x}, t)\, d\mathbf{x} = \mathbf{0}. \tag{3.3}$$

It can be easily shown that, due to the incompressibility condition and the boundary conditions, the integral of $(\mathbf{u} \cdot \nabla)\mathbf{u}$ over Ω vanishes so that its $\mathbf{k} = \mathbf{0}$ Fourier coefficient vanishes for all $t > 0$, and a necessary condition for (3.3) to be valid for all time $t > 0$ is that $\hat{\mathbf{f}}(\mathbf{0}, t) = \mathbf{0}$, for all $t > 0$.

The Fourier expansions above can then be rewritten as

$$\mathbf{u}(\mathbf{x}, t) = \sum_{\mathbf{k} \neq 0} \hat{\mathbf{u}}(\mathbf{k}, t)\, e^{\mathrm{i}(2\pi/L_1)\mathbf{k}\cdot\mathbf{x}} \quad \text{and} \tag{3.4}$$

$$\mathbf{f}(\mathbf{x}, t) = \sum_{\mathbf{k} \neq 0} \hat{\mathbf{f}}(\mathbf{k}, t)\, e^{\mathrm{i}(2\pi/L_1)\mathbf{k}\cdot\mathbf{x}}.$$

Taking the $[L^2(\Omega)]^n$ inner product (see Chapter 1, Section 1.4) with the basis functions $e^{-\mathbf{i}(2\pi/L_1)\mathbf{k}\cdot\mathbf{x}}$, we obtain the spectral form of the system (3.2):

$$\begin{aligned}
&\frac{\partial \hat{\mathbf{u}}(\mathbf{k})}{\partial t} + \nu \frac{4\pi^2}{L_1^2} k^2\, \hat{\mathbf{u}}(\mathbf{k}) \\
&\quad + \widehat{(\omega \times \mathbf{u})}(\mathbf{k}) + \mathbf{i}\frac{2\pi}{L_1}\mathbf{k}\left(\hat{p}(\mathbf{k}) + \frac{1}{2}\widehat{|\mathbf{u}|^2}(\mathbf{k})\right) = \hat{\mathbf{f}}(\mathbf{k}), \\
&\qquad \mathbf{i}\mathbf{k}\cdot\hat{\mathbf{u}}(\mathbf{k}) = 0, \\
&\qquad \hat{\mathbf{u}}(\mathbf{k}, t=0) = \hat{\mathbf{u}}_0(\mathbf{k}) \qquad \text{for all} \quad \mathbf{k} \neq \mathbf{0}, \\
&\qquad \hat{\mathbf{u}}(\mathbf{k}=\mathbf{0}, t=0) = \mathbf{0},
\end{aligned} \tag{3.5}$$

where k stands for $|\mathbf{k}| = (\sum_{i=1}^{n} k_i^2)^{1/2}$. The spectral form of the nonlinear term in (3.5) can be derived from the Fourier expansion (3.4), namely

$$\widehat{(\omega \times \mathbf{u})}_j(\mathbf{k}, t) = \frac{2\pi}{L_1}\mathbf{i}\sum_{\ell \neq j}\ \sum_{\mathbf{m}+\mathbf{r}=\mathbf{k}} \left[m_\ell \hat{u}_\ell(\mathbf{r})\hat{u}_j(\mathbf{m}) - \frac{1}{2}k_j \hat{u}_\ell(\mathbf{r})\hat{u}_\ell(\mathbf{m})\right].$$

The Helmholtz Decomposition

In the particular case of periodic boundary conditions, the operator P introduced in Chapter 1 can be easily implemented.

Taking the dot product in $\mathbb{R}^n$ of the momentum equation above with the wavenumber $2\pi\mathbf{k}/L_1$, and recalling the incompressibility condition, the pressure can be expressed in terms of the velocity field:

$$\hat{p}(\mathbf{k}) + \frac{1}{2}\widehat{|\mathbf{u}|^2}(\mathbf{k}) = -\frac{\mathbf{i}}{k^2}\frac{L_1}{2\pi}\left\{\mathbf{k}\cdot\left[\hat{\mathbf{f}}(\mathbf{k}) - \widehat{(\omega \times \mathbf{u})}(\mathbf{k})\right]\right\} \tag{3.6}$$

for all $\mathbf{k} \in \mathbb{Z}^n$, $\mathbf{k} \neq \mathbf{0}$. Substituting (3.6) into the momentum equation in (3.5), we obtain the spectral representation of the abstract evolution equation (1.30)

$$\frac{\partial \hat{\mathbf{u}}(\mathbf{k})}{\partial t} + \nu\frac{4\pi^2}{L_1^2}k^2\, \hat{\mathbf{u}}(\mathbf{k}) + \hat{B}_{\mathbf{k}}(\mathbf{u}, \mathbf{u}) = \hat{\mathbf{g}}(\mathbf{k}), \tag{3.7}$$

for all $\mathbf{k} \in \mathbb{Z}^n$, $\mathbf{k} \neq \mathbf{0}$, where

$$\begin{aligned}
\hat{B}_{\mathbf{k}}(\mathbf{u}, \mathbf{u}) &= \left[\widehat{(\omega \times \mathbf{u})}(\mathbf{k}) - \frac{\mathbf{k}}{k^2}\left[\mathbf{k}\cdot\widehat{(\omega \times \mathbf{u})}(\mathbf{k})\right]\right], \\
\hat{\mathbf{g}}(\mathbf{k}) &= \left[\hat{\mathbf{f}}(\mathbf{k}) - \frac{\mathbf{k}}{k^2}\left[\mathbf{k}\cdot\hat{\mathbf{f}}(\mathbf{k})\right]\right].
\end{aligned} \tag{3.8}$$

We denote by $L^2_{\text{per}}(\Omega)$ the subspace of periodic functions in $\mathbb{R}^n$ that are locally L^2.

The operator P transforming any field $\varphi \in [L^2_{\text{per}}(\Omega)]^n$ such that $\hat{\varphi}(\mathbf{0}) = \mathbf{0}$ into a free divergence field of $[L^2_{\text{per}}(\Omega)]^n$ is then defined by setting

$$P\varphi(\mathbf{x}) = \sum_{\mathbf{k}\neq 0} \left[\hat{\varphi}(\mathbf{k}) - \frac{\mathbf{k}}{k^2}\,[\mathbf{k}\cdot\hat{\varphi}(\mathbf{k})]\right] e^{\mathrm{i}(2\pi/L_1)\mathbf{k}\cdot\mathbf{x}}. \tag{3.9}$$

This operator is called the Leray–Helmholtz projector for the problem. Hence, with the help of (3.9), (3.8) becomes

$$\hat{B}_{\mathbf{k}}(\mathbf{u},\mathbf{u}) = P(\widehat{\omega\times \mathbf{u}})(\mathbf{k}) \text{ and } \hat{\mathbf{g}}(\mathbf{k}) = \widehat{P\mathbf{f}}(\mathbf{k}).$$

We associate with the coefficients $\hat{B}_{\mathbf{k}}(\mathbf{u},\mathbf{u})$ the bilinear form

$$B(\mathbf{u},\mathbf{u}) = \sum_{\mathbf{k}\in\mathbb{Z}^n,\ \mathbf{k}\neq 0} \hat{B}_{\mathbf{k}}(\mathbf{u},\mathbf{u})\, e^{\mathrm{i}(2\pi/L_1)\mathbf{k}\cdot\mathbf{x}}, \tag{3.10}$$

and we recover the evolution equation (1.30); thus, (3.7) is equivalent to

$$\frac{\partial \mathbf{u}}{\partial t} + \nu A\mathbf{u} + B(\mathbf{u},\mathbf{u}) = P\mathbf{f}.$$

Fourier–Galerkin Approximation

We now introduce P_N, the orthogonal projector from $[L^2_{\text{per}}(\Omega)]^n$, onto the space V_N of trigonometric polynomials of degree $\leq N$. The spectral Galerkin method consists in looking for

$$\mathbf{u}_N(\mathbf{x},t) = \sum_{\mathbf{k}\in S_N} \tilde{\mathbf{u}}(\mathbf{k},t)\, e^{\mathrm{i}(2\pi/L_1)\mathbf{k}\cdot\mathbf{x}}, \tag{3.11}$$

where $S_N = \{\mathbf{k}\in\mathbb{Z}^n,\ \mathbf{k}\neq\mathbf{0} : |k_j| \leq N/2-1,\ j=1,\ldots,n\}$, which is solution of

$$\frac{\partial \mathbf{u}_N}{\partial t} + \nu A\mathbf{u}_N + P_N B(\mathbf{u}_N,\mathbf{u}_N) = \mathbf{g}_N. \tag{3.12}$$

According to (3.7)–(3.10), (3.12) is equivalent to the following set of equations:

$$\frac{\partial \tilde{\mathbf{u}}}{\partial t}(\mathbf{k},t) + \nu\frac{4\pi^2}{L_1^2}k^2\,\tilde{\mathbf{u}}(\mathbf{k},t) + \hat{B}_{\mathbf{k}}(\mathbf{u}_N,\mathbf{u}_N) = \hat{\mathbf{g}}(\mathbf{k},t) \tag{3.13}$$

for any $\mathbf{k}\in S_N \cap \{k_n \geq 0\}$.[1]

[1] As the velocity field is real, its Fourier coefficients enjoy a symmetry property: $\tilde{u}_j(\mathbf{k}) = \tilde{u}_j^*(-\mathbf{k})$; hence, only the coefficients with wavenumber $2\pi\mathbf{k}/L_1$ satisfying $k_n \geq 0$ need to be computed.

Time Discretization

By multiplying (3.13) by the exponential operator $\exp(4\pi^2 \nu k^2 t/L_1^2)$, we obtain

$$\frac{d}{dt}\left(e^{\nu(4\pi^2/L_1^2)k^2 t}\tilde{\mathbf{u}}(\mathbf{k},t)\right) = e^{\nu(4\pi^2/L_1^2)k^2 t}\,\hat{\mathbf{h}}_{\mathbf{k}}(\mathbf{u}_N,\mathbf{u}_N,t) \tag{3.14}$$

for any $\mathbf{k} \in S_N \cap \{k_n \geq 0\}$, where we have set $\hat{\mathbf{h}}_{\mathbf{k}}(\mathbf{u}_N,\mathbf{u}_N,t) = (\hat{\mathbf{g}}(\mathbf{k},t) - \hat{B}_{\mathbf{k}}(\mathbf{u}_N(t),\mathbf{u}_N(t)))$. With the form (3.14) of the equations, the linear (Stokes) operator can be integrated exactly, avoiding stability and accuracy restrictions due to the time discretization. At this time, the only difficulties in the numerical approximation of (3.14) reside in the computation of the nonlinear terms and in the time integration of the right-hand side.

Now, by integrating the system of ordinary differential equations (ODEs) (3.14) over a time interval $[t_m, t_{m+1}]$, where $t_m = m\,\Delta t$ (Δt is a given time step), we obtain

$$\begin{aligned}\tilde{\mathbf{u}}(\mathbf{k},t_{m+1}) &= e^{-\nu(4\pi^2/L_1^2)k^2\Delta t}\,\tilde{\mathbf{u}}(\mathbf{k},t_m) \\ &\quad + \int_{t_m}^{t_{m+1}} e^{\nu(4\pi^2/L_1^2)k^2(t-t_{m+1})}\,\hat{\mathbf{h}}_{\mathbf{k}}(\mathbf{u}_N,\mathbf{u}_N,t)\,dt.\end{aligned} \tag{3.15}$$

When one-step explicit schemes of Runge–Kutta type are used for the approximation of the integral in the right-hand side of the dynamical system (3.15), the following fully discretized system is obtained: Given $\mathbf{u}_N^0 = P_N\mathbf{u}_0$, compute

$$\mathbf{u}_N^{m+1}(\mathbf{x}) = \sum_{\mathbf{k}\in S_N} \tilde{\mathbf{u}}^{m+1}(\mathbf{k})\,e^{\mathrm{i}(2\pi/L_1)\mathbf{k}\cdot\mathbf{x}} \in V_N,$$

such that

$$\begin{aligned}\tilde{\mathbf{u}}^{m,0}(\mathbf{k}) &= \tilde{\mathbf{u}}^m(\mathbf{k}),\\ \tilde{\mathbf{u}}^{m,j}(\mathbf{k}) &= e^{-\nu(4\pi^2/L_1^2)k^2\alpha_j\Delta t}\,\tilde{\mathbf{u}}^{m,j-1}(\mathbf{k})\\ &\quad + \Delta t\,e^{-\nu(4\pi^2/L_1^2)k^2\alpha_j\Delta t}\sum_{l=0}^{j-1} b_{j,l}\,\hat{\mathbf{h}}_{\mathbf{k}}\left(\mathbf{u}_N^{m,l},\mathbf{u}_N^{m,l},t_{m,l}\right), \qquad \text{for } j=1,\dots,s,\\ \tilde{\mathbf{u}}^{m+1}(\mathbf{k}) &= \tilde{\mathbf{u}}^{m,s}(\mathbf{k}),\end{aligned} \tag{3.16}$$

for any $\mathbf{k} \in S_N \cap \{k_n \geq 0\}$, where $t_{m,j} = t_m + c_j\,\Delta t$ are intermediate times in the interval $[t_m, t_{m+1})$, with $t_{m,0} = t_m$, $t_{m,s} = t_{m+1}$, so that $[t_m, t_{m+1}) = \bigcup_{j=0}^{s-1}[t_{m,j}, t_{m,j+1})$. We have set $\alpha_j = c_j - c_{j-1}$, so that $t_{m,j} = t_{m,j-1} + \alpha_j\,\Delta t$ in (3.16). The coefficients $b_{j,l}$, $l = 0,\dots,j-1$, are real. Therefore $\mathbf{u}_N^{m,j}$ is an approximation of $\mathbf{u}_N$ at time $t_{m,j}$.

Note that $\mathbf{u}_N^{m+1}(\mathbf{x})$ defined by (3.16) satisfies the continuity equation $\nabla \cdot \mathbf{u}_N^{m+1} = 0$ if and only if the initial condition satisfies also the continuity equation; this condition was required in the existence and uniqueness theorem stated in Chapter 1.

Computational Complexity

The computation of $\tilde{\mathbf{u}}^{m+1}(\mathbf{k})$ is then reduced to s evaluations of the right-hand side of (3.14) at different discrete times. The pseudospectral method (see Canuto et al. (1988) for instance) based on the fast Fourier transform (FFT) is used to compute the truncated nonlinear term $P_N B(\mathbf{u}_N, \mathbf{u}_N)$; this requires $(4n-3) \times s$ calls to the FFT per time step, which correspond to $(4n-3)\ s \times c_2\ N^n \log_2(N)$ operations, where $c_2 \simeq 7.3$ in our codes. In all the simulations presented in Chapter 10, an explicit (low-storage) Runge–Kutta method of third order has been used (see Canuto et al. (1988), Dubois, Jauberteau, and Temam (1993)).

When computing the convolution sums occurring in the nonlinear terms via a pseudospectral method, aliasing errors are introduced, which may deteriorate the accuracy of the Galerkin approximation. When the resolution is high enough, these errors are very small and can be neglected (see Kerr (1985), for instance, and Jiménez (1994)). Aliasing errors can be removed by using the so-called $\frac{3}{2}$ rule (see Canuto et al. (1988)). In such a case, the nonlinear terms are evaluated by performing FFTs on a finer grid, with $3N/2$ instead of N modes in each spatial direction. This leads to a considerable increase of the number of operations per time step, to $(4n-3)\ s \times c_2(3N/2)^n \log_2 (3N/2)$.

For the sake of simplicity, we write hereafter $\hat{h}_j(\mathbf{k}) = \hat{\mathbf{h}}_\mathbf{k}(\mathbf{u}_N, \mathbf{u}_N) \cdot \mathbf{e}_j$, $j = 1, \dots, n$, where $\mathbf{e}_j$ is the jth unit vector in $\mathbb{R}^n$. Note that the time dependence is omitted. By definition, the right-hand side of (3.16) is orthogonal to the wavenumber $\mathbf{k}$, so that the following relation holds:

$$\hat{h}_n(\mathbf{k}) = -\sum_{\ell=1}^{n-1} k_\ell \hat{h}_\ell(\mathbf{k})/k_n \qquad \text{for} \quad \mathbf{k} \in S_N, \ \ k_n > 0. \tag{3.17}$$

The computation of $\{\hat{h}_\ell(\mathbf{k})\}_{\ell=1,n}$ requires the computation of the n components of $\tilde{B}_\mathbf{k}(\mathbf{u}_N, \mathbf{u}_N)$. However, due to (3.17), we only need to keep in memory the first $n-1$ components of the right-hand side $\hat{\mathbf{h}}$ and $\hat{h}_n(\mathbf{k})$ for $k_n = 0$; indeed, $\hat{h}_n(\mathbf{k})$ for $k_n > 0$ can be recovered at any time via (3.17). Similarly, for any vector field $\hat{\varphi}(\mathbf{k})$ satisfying the continuity equation, we have stored in our codes the following information: $\{\hat{\varphi}_j(\mathbf{k})\}_{j=1}^{n-1}$ for all $\mathbf{k} \in S_N$ and $\hat{\varphi}_n(\mathbf{k})$ for $k_n = 0$ only. This represents $2(N/2+1)[(n-1)N^{n-1} + N^{n-2}]$ real numbers (remember that the Fourier coefficients are complex numbers and that $k_n \geq 0$). Then the

Table 3.1. *Estimation of the memory size of three-dimensional Galerkin pseudospectral code (real quantities are assumed to be stored in double accuracy)*

Resolution	32^3	64^3	128^3	256^3	512^3	1024^3
Memory size (in Mega–Octets)	2.62	21	168	1,341	10,731	85,852

storage of the spectral coefficients of the velocity and of the nonlinear term represents $(N+2)[(n-1)N^{n-1}+N^{n-2}]$ real numbers; moreover, $2n-1$ additional (work) arrays of $(N+2)N^{n-1}$ real numbers have been used in order to perform the pseudospectral evaluation of the nonlinear term $\omega_N \times \mathbf{u}_N$. The memory size of our numerical code is then roughly $(N+2)[(4n-3)N^{n-1}+2N^{n-2}]$. When the $\frac{3}{2}$ rule is used for dealiasing, the size of the work arrays has to be increased, and it corresponds to $[3(N+2)/2](3N/2)^{n-1}$ real numbers.

It is important to note that the use of the relation (3.17) to reduce the amount of storage induces an overhead in terms of the number of operations. However, this overhead is small by comparison with the total number of operations required to solve (3.16). Moreover, the modifications induced by the implementation of (3.17) do not imply any complexity or loss of performance in terms of vectorization or multitasking.

In Table 3.1, we have listed estimated values of the memory used by the three-dimensional Galerkin code at different resolutions. Note that simulations with a resolution below 256^3 can be achieved on (small) computers such as the Cray-YMP EL or Cray-JEDI. Finer resolutions, namely 512^3 modes, have been reported by Chen et al. (1993) and Jiménez et al. (1993). These simulations were obtained on massively parallel computers. More recently, a 1024^3 simulation was done on a cluster of workstations (Woodward et al. (1995)).

Computational Efficiency

The simulations reported in Chapter 10 have been performed on a Cray YMP-EL (4 processors) and on a Cray C90 (8 processors). A particular effort has been made to optimize the codes; multitasked versions have been developed. The average performances of the codes are of the order of 485 megaflops per processor for the 128^3 and 256^3 simulations on the Cray C90. Moreover, the average number of processors used in the multitasking mode is of the order of 4.5 when 6 processors are requested, so that 2.18 gigaflops is reached.

Remark 3.1. Finally, we want to mention that the projector P can be implemented in different manners. Indeed, noting that, in the Fourier space, the vector $\hat{\mathbf{u}}(\mathbf{k})$ is orthogonal to the wavevector $\mathbf{k}$ (continuity equation), we see that $\hat{\mathbf{u}}(\mathbf{k})$

can be expressed in a basis orthogonal to $\mathbf{k}$. Such a decomposition is called the Craya decomposition; the reader is referred to Lesieur (1990) for details.

For instance, in the two-dimensional case, the Fourier expansion (3.4) can be rewritten as

$$\mathbf{u}(\mathbf{x}, t) = \sum_{\mathbf{k} \neq 0} \frac{\mathbf{k}^{\perp}}{k} \tilde{u}(\mathbf{k}, t)\, e^{i\mathbf{k}_L \cdot \mathbf{x}}, \tag{3.18}$$

where $\mathbf{k}^{\perp} = (k_2, -k_1)$ and $\tilde{u}(\mathbf{k})$ is a complex scalar. An evolution equation for $\tilde{u}(\mathbf{k})$ can be easily derived; see Jolly and Xiong (1995) and the references therein. The implementation of this decomposition is simple and efficient in two dimensions, but is more complicated in three dimensions. Hence, in our three-dimensional codes we have preferred, for its simplicity, the procedure based on the relation (3.17) described above.

Note that both methods reduce the dimensionality of the problem and lead to the computation of approximately $(n - 1)N^n$ unknowns. □

3.2. Flows in an Infinite Channel: Nonhomogeneous Turbulence

In this section, we consider wall-bounded flows – more precisely, flows in the domain $\Omega = (0, L_1) \times (-h, h) \times (0, L_3)$, where h is the half width of the channel and L_1 (L_3) is the length in the streamwise (spanwise) direction. The boundary conditions correspond to the problem ($\mathcal{P}_{\text{ch}}$) introduced in Chapter 1. In such a case, the projection operator P cannot be implemented as easily as in the fully periodic case. Therefore several numerical methods that we describe hereafter have been proposed in order to overcome this difficulty and to impose the continuity equation as well as the boundary conditions. Note that in Section 3.2.4 below, a new algorithm is presented; this is an improved version of the Galerkin approximation applied to flows in a plane channel and to Taylor–Couette flows by Moser, Moin, and Leonard (1983).

3.2.1. Preliminary

We assume here that the flow is sustained by a mean pressure gradient K_p applied in the streamwise direction. The velocity field $\mathbf{u}(\mathbf{x}, t)$ and the pressure $p(\mathbf{x}, t)$ then satisfy

$$\begin{aligned} \frac{\partial \mathbf{u}}{\partial t} - \nu \Delta \mathbf{u} + \omega \times \mathbf{u} + \nabla \left(p + \frac{1}{2}|\mathbf{u}|^2 \right) + K_p\, \mathbf{e}_1 = \mathbf{0}, \\ \nabla \cdot \mathbf{u} = \mathbf{0}, \\ \mathbf{u}(\mathbf{x}, t = 0) = \mathbf{u}_0(\mathbf{x}), \end{aligned} \tag{3.19}$$

where $\mathbf{e}_1 = (1, 0, 0)^T$ is the unit vector in the streamwise direction; in what follows, we will denote by $\mathbf{e}_j$ the unit vector in the j-direction in $\mathbb{R}^n$. The system (3.19) is supplemented with the following boundary conditions:

$$\begin{gathered} \mathbf{u}(x_1, \pm h, x_3, t) = \mathbf{0} \qquad \text{for any} \quad (x_1, x_3) \in Q = (0, L_1) \times (0, L_3), \\ \mathbf{u}(\mathbf{x} + L_j \mathbf{e}_j, t) = \mathbf{u}(\mathbf{x}, t), \qquad p(\mathbf{x} + L_j \mathbf{e}_j, t) = p(\mathbf{x}, t) \\ \text{for } j = 1, 3 \text{ and for any } \mathbf{x} \in \mathbb{R} \times (-h, h) \times \mathbb{R}. \end{gathered} \tag{3.20}$$

We set $q(\mathbf{x}, t) = p(\mathbf{x}, t) + \frac{1}{2}|\mathbf{u}|^2$ and rewrite the momentum equation in (3.19) in the form

$$\frac{\partial \mathbf{u}}{\partial t} - \nu\, \Delta \mathbf{u} + \boldsymbol{\omega} \times \mathbf{u} + \nabla q + K_p\, \mathbf{e}_1 = \mathbf{0}. \tag{3.21}$$

By definition, and due to (3.20), the pressure $q(\mathbf{x}, t)$ with zero mean gradient is periodic in the streamwise and spanwise directions.

A Velocity-Vorticity Formulation of the Navier–Stokes Equations

Instead of solving (3.19)–(3.20), the vorticity formulation can be considered. This problem is simpler, since no pressure term appears in the vorticity equation. However, boundary conditions for the vorticity are not provided, so that the problem is not well posed. Nevertheless, with the particular set of boundary conditions (3.20), we show hereafter that (3.19)–(3.20) can be reformulated as a system of equations involving only the velocity field $\mathbf{u}$, that is, the pressure can be removed. Moreover, boundary conditions for this system can be directly derived from (3.19)–(3.20). This velocity formulation has been used by Patera and Orzsag (1980) and Kim, Moin, and Moser (1987).

As a consequence of the boundary conditions (3.20) above, the normal component of the vorticity field, $\omega_2 = (\nabla \times \mathbf{u}) \cdot \mathbf{e}_2$, vanishes at the wall and is periodic in the streamwise and spanwise directions. Hence ω_2 is fully determined by the following evolution equation:

$$\begin{gathered} \frac{\partial \omega_2}{\partial t} - \nu \Delta \omega_2 + \frac{\partial H_1}{\partial x_3} - \frac{\partial H_3}{\partial x_1} = 0, \\ \omega_2(\mathbf{x}, t = 0) = (\nabla \times \mathbf{u}_0) \cdot \mathbf{e}_2, \end{gathered} \tag{3.22}$$

where we have set $\mathbf{H} = \boldsymbol{\omega} \times \mathbf{u}$. Note that $\mathbf{H}$ is a bilinear form, that is,

$$\mathbf{H}(\varphi, \psi) = (\nabla \times \varphi) \times \psi,$$

for any flow variables φ, ψ. For the sake of simplicity, unless it is necessary, this bilinear dependence will be omitted in what follows.

Assuming that the pressure satisfies the Poisson equation (1.21), and applying the Laplacian operator to the normal momentum equation, we formally derive a fourth-order evolution equation for the normal velocity $u_2(\mathbf{x}, t)$, namely,

$$\frac{\partial \Delta u_2}{\partial t} - \nu \Delta^2 u_2 + \Delta H_2 - \frac{\partial}{\partial x_2}(\nabla \cdot \mathbf{H}) = 0,$$
$$u_2(\mathbf{x}, t = 0) = \mathbf{u}_0(\mathbf{x}) \cdot \mathbf{e}_2. \tag{3.23}$$

Now, by assuming that the continuity equation holds at the wall, we deduce that $\partial u_2/\partial x_2$, the normal derivative of u_2, vanishes at the wall. Hence, (3.23) is supplemented with the boundary conditions

$$u_2(x_1, \pm h, x_3, t) = \frac{\partial u_2}{\partial x_2}(x_1, \pm h, x_3, t) = 0 \tag{3.24}$$

for any $(x_1, x_3) \in Q$ and $t \geq 0$.

Remark 3.2. Let (**u**, q) be a classical solution of (3.19)–(3.21) satisfying the Poisson equation (1.21), that is,

$$\Delta q = -\nabla \cdot \mathbf{H}.$$

Then a necessary condition for **u** to satisfy the continuity equation is that $\nabla \cdot \mathbf{u}$ vanishes at the wall. Indeed, by taking the dot product of the momentum equation with the ∇ operator we obtain

$$\frac{\partial}{\partial t} \nabla \cdot \mathbf{u} - \nu\, \Delta(\nabla \cdot \mathbf{u}) = 0 \quad \text{in } \Omega.$$

Hence, $\nabla \cdot \mathbf{u} = 0$ in Ω if it vanishes at the wall. □

We now introduce the operator

$$\langle \varphi \rangle = \frac{1}{L_1 L_3} \int_0^{L_1} \int_0^{L_3} \varphi(\mathbf{x})\, dx_1\, dx_3,$$

for any vector field φ in $[L^2(\Omega)]^n$, for instance. By definition, $\langle \varphi \rangle$ is a function of the real variable x_2 with values in $\mathbb{R}^n$. We write

$$\mathbf{U}(x_2, t) = \langle \mathbf{u}(\mathbf{x}, t) \rangle \qquad \text{for} \quad t \geq 0,\ x_2 \in (-h, h).$$

Since **u** satisfies the no-slip boundary condition at the walls, we have $\mathbf{U}(\pm h, t) = \mathbf{0}$.

The fluctuating part of the normal velocity $u_2 - U_2$ is fully determined by (3.23)–(3.24). It can be easily shown that U_2 cannot be obtained from (3.23). Moreover, U_2 can be obtained by imposing the continuity equation. Indeed, by averaging with respect to $\langle\cdot\rangle$ the continuity equation and by invoking the periodic boundary conditions for u_1 and u_3 we deduce that

$$\frac{\partial U_2}{\partial x_2}(\mathbf{x}, t) = 0.$$

As U_2 vanishes at the wall, we have that

$$U_2(x_2, t) = 0 \qquad \text{for any } x_2 \in (-h, h) \quad \text{and} \quad t \geq 0. \tag{3.25}$$

Knowing ω_2 and u_2 respectively from (3.22) and (3.23), we can easily deduce from the continuity equation the streamwise and spanwise components of the fluctuating velocity, $u_j - U_j$ for $j = 1, 3$. In order to be able to recover the whole velocity field, additional equations are needed for $U_j,\ j = 1, 3$. By averaging the streamwise and spanwise momentum equations and using the continuity equation, we find

$$\begin{gathered}\frac{\partial U_j}{\partial t} - \nu \frac{\partial^2 U_j}{\partial x_2^2} + \frac{\partial}{\partial x_2}\langle u_j\, u_2\rangle + K_p \mathbf{e}_1 = 0,\\ U_j(\pm h, t) = 0, \quad U_j(x_2, t = 0) = \langle \mathbf{u}_0\rangle \cdot \mathbf{e}_j \quad \text{for} \quad j = 1, 3.\end{gathered} \tag{3.26}$$

Note that, due to the periodic boundary conditions in the streamwise and spanwise directions, the pressure is no longer present in (3.26).

Finally, combining the equations (3.22) to (3.26), we obtain the following system:

$$\begin{gathered}\frac{\partial \omega_2}{\partial t} - \nu\, \Delta\omega_2 + \frac{\partial H_1}{\partial x_3} - \frac{\partial H_3}{\partial x_1} = 0,\\ \frac{\partial \Delta u_2}{\partial t} - \nu\, \Delta^2 u_2 + \Delta H_2 - \frac{\partial}{\partial x_2}(\boldsymbol{\nabla}\cdot\mathbf{H}) = 0,\\ \boldsymbol{\nabla}\cdot\mathbf{u} = 0,\\ \frac{\partial U_j}{\partial t} - \nu \frac{\partial^2 U_j}{\partial x_2^2} + \frac{\partial}{\partial x_2}\langle u_j\, u_2\rangle + K_p \mathbf{e}_1 = 0 \qquad \text{for} \quad j = 1, 3,\\ u_2(\mathbf{x}, t = 0) = \mathbf{u}_0(\mathbf{x}) \cdot \mathbf{e}_2, \qquad \omega_2(\mathbf{x}, t = 0) = (\boldsymbol{\nabla} \times \mathbf{u}_0) \cdot \mathbf{e}_2,\\ U_j(x_2, t = 0) = \langle \mathbf{u}_0\rangle \cdot \mathbf{e}_j \qquad \text{for} \quad j = 1, 3,\end{gathered} \tag{3.27}$$

which we supplement with the following boundary conditions:

$$\begin{aligned} &\omega_2(x_1, \pm h, x_3, t) = u_2(x_1, \pm h, x_3, t) = \frac{\partial u_2}{\partial x_2}(x_1, \pm h, x_3, t) = 0 \\ &\qquad\qquad\qquad\qquad\qquad\qquad \text{for any} \quad (x_1, x_3) \in Q, \\ &\omega_2(\mathbf{x} + L_j \mathbf{e}_j, t) = \omega_2(\mathbf{x}, t), \qquad u_2(\mathbf{x} + L_j \mathbf{e}_j, t) = u_2(\mathbf{x}, t), \\ &U_j(\pm h, t) = 0 \qquad \text{for} \quad j = 1, 3 \text{ and for any } \mathbf{x} \in \mathbb{R} \times (-h, h) \times \mathbb{R}. \end{aligned} \tag{3.28}$$

The existence and uniqueness of solutions for the problem (3.27)–(3.28) have been studied in Dubois and Miranville (1994). The system (3.27)–(3.28) possesses a solution that is unique only on (small) finite time intervals and for solutions with enough regularity. In this case, the problems (3.19)–(3.20) and (3.27)–(3.28) are equivalent.

3.2.2. Fourier–Galerkin Approximations and Time Discretizations

Fourier–Galerkin Approximations in the Streamwise and Spanwise Directions

We first derive the spectral form of (3.19)–(3.20). The velocity field as well as the pressure being periodic in the streamwise and spanwise directions, we expand them in Fourier series:

$$\mathbf{u}(\mathbf{x}, t) = \sum_{\mathbf{k} \in \mathbb{Z}^2} \hat{\mathbf{u}}(\mathbf{k}, x_2, t)\, e^{\mathbf{i}\underline{\mathbf{k}}\cdot\mathbf{x}} \quad \text{and} \quad q(\mathbf{x}, t) = \sum_{\mathbf{k} \in \mathbb{Z}^2} \hat{q}(\mathbf{k}, x_2, t)\, e^{\mathbf{i}\underline{\mathbf{k}}\cdot\mathbf{x}}.$$

Here $\underline{\mathbf{k}} = (2\pi k_1/L_1, 0, 2\pi k_3/L_3)$ and $\mathbf{k} = (k_1, k_3)$. By multiplying (3.19) by $e^{-\mathbf{i}\underline{\mathbf{k}}\cdot\mathbf{x}}$ and by averaging over the plane $Q = (0, L_1) \times (0, L_3)$, we obtain the following spectral form:

$$\begin{aligned} &\frac{\partial \hat{\mathbf{u}}(\mathbf{k})}{\partial t} - \nu D_{\mathbf{k}}^2 \hat{\mathbf{u}}(\mathbf{k}) + \hat{\mathbf{H}}_{\mathbf{k}}(\mathbf{u}, \mathbf{u}) + D_{\mathbf{k}} \hat{q}(\mathbf{k}) + K_p \delta_{\mathbf{k},\mathbf{0}}\, \mathbf{e}_1 = \mathbf{0}, \\ &\qquad\qquad\qquad\qquad\qquad\qquad D_{\mathbf{k}} \cdot \hat{\mathbf{u}}(\mathbf{k}) = 0, \\ &\hat{\mathbf{u}}(\mathbf{k}, t = 0) = \hat{\mathbf{u}}_0(\mathbf{k}), \\ &\hat{\mathbf{u}}(\mathbf{k}, \pm h, t) = \mathbf{0} \qquad \text{for all} \quad \mathbf{k} \in \mathbb{Z}^2, \end{aligned} \tag{3.29}$$

where $D_{\mathbf{k}}$ is the spectral form of ∇, that is, $D_{\mathbf{k}} = (\mathbf{i}2\pi k_1/L_1, \partial/\partial x_2, \mathbf{i}2\pi k_3/L_3)^T$; $D_{\mathbf{k}}^2$ is the spectral form of Δ, that is, $D_{\mathbf{k}}^2 = D_{\mathbf{k}}^T D_{\mathbf{k}} = (-\underline{k}^2 + \partial^2/\partial x_2^2)$; and $\underline{k}^2 = |\underline{\mathbf{k}}|^2 = 4\pi^2(k_1^2/L_1^2 + k_3^2/L_3^2)$. For the sake of simplicity, the dependences on x_2 and t have been omitted in (3.29).

The Galerkin approximation of (3.29) consists in looking for

$$\mathbf{u}_N(\mathbf{x},t)=\sum_{\mathbf{k}\in S_N}\tilde{\mathbf{u}}(\mathbf{k},x_2,t)\,e^{\mathrm{i}\underline{\mathbf{k}}\cdot\mathbf{x}}\quad\text{and}\quad q_N(\mathbf{x},t)=\sum_{\mathbf{k}\in S_N}\tilde{q}(\mathbf{k},x_2,t)\,e^{\mathrm{i}\underline{\mathbf{k}}\cdot\mathbf{x}}$$

where[2] $S_N=\{\mathbf{k}\in\mathbb{Z}^2;\ |k_j|\le\frac{N}{2}-1,\ j=1,3\}$, which is solution of

$$\begin{aligned}
\frac{\partial\tilde{\mathbf{u}}(\mathbf{k})}{\partial t}-\nu D_{\mathbf{k}}^2\tilde{\mathbf{u}}(\mathbf{k})+\hat{\mathbf{H}}_{\mathbf{k}}(\mathbf{u}_N,\mathbf{u}_N)+D_{\mathbf{k}}\tilde{q}(\mathbf{k})+K_p\delta_{\mathbf{k},\mathbf{0}}\,\mathbf{e}_1&=\mathbf{0},\\
D_{\mathbf{k}}\cdot\tilde{\mathbf{u}}(\mathbf{k})&=0,\\
\tilde{\mathbf{u}}(\mathbf{k},t=0)=\hat{\mathbf{u}}_0(\mathbf{k}),&\\
\tilde{\mathbf{u}}(\mathbf{k},\pm h,t)=\mathbf{0}\qquad\text{for all}\quad\mathbf{k}\in S_N\cap\{k_3\ge 0\}.&
\end{aligned}\tag{3.30}$$

Similarly, the Fourier–Galerkin approximation of (3.27) consists in looking for $\mathbf{u}_N(\mathbf{x},t)$ expanded as above, which is solution of

$$\begin{aligned}
\frac{\partial\tilde{\omega}_2(\mathbf{k})}{\partial t}-\nu D_{\mathbf{k}}^2\tilde{\omega}_2(\mathbf{k})+\mathrm{i}\underline{\mathbf{k}}^{\perp}\cdot\hat{\mathbf{H}}_{\mathbf{k}}(\mathbf{u}_N,\mathbf{u}_N)&=0,\\
\frac{\partial}{\partial t}D_{\mathbf{k}}^2\tilde{u}_2(\mathbf{k})-\nu D_{\mathbf{k}}^4\tilde{u}_2(\mathbf{k})+\hat{h}_{\mathbf{k}}(\mathbf{u}_N,\mathbf{u}_N)&=0,\\
D_{\mathbf{k}}\cdot\tilde{\mathbf{u}}(\mathbf{k})&=0,
\end{aligned}$$

$$\begin{aligned}
\tilde{\mathbf{u}}(\mathbf{k},x_2,t=0)=\hat{\mathbf{u}}_0(\mathbf{k},x_2)\qquad&\text{for all}\quad\mathbf{k}\in S_N\cap\{k_3\ge0\},\ \mathbf{k}\ne\mathbf{0},\\
\frac{\partial\tilde{u}_j(\mathbf{0})}{\partial t}-\nu\frac{\partial^2\tilde{u}_j}{\partial x_2^2}(\mathbf{0})+\hat{H}_{j,\mathbf{0}}(\mathbf{u}_N,\mathbf{u}_N)+K_p\mathbf{e}_1=0,\qquad&\text{for}\quad j=1,3,\\
\tilde{u}_j(\mathbf{0},x_2,t=0)=\langle\mathbf{u}_0\rangle\cdot\mathbf{e}_j\qquad&\text{for}\quad j=1,3,\\
\tilde{u}_2(\mathbf{0},x_2,t)=0\qquad&\text{for any}\quad x_2\in(-h,h),
\end{aligned}\tag{3.31}$$

where

$$\begin{aligned}
\tilde{\omega}_2(\mathbf{k},x_2)&=(\widetilde{\nabla\times\mathbf{u}_N})(\mathbf{k},x_2)=\mathrm{i}\underline{\mathbf{k}}^{\perp}\cdot\tilde{\mathbf{u}}(\mathbf{k},x_2),\\
\underline{\mathbf{k}}^{\perp}&=\left(\frac{2\pi}{L_3}k_3,0,-\frac{2\pi}{L_1}k_1\right),\\
D_{\mathbf{k}}^4&=D_{\mathbf{k}}^2D_{\mathbf{k}}^2.
\end{aligned}$$

[2] For the sake of simplicity, we have restricted ourselves to the particular case where the same number of modes are used in the streamwise and spanwise directions. However, in numerical simulations, different values are often considered.

We have also set in (3.31)

$$\hat{h}_{\mathbf{k}}(\mathbf{u}_N, \mathbf{u}_N) = D_{\mathbf{k}}^2 \hat{H}_{2,\mathbf{k}}(\mathbf{u}_N, \mathbf{u}_N) - \frac{\partial}{\partial x_2}\left(D_{\mathbf{k}} \cdot \hat{\mathbf{H}}_{\mathbf{k}}(\mathbf{u}_N, \mathbf{u}_N)\right).$$

The boundary conditions (3.28) can be rewritten as

$$\begin{aligned} \tilde{\omega}_2(\mathbf{k}, \pm h, t) = \tilde{u}_2(\mathbf{k}, \pm h, t) &= \frac{\partial \tilde{u}_2}{\partial x_2}(\mathbf{k}, \pm h, t) = 0 \\ \text{for any} \quad &\mathbf{k} \in S_N \cap \{k_3 \geq 0\}\ \mathbf{k} \neq \mathbf{0}, \\ \tilde{u}_j(\mathbf{0}, \pm h, t) = 0, \qquad &\text{for} \quad j = 1, 3. \end{aligned} \tag{3.32}$$

Computation of the Pressure

As was mentioned in Chapter 1, the pressure can be obtained from the velocity by solving the Poisson equation (1.21) with the following boundary conditions:

$$\begin{aligned} &\frac{\partial q_N}{\partial x_2}(x_1, \pm h, x_3, t) = \nu \frac{\partial^2 \mathbf{u}_N}{\partial x_2^2}(x_1, \pm h, x_3, t) \cdot \mathbf{e}_2, \\ &q_N(\mathbf{x} + L_j \mathbf{e}_j, t) = q_N(\mathbf{x}, t) \qquad \text{for} \quad j = 1, 3 \text{ and } \mathbf{x} \in \mathbb{R} \times (-h, h) \times \mathbb{R}. \end{aligned} \tag{3.33}$$

Let us assume that the pressure is a solution of (1.21) and (3.33), and that the normal velocity is a solution of the fourth-order evolution equation in (3.31); then the normal momentum equation holds:

$$\frac{\partial \tilde{u}_2}{\partial t}(\mathbf{k}) - \nu D_{\mathbf{k}}^2 \tilde{u}_2(\mathbf{k}) + \hat{H}_{2,\mathbf{k}}(\mathbf{u}_N, \mathbf{u}_N) + \frac{\partial \tilde{q}}{\partial x_2}(\mathbf{k}) = 0$$

for all $\mathbf{k} \in S_N \cap \{k_3 \geq 0\}$. By differentiating the above equation with respect to x_2 and using the continuity and the Poisson equations, we deduce that $\tilde{q}(\mathbf{k}, x_2)$ satisfies

$$\begin{aligned} \tilde{q}(\mathbf{k}, x_2, t) = \frac{1}{\underline{k}^2}\left[\frac{\partial \tilde{f}}{\partial t}(\mathbf{k}, x_2) - \nu D_{\mathbf{k}}^2 \tilde{f}(\mathbf{k}, x_2) + \mathbf{i}\, \underline{\mathbf{k}} \cdot \hat{\mathbf{H}}_{\mathbf{k}}(\mathbf{u}_N, \mathbf{u}_N)\right] \\ \text{for all} \quad x_2 \in (-h, h), \quad t > 0, \quad \mathbf{k} \in S_N \cap \{k_3 \geq 0\}, \quad \mathbf{k} \neq \mathbf{0}, \end{aligned} \tag{3.34}$$

where we have set $\tilde{f}(\mathbf{k}, x_2) = \mathbf{i}\, \underline{\mathbf{k}} \cdot \tilde{\mathbf{u}}(\mathbf{k}, x_2)$. Therefore, the pressure satisfies the following Dirichlet boundary conditions:

$$\tilde{q}(\mathbf{k}, \pm h, t) = -\frac{\nu}{\underline{k}^2} D_{\mathbf{k}}^2 \tilde{f}(\mathbf{k}, \pm h) \qquad \text{for all} \quad \mathbf{k} \in S_N \cap \{k_3 \geq 0\}, \quad \mathbf{k} \neq \mathbf{0}. \tag{3.35}$$

Now, we assume that the continuity equation holds, namely

$$\tilde{f}(\mathbf{k}, x_2) = -\frac{\partial \tilde{u}_2}{\partial x_2}(\mathbf{k}, x_2), \qquad \mathbf{k} \in S_N \cap \{k_3 \geq 0\},$$

and that the normal velocity u_2 is a solution of the fourth-order evolution equation in (3.31). If the pressure is given by (3.34), then the normal momentum equation is satisfied. In this case, $\tilde{q}(\mathbf{k}, x_2)$ satisfies the Neumann boundary conditions (3.33) and the Poisson equation.

In summary, computing the pressure from (1.21) and (3.33) or from (3.34) is equivalent. In both cases, the boundary conditions (3.33) as well as (3.35) are satisfied.

Time Discretization

We now briefly describe the time discretization of the systems (3.30) and (3.31). As in Section 3.1, one-step explicit schemes of Runge–Kutta type are used for the approximation of the nonlinear terms, while a Crank–Nicholson scheme is retained for the diffusive (linear) part as well as the pressure term in (3.30). As usual, we denote the time step by Δt. The discretization of (3.30) is as follows: Given $\mathbf{u}_N^0(\mathbf{x}) = \sum_{\mathbf{k} \in S_N} \hat{\mathbf{u}}_0(\mathbf{k}, x_2)\, e^{\underline{\mathbf{i}\mathbf{k}} \cdot \mathbf{x}}$, compute

$$\mathbf{u}_N^{m+1}(\mathbf{x}) = \sum_{\mathbf{k} \in S_N} \tilde{\mathbf{u}}^{m+1}(\mathbf{k}, x_2)\, e^{\underline{\mathbf{i}\mathbf{k}} \cdot \mathbf{x}},$$

such that, for any $\mathbf{k} \in S_N \cap \{k_3 \geq 0\}$,

$$\begin{aligned}
&\tilde{\mathbf{u}}^{m,0}(\mathbf{k}) = \tilde{\mathbf{u}}^{m}(\mathbf{k}), \\
&\left(1 - \frac{\nu \alpha_j\, \Delta t}{2} D_{\mathbf{k}}^2\right) \tilde{\mathbf{u}}^{m,j}(\mathbf{k}) + \frac{\alpha_j}{2}\, \Delta t\, D_{\mathbf{k}} \tilde{q}^{m,j}(\mathbf{k}) \\
&\quad = -\frac{\alpha_j}{2}\, \Delta t\, D_{\mathbf{k}} \tilde{q}^{m,j-1}(\mathbf{k}) + \left(1 + \frac{\nu \alpha_j\, \Delta t}{2} D_{\mathbf{k}}^2\right) \tilde{\mathbf{u}}^{m,j-1}(\mathbf{k}) \\
&\quad\quad - \Delta t \sum_{l=0}^{j-1} b_{j,l} \left[\hat{\mathbf{H}}_{\mathbf{k}}\left(\mathbf{u}_N^{m,l}, \mathbf{u}_N^{m,l}\right) + K_p \delta_{\mathbf{k},\mathbf{0}}\, \mathbf{e}_1\right], \\
&D_{\mathbf{k}} \cdot \tilde{\mathbf{u}}^{m,j} = 0, \quad \tilde{\mathbf{u}}^{m,j}(\mathbf{k}, \pm h) = \mathbf{0} \qquad \text{for} \quad j = 1, \ldots, s, \\
&\tilde{\mathbf{u}}^{m+1}(\mathbf{k}) = \tilde{\mathbf{u}}^{m,s}(\mathbf{k}),
\end{aligned} \tag{3.36}$$

where $t_{m,j} = t_m + c_j\, \Delta t$ are intermediate times in the interval $[t_m, t_{m+1})$, with

$t_{m,0} = t_m$, $t_{m,s} = t_{m+1}$, so that $[t_m, t_{m+1}) = \bigcup_{j=0}^{s-1}[t_{m,j}, t_{m,j+1})$. We have set $\alpha_j = c_j - c_{j-1}$ so that $t_{m,j} = t_{m,j-1} + \alpha_j \,\Delta t$. The coefficients $b_{j,l}$, $l = 0, \ldots, j-1$, are real. Therefore $\mathbf{u}_N^{m,j}$ is an approximation of $\mathbf{u}_N$ at the intermediate times $t_{m,j}$. Note that here the operator $D_{\mathbf{k}}^2$ is equal to $-\underline{k}^2 + d^2/dx_2^2$, that is, the partial normal derivatives become standard normal derivatives. For the sake of simplicity the same notation $D_{\mathbf{k}}^2$ as before is used.

A similar time approximation of the velocity formulation (3.31) leads to the following sub-time-step computations:

$$\left(1 - \frac{\nu\alpha_j\,\Delta t}{2} D_{\mathbf{k}}^2\right) \tilde{\omega}_2^{m,j}(\mathbf{k}) = \left(1 + \frac{\nu\alpha_j\,\Delta t}{2} D_{\mathbf{k}}^2\right) \tilde{\omega}_2^{m,j-1}(\mathbf{k}) - \Delta t \sum_{l=0}^{j-1} b_{j,l}\, \mathrm{i}\underline{\mathbf{k}}^{\perp} \cdot \hat{\mathbf{H}}_{\mathbf{k}} \left(\mathbf{u}_N^{m,j-1}, \mathbf{u}_N^{m,j-1}\right),$$

$$\left(1 - \frac{\nu\alpha_j\,\Delta t}{2} D_{\mathbf{k}}^2\right) D_{\mathbf{k}}^2 \tilde{u}_2^{m,j}(\mathbf{k}) = \left(1 + \frac{\nu\alpha_j\,\Delta t}{2} D_{\mathbf{k}}^2\right) D_{\mathbf{k}}^2 \tilde{u}_2^{m,j-1}(\mathbf{k}) - \Delta t \sum_{l=0}^{j-1} b_{j,l}\, \hat{h}_{\mathbf{k}} \left(\mathbf{u}_N^{m,j-1}, \mathbf{u}_N^{m,j-1}\right), \quad (3.37)$$

$$D_{\mathbf{k}} \cdot \tilde{\mathbf{u}}^{m,j}(\mathbf{k}) = 0, \qquad \text{for any} \quad \mathbf{k} \in S_N \cap \{k_3 \geq 0\},\ \mathbf{k} \neq \mathbf{0},$$

$$\left(1 - \frac{\nu\alpha_j\,\Delta t}{2} \frac{d^2}{dx_2^2}\right) \tilde{u}_i^{m,j}(\mathbf{0}) = \left(1 + \frac{\nu\alpha_j\,\Delta t}{2} \frac{d^2}{dx_2^2}\right) \tilde{u}_i^{m,j}(\mathbf{0}) - \Delta t \sum_{l=0}^{j-1} b_{j,l} \left[\hat{H}_{i,\mathbf{0}} \left(\mathbf{u}_N^{m,j-1}, \mathbf{u}_N^{m,j-1}\right) + K_p \mathbf{e}_1\right], \qquad \text{for} \quad i = 1, 3,$$

$$\tilde{u}_2^{m,j}(\mathbf{0}) = 0,$$

which is supplemented with the following boundary conditions:

$$\tilde{\omega}_2^{m,j}(\mathbf{k}, \pm h) = \tilde{u}_2^{m,j}(\mathbf{k}, \pm h) = \frac{d\tilde{u}_2^{m,j}}{dx_2}(\mathbf{k}, \pm h) = 0 \qquad \text{for any} \quad \mathbf{k} \in S_N \cap \{k_3 \geq 0\}, \quad \mathbf{k} \neq \mathbf{0}, \qquad (3.38)$$

$$\tilde{u}_i^{m,j}(\mathbf{0}, \pm h) = 0 \qquad \text{for} \quad i = 1, 3, \quad j = 1, \ldots, s.$$

Remark 3.3. Note that in (3.37), the operators involved in the computation of $\tilde{\omega}_2^{m,j}(\mathbf{k})$ and $\tilde{u}_2^{m,j}(\mathbf{k})$ depend on the wavenumber $\mathbf{k}$ via the modulus of $\underline{\mathbf{k}}$. After discretization of the normal derivatives, this will result in the computation and storage of roughly $2 \times (N/2+1)^2$ complex matrices of size $(M+1)^2$ if $M+1$ points are used in the normal direction. However, this extra computations can be reduced. Indeed, by multiplying the $\tilde{\omega}_2(\mathbf{k})$ evolution equation by $\exp(\nu \underline{k}^2 t)$, we obtain

$$\frac{\partial}{\partial t}\left[e^{\nu \underline{k}^2 t}\tilde{\omega}_2(\mathbf{k})\right] - \nu\, e^{\nu \underline{k}^2 t}\,\frac{\partial^2 \tilde{\omega}_2}{\partial x_2^2}(\mathbf{k}) + e^{\nu \underline{k}^2 t}\,\mathbf{i}\underline{\mathbf{k}}^{\perp}\cdot \hat{\mathbf{H}}_{\mathbf{k}}(\mathbf{u}_N, \mathbf{u}_N) = 0.$$

After time discretization, we find

$$\left(1 - \frac{\nu\alpha_j\,\Delta t}{2}\frac{d^2}{dx_2^2}\right)\tilde{\omega}_2^{m,j}(\mathbf{k}) \tag{3.39}$$

$$= e^{-\nu \underline{k}^2 \alpha_j \Delta t}\left(1 + \frac{\nu\alpha_j\,\Delta t}{2}\frac{d^2}{dx_2^2}\right)\tilde{\omega}_2^{m,j}(\mathbf{k})$$

$$- \Delta t\, e^{-\nu \underline{k}^2 \alpha_j \Delta t}\sum_{l=0}^{j-1} b_{j,l}\,\mathbf{i}\underline{\mathbf{k}}^{\perp}\cdot \hat{\mathbf{H}}_{\mathbf{k}}\left(\mathbf{u}_N^{m,j-1}, \mathbf{u}_N^{m,j-1}\right).$$

Therefore, the operator in the left-hand side of (3.39) is independent of the wavenumber and is the same as the one appearing in the discretized equation for $\tilde{u}_i^{m,j}(\mathbf{0})$ in (3.37). □

3.2.3. Methods Based on Pressure Solvers

When considering the problem in the primitive variable form (3.36), a coupled system of ordinary differential equations (ODEs), involving the velocity field and the pressure, has to be solved at each discrete time $t_{m,j},\ j = 1, \ldots, s$, namely,

$$\left(\beta(j,\mathbf{k}) - \frac{\nu}{2}\frac{d^2}{dx_2^2}\right)\tilde{\mathbf{u}}^{m,j}(\mathbf{k}, x_2) + \frac{1}{2}\,D_{\mathbf{k}}\tilde{q}^{m,j}(\mathbf{k}, x_2) = \hat{\mathbf{R}}^{m,j}(\mathbf{k}, x_2),$$

$$D_{\mathbf{k}}\cdot \tilde{\mathbf{u}}^{m,j}(\mathbf{k}, x_2) = 0, \tag{3.40}$$

$$\tilde{\mathbf{u}}^{m,j}(\mathbf{k}, \pm h) = \mathbf{0} \qquad \text{for} \quad \mathbf{k} \in S_N \cap \{k_3 \geq 0\},$$

where $\beta(j,\mathbf{k}) = 1/(\alpha_j\,\Delta t) + (\nu/2)\underline{k}^2$ and $\hat{\mathbf{R}}^{m,j}(\mathbf{k}, x_2)$ denotes the right-hand side in (3.36) divided by $\alpha_j\,\Delta t$.

Taking the inner product of the first equation in (3.40) with the discrete gradient operator $D_{\mathbf{k}}$ leads to

$$\left(\underline{k}^2 - \frac{d^2}{dx_2^2}\right) \tilde{q}^{m,j}(\mathbf{k}, x_2) = -2\, D_{\mathbf{k}} \cdot \hat{\mathbf{R}}^{m,j}(\mathbf{k}, x_2), \quad \mathbf{k} \in S_N \cap \{k_3 \geq 0\}.$$

By imposing the above Poisson equation, the continuity equation holds if the boundary condition $(d\tilde{u}_2^{m,j}/dx_2)(\mathbf{k}, \pm h) = 0$ is satisfied. Therefore (3.40) can be equivalently rewritten as

$$\begin{aligned}
&\left(\beta(j,\mathbf{k}) - \frac{\nu}{2}\frac{d^2}{dx_2^2}\right) \tilde{\mathbf{u}}^{m,j}(\mathbf{k}, x_2) + \frac{1}{2}\, D_{\mathbf{k}}\tilde{q}^{m,j}(\mathbf{k}, x_2) = \hat{\mathbf{R}}^{m,j}(\mathbf{k}, x_2),\\
&\left(\underline{k}^2 - \frac{d^2}{dx_2^2}\right) \tilde{q}^{m,j}(\mathbf{k}, x_2) = -2\, D_{\mathbf{k}} \cdot \hat{\mathbf{R}}^{m,j}(\mathbf{k}, x_2),\\
&\tilde{\mathbf{u}}^{m,j}(\mathbf{k}, \pm h) = \mathbf{0},\\
&\frac{d\tilde{u}_2^{m,j}}{dx_2}(\mathbf{k}, \pm h) = 0 \qquad \text{for} \quad \mathbf{k} \in S_N \cap \{k_3 \geq 0\}.
\end{aligned} \tag{3.41}$$

Note that once the pressure is known, the streamwise and spanwise velocity components are easily obtained by solving

$$\begin{aligned}
&\left(\beta(j,\mathbf{k}) - \frac{\nu}{2}\frac{d^2}{dx_2^2}\right) \tilde{u}_l^{m,j}(\mathbf{k}, x_2) = \hat{R}_l^{m,j}(\mathbf{k}, x_2) - \mathbf{i}\frac{2\pi}{L_l} k_l\, \tilde{q}^{m,j}(\mathbf{k}, x_2),\\
&\tilde{u}_l^{m,j}(\mathbf{k}, \pm h) = \mathbf{0}, \qquad l = 1, 3, \quad \mathbf{k} \in S_N \cap \{k_3 \geq 0\}.
\end{aligned} \tag{3.42}$$

For instance, the Chebyshev-tau method can be used to impose the boundary conditions in (3.42). The normal velocity and the pressure then satisfy the following coupled one-dimensional Helmholtz equations:

$$\begin{aligned}
&\left(\beta(j,\mathbf{k}) - \frac{\nu}{2}\frac{d^2}{dx_2^2}\right) \tilde{u}_2^{m,j}(\mathbf{k}, x_2) + \frac{1}{2}\frac{d\tilde{q}}{dx_2}^{m,j}(\mathbf{k}, x_2) = \hat{R}_2^{m,j}(\mathbf{k}, x_2),\\
&\left(\underline{k}^2 - \frac{d^2}{dx_2^2}\right) \tilde{q}^{m,j}(\mathbf{k}, x_2) = -2\, D_{\mathbf{k}} \cdot \hat{\mathbf{R}}^{m,j}(\mathbf{k}, x_2),\\
&\tilde{u}_2^{m,j}(\mathbf{k}, \pm h) = \frac{d\tilde{u}_2}{dx_2}^{m,j}(\mathbf{k}, \pm h) = 0, \qquad \text{for} \quad \mathbf{k} \in S_N \cap \{k_3 \geq 0\}.
\end{aligned} \tag{3.43}$$

Using the linearity of this ODE system, Kleiser and Schuman (1980) proposed to use an influence matrix in order to impose the boundary conditions in (3.43). This method consists in solving the following steps:

Step 1: Compute $\{\tilde{u}_{2,0}^{m,j}(\mathbf{k}), \tilde{q}_0^{m,j}(\mathbf{k})\}_{\mathbf{k}\in S_N\cap\{k_3\geq 0\}}$ by solving the following Helmholtz equations

$$\begin{gathered}
\left(\underline{k}^2 - \frac{d^2}{dx_2^2}\right)\tilde{q}_0^{m,j}(\mathbf{k}) = -2\,D_{\mathbf{k}}\cdot\hat{\mathbf{R}}^{m,j}(\mathbf{k}),\\
\tilde{q}_0^{m,j}(\mathbf{k},\pm h) = 0,\\
\left(\beta(j,\mathbf{k}) - \frac{\nu}{2}\frac{d^2}{dx_2^2}\right)\tilde{u}_{2,0}^{m,j}(\mathbf{k}) = \hat{R}_2^{m,j}(\mathbf{k},x_2) - \frac{1}{2}\frac{d\tilde{q}_0^{m,j}}{dx_2}(\mathbf{k}),\\
\tilde{u}_{2,0}(\mathbf{k},\pm h) = 0, \qquad \mathbf{k}\in S_N\cap\{k_3\geq 0\}.
\end{gathered} \tag{3.44}$$

Step 2: Compute $\{\tilde{u}_{2,i}(\mathbf{k}), \tilde{q}_i(\mathbf{k})\}_{\mathbf{k}\in S_N, i=1,2}$ by solving

$$\begin{gathered}
\left(\underline{k}^2 - \frac{d^2}{dx_2^2}\right)\tilde{q}_i(\mathbf{k}) = 0,\\
\tilde{q}_i(\mathbf{k},-h) = i-1, \qquad \tilde{q}_i(\mathbf{k},h) = 2-i,\\
\left(\beta(j,\mathbf{k}) - \frac{\nu}{2}\frac{d^2}{dx_2^2}\right)\tilde{u}_{2,i}(\mathbf{k}) = -\frac{1}{2}\frac{d\tilde{q}_i}{dx_2}(\mathbf{k}),\\
\tilde{u}_{2,i}(\mathbf{k},\pm h) = 0, \qquad \mathbf{k}\in S_N\cap\{k_3\geq 0\}.
\end{gathered} \tag{3.45}$$

Step 3: The normal velocity is finally formed as the linear combination

$$\tilde{u}_2^{m,j}(\mathbf{k},x_2) = \tilde{u}_{2,0}^{m,j}(\mathbf{k},x_2) + \sum_{i=1}^{2}\gamma_i^{m,j}\tilde{u}_{2,i}(\mathbf{k},x_2),$$

where the coefficients $\gamma_i^{m,j}$ are computed by imposing

$$\frac{d\tilde{u}_2^{m,j}}{dx_2}(\mathbf{k},\pm h) = 0,$$

which involves the resolution of the 2×2 influence matrix problem

$$\begin{pmatrix} \dfrac{d\tilde{u}_{2,1}}{dx_2}(\mathbf{k},-h) & \dfrac{d\tilde{u}_{2,2}}{dx_2}(\mathbf{k},-h)\\ \dfrac{d\tilde{u}_{2,1}}{dx_2}(\mathbf{k},h) & \dfrac{d\tilde{u}_{2,2}}{dx_2}(\mathbf{k},h)\end{pmatrix}\begin{pmatrix}\gamma_1^{m,j}\\ \gamma_2^{m,j}\end{pmatrix} = -\begin{pmatrix}\dfrac{d\tilde{u}_{2,0}^{m,j}}{dx_2}(\mathbf{k},-h)\\ \dfrac{d\tilde{u}_{2,0}^{m,j}}{dx_2}(\mathbf{k},h)\end{pmatrix}$$

for all $\mathbf{k}\in S_N\cap\{k_3\geq 0\}$.

Note that the systems (3.45) are time-independent and thus they can be solved once at the beginning of the time iterations and stored. A similar method has been implemented by Jiménez (1990) for solving the vorticity formulation for a flow in a two-dimensional channel.

Remark 3.4. The resolution of (3.44)–(3.45) can now be easily handled in the Chebyshev representation and the boundary conditions can be applied by using a Chebyshev-tau method. However, in order to impose the boundary conditions, the tau method modifies the equations, and tau-corrector terms appear acting only on the two highest modes of the Chebyshev representation. Due to the coupling between the pressure and the normal velocity, these tau terms will propagate and contaminate the approximation of all modes. The numerical effect is that the continuity equation will not be satisfied. Corrections to the influence matrix technique have to be made in order to recover a divergence-free velocity field. Such modified methods were proposed by Kleiser and Schumann (1980) and more recently by Werne (1995). The reader is referred to these articles for the details. □

3.2.4. Spectral Approximations of the Velocity Formulation

We now consider the velocity-vorticity formulation (3.27)–(3.28), which reduces to (3.37)–(3.38) after Fourier–Galerkin approximation and time discretization. The boundary conditions can be applied to the normal vorticity and velocity equations in (3.37) by using either a Chebyshev-tau method or a (Legendre or Chebyshev) Galerkin method. Note that the collocation method based on the Chebyshev–Gauss–Lobatto points $x_{2,j} = h\cos(j\pi/M)$, $j = 0, \ldots, M$, can also be used. However, this method is not described here, and the interested reader is referred to Gottlieb and Orzag (1977) and Canuto et al. (1988).

Chebyshev-Tau Method

The normal vorticity $\tilde{\omega}_2^{m,j}(\mathbf{k})$ satisfies a Dirichlet boundary value problem, namely

$$\begin{aligned} \left(\beta(j,\mathbf{k}) - \frac{\nu}{2}\frac{d^2}{dx_2^2}\right)\tilde{\omega}_2^{m,j}(\mathbf{k},x_2) &= \hat{R}_{\omega}^{m,j}(\mathbf{k}), \\ \tilde{\omega}_2^{m,j}(\mathbf{k},\pm h) &= 0, \end{aligned} \tag{3.46}$$

where $\hat{R}_{\omega}^{m,j}(\mathbf{k})$ denotes the right-hand side of the normal vorticity equation in (3.37) divided by $\alpha_j\,\Delta t$. The problem (3.46) can be easily approximated with the Chebyshev-tau method. Similar equations are satisfied by the average streamwise and spanwise velocity components. By noting that these averages are real and recalling that $\tilde{\omega}_2(\mathbf{0}, x_2, t) = 0$ for all time $t \geq 0$, we can set

$$\tilde{\mathrm{w}}^0(\mathbf{k}, x_2) = \begin{cases} \hat{\omega}_2^0(\mathbf{k}, x_2), & \mathbf{k} \neq \mathbf{0}, \\ \hat{u}_1^0(\mathbf{0}, x_2) + \mathbf{i}\hat{u}_3^0(\mathbf{0}, x_2), & \mathbf{k} = 0. \end{cases}$$

Moreover, by setting

$$\hat{\mathcal{R}}^{m,j}(\mathbf{k}) = \hat{R}_\omega^{m,j}(\mathbf{k}), \quad \mathbf{k} \neq \mathbf{0},$$

and

$$\hat{\mathcal{R}}^{m,j}(\mathbf{0}) = \left(1 + \frac{\nu\alpha_j \, \Delta t}{2} \frac{d^2}{dx_2^2}\right) \tilde{\omega}_2^{m,j}(\mathbf{0}) - \Delta t \sum_{l=0}^{j-1} b_{j,l}$$
$$\left[\hat{H}_{1,0}\left(\mathbf{u}_N^{m,j-1}, \mathbf{u}_N^{m,j-1}\right) + \mathbf{i}\hat{H}_{3,0}\left(\mathbf{u}_N^{m,j-1}, \mathbf{u}_N^{m,j-1}\right) + K_p \mathbf{e}_1\right],$$

we obtain

$$\tilde{\mathrm{w}}^{m,j}(\mathbf{k}, x_2) = \begin{cases} \hat{\omega}_2^{m,j}(\mathbf{k}, x_2), & \mathbf{k} \neq \mathbf{0}, \\ \hat{u}_1^{m,j}(\mathbf{0}, x_2) + \mathbf{i}\, \hat{u}_3^{m,j}(\mathbf{0}, x_2), & \mathbf{k} = \mathbf{0}, \end{cases}$$

which is solution of

$$\left(\beta(j, \mathbf{k}) - \frac{\nu}{2} \frac{d^2}{dx_2^2}\right) \tilde{\mathrm{w}}^{m,j}(\mathbf{k}, x_2) = \hat{\mathcal{R}}^{m,j}(\mathbf{k}),$$
$$\tilde{\mathrm{w}}^{m,j}(\mathbf{k}, \pm h) = 0,$$

for all $m > 0$ and $j = 1, \ldots, s$.

However, one must be careful when approximating the normal velocity $\tilde{u}_2^{m,j}(\mathbf{k})$, which satisfies a fourth-order differential equation:

$$\left(\beta(j, \mathbf{k}) - \frac{d^2}{dx_2^2}\right) D_{\mathbf{k}}^2 \tilde{u}_2^{m,j}(\mathbf{k}, x_2) = \hat{R}_u^{m,j}(\mathbf{k}, x_2),$$
$$\tilde{u}_2^{m,j}(\mathbf{k}, \pm h) = \frac{d\tilde{u}_2^{m,j}}{dx_2}(\mathbf{k}, \pm h) = 0, \quad \mathbf{k} \in S_N \cap \{k_3 \geq 0\}, \tag{3.47}$$

where $\beta(j, \mathbf{k}) = (2/(\nu\alpha_j \, \Delta t)) + \underline{k}^2$ and $\hat{R}_u^{m,j}(\mathbf{k})$ denotes the right-hand side of the normal velocity equation in (3.37) divided by $\nu\alpha_j \, \Delta t/2$.

It is well known (see for instance Gottlieb and Orzag (1977)) that the direct application of the tau method to fourth-order equations leads to an ill-conditioned system that is numerically unstable. In order to overcome this difficulty, Kim, Moin, and Moser (1987) used a split-step integration scheme and applied the boundary conditions at the end of the last partial step by using an influence matrix technique similar to that described in the previous section.

The whole scheme consists of the following steps:

Step 1: Compute $\{\tilde{u}_{2,0}^{m,j}(\mathbf{k})\}_{\mathbf{k}\in S_N\cap\{k_3\geq 0\}}$ by solving the following system of Helmholtz equations:

$$\begin{aligned}
&\left(\beta(j,\mathbf{k})-\frac{d^2}{dx_2^2}\right)\tilde{\Phi}_0^{m,j}(\mathbf{k})=\hat{R}_u^{m,j}(\mathbf{k}),\\
&\tilde{\Phi}_0^{m,j}(\mathbf{k},\pm h)=0,\\
&\left(-\underline{k}^2+\frac{d^2}{dx_2^2}\right)\tilde{u}_{2,0}^{m,j}(\mathbf{k})=\tilde{\Phi}_0^{m,j}(\mathbf{k}),\\
&\tilde{u}_{2,0}^{m,j}(\mathbf{k},\pm h)=0 \qquad \text{for all}\quad \mathbf{k}\in S_N\cap\{k_3\geq 0\}.
\end{aligned} \tag{3.48}$$

Step 2: Compute $\{\tilde{u}_{2,i}(\mathbf{k})\}_{\mathbf{k}\in S_N\cap\{k_3\geq 0\},i=1,2}$ by solving the following systems:

$$\begin{aligned}
&\left(\beta(j,\mathbf{k})-\frac{d^2}{dx_2^2}\right)\tilde{\Phi}_i(\mathbf{k})=0,\\
&\tilde{\Phi}_i(\mathbf{k},-h)=i-1,\qquad \tilde{\Phi}_i(\mathbf{k},+h)=2-i,\\
&\left(-\underline{k}^2+\frac{d^2}{dx_2^2}\right)\tilde{u}_{2,i}(\mathbf{k})=\tilde{\Phi}_i(\mathbf{k}),\\
&\tilde{u}_{2,i}(\mathbf{k},\pm h)=0 \qquad \text{for all}\quad \mathbf{k}\in S_N\cap\{k_3\geq 0\}.
\end{aligned} \tag{3.49}$$

Step 3: The component of the normal velocity $\tilde{u}_2^{m,j}(\mathbf{k},x_2)$ is obtained as in the previous section after solving a 2×2 influence matrix problem.

Once $\tilde{\omega}_2^{m,j}(\mathbf{k})$ and $\tilde{u}_2^{m,j}(\mathbf{k})$ have been computed for $\mathbf{k}\in S_N,\ \mathbf{k}\neq\mathbf{0}$, we obtain the streamwise and spanwise velocity components by solving

$$\begin{aligned}
&\underline{\mathbf{k}}^{\perp}\cdot\tilde{\mathbf{u}}^{m,j}(\mathbf{k})=\tilde{\omega}_2^{m,j}(\mathbf{k}),\\
&\underline{\mathbf{k}}\cdot\tilde{\mathbf{u}}^{m,j}(\mathbf{k})=-\frac{d\tilde{u}_2^{m,j}}{dx_2}(\mathbf{k}),\qquad \mathbf{k}\in S_N,\ \mathbf{k}\neq\mathbf{0}.
\end{aligned} \tag{3.50}$$

Therefore, with this method the continuity equation is always satisfied.

Remark 3.5. In Section 3.2.2, we have noted that imposing the Poisson equation supplemented by the Neumann boundary condition or imposing the evolution equation for $\mathbf{i}\underline{\mathbf{k}}\cdot\tilde{\mathbf{u}}^{m,j}(\mathbf{k})$ can be equivalently used for computing the pressure. However this is no longer the case at the fully discrete level. Indeed, when the tau method is used to approximate the problems (3.48)–(3.50), the fourth-order equation (3.47) is not satisfied by the approximate normal velocity. Tau corrector terms appear on the highest modes of the approximated equations, and they propagate to all modes under the effect of discrete normal derivative operators. □

Legendre– and Chebyshev–Galerkin Approximations

We introduce the polynomial spaces

$$V_M = \{\varphi \in Y_M;\ \ \varphi(\pm h) = 0\},$$

and

$$W_M = \left\{\varphi \in Y_M;\ \ \varphi(\pm h) = \frac{d\varphi}{dx_2}(\pm h) = 0\right\},$$

where Y_M is the space of polynomials of degree $\leq M$. We denote by $(\cdot, \cdot)_w$ the following inner product:

$$(\varphi_1, \varphi_2)_w = \int_{-h}^{+h} \varphi_1(x_2)\varphi_2(x_2)w(x_2)\, dx_2,$$

where $w(x_2)$ is the weight function; $w(x_2) = (1 - x_2^2)^{-1/2}$ in the case of the Chebyshev polynomials, and $w(x_2) = 1$ in the case of the Legendre polynomials (see Gottlieb and Orzag (1977) and Canuto et al. (1988)).

The space Y_M is spanned by the $M + 1$ first Chebyshev or Legendre polynomials, that is,

$$Y_M = \text{span}\{T_0, T_1, \ldots, T_M\} = \text{span}\{L_0, L_1, \ldots, L_M\},$$

so that

$$V_M = \text{span}\{\phi_0, \phi_1, \ldots, \phi_{M-2}\},$$

where $\phi_j(x_2) = T_{j+2}(x_2) - T_j(x_2)$ in the Chebyshev case and $\phi_j(x_2) = [L_{j+2}(x_2) - L_j(x_2)]/\sqrt{4j+6}$ in the Legendre case. Moreover, following Shen (1994, 1995), we have

$$W_M = \text{span}\{\psi_0, \psi_1, \ldots, \psi_{M-4}\},$$

where

$$\psi_j(x_2) = T_j(x_2) - \frac{2j+4}{j+3}T_{j+2}(x_2) + \frac{j+1}{j+3}T_{j+4}(x_2),$$

or

$$\psi_j(x_2) = \frac{1}{(2j+3)\sqrt{4j+10}} L_j(x_2) - \frac{\sqrt{4j+10}}{(2j+3)(2j+7)} L_{j+2}(x_2) + \frac{1}{(2j+7)\sqrt{4j+10}} L_{j+4}(x_2).$$

Then the standard Galerkin approximation of (3.46)–(3.47) consists in the following: Find $\mathbf{u}_{NM}^{m,j}(\mathbf{x}) = \sum_{\mathbf{k}\in S_N} \tilde{\mathbf{u}}_M^{m,j}(\mathbf{k})\, e^{i\underline{\mathbf{k}}\cdot\mathbf{x}}$, where $\tilde{\mathbf{u}}_M^{m,j}(\mathbf{k}) \in \mathcal{V}_M \times \mathcal{W}_M \times \mathcal{V}_M$, such that for all $(\phi, \psi) \in V_M \times W_M$

$$\begin{aligned} &\beta(j,\mathbf{k}) \left(\tilde{\omega}_{2,M}^{m,j}(\mathbf{k}), \phi\right)_w - \left(\frac{d^2\tilde{\omega}_{2,M}^{m,j}}{dx_2^2}(\mathbf{k}), \phi\right)_w = \left(\hat{R}_\omega^{m,j}(\mathbf{k}), \phi\right)_w, \\ &-\underline{k}^2\beta(j,\mathbf{k}) \left(\tilde{u}_{2,M}^{m,j}(\mathbf{k}), \psi\right)_w + [\underline{k}^2 + \beta(j,\mathbf{k})] \left(\frac{d^2\tilde{u}_{2,M}^{m,j}}{dx_2^2}(\mathbf{k}), \psi\right)_w \\ &- \left(\frac{d^4\tilde{u}_{2,M}^{m,j}}{dx_2^4}(\mathbf{k}), \psi\right)_w = \left(\hat{R}_u^{m,j}(\mathbf{k}), \psi\right)_w \end{aligned} \tag{3.51}$$

for all $\mathbf{k} \in S_N \cap \{k_3 \geq 0\}$.
We have denoted by $\mathcal{V}_M$ ($\mathcal{W}_M$) the space of complex-valued polynomials of degree M with real and imaginary parts in V_M (W_M).

It is well known that a Galerkin approximation achieves the optimal convergence rate (see Canuto et al. (1988) and Maday and Métivet (1987)); however, its efficiency depends on the basis of the polynomial spaces V_M and W_M that is used (i.e., that one is able to derive). Several such attempts have been made in the past in Gottlieb and Orszag (1977) and in Moser et al. (1983). The basis proposed in the former reference led to linear systems with full matrices so that their cost is prohibitive. In the latter reference, more efficient bases leading to sparse (banded) matrices are used. However, as the authors were looking for divergence-free functions formed as combinations of Chebyshev polynomials, their bases were quite complicated. In Shen (1994, 1995), the author determined the bases given above for V_M and W_M that lead to easily invertible linear systems, namely, the solution can be computed in $O(M)$ operations with a standard direct method such as the LU decomposition.

The system (3.51) can be rewritten as

$$\beta(j,\mathbf{k}) \left(\tilde{\omega}_{2,M}^{m,j}(\mathbf{k}), \phi_\ell\right)_w - \left(\frac{d^2\tilde{\omega}_{2,M}^{m,j}}{dx_2^2}(\mathbf{k}), \phi_\ell\right)_w = \left(\hat{R}_\omega^{m,j}(\mathbf{k}), \phi_\ell\right)_w \quad \text{for } \ell = 0, \ldots, M-2,$$

$$-\underline{k}^2\beta(j,\mathbf{k})\left(\tilde{u}_{2,M}^{m,j}(\mathbf{k}),\psi_\ell\right)_w+(\underline{k}^2+\beta(j,\mathbf{k}))\left(\frac{d^2\tilde{u}_{2,M}^{m,j}}{dx_2^2}(\mathbf{k}),\psi_\ell\right)_w \tag{3.52}$$
$$-\left(\frac{d^4\tilde{u}_{2,M}^{m,j}}{dx_2^4}(\mathbf{k}),\psi_\ell\right)_w=\left(\hat{R}_u^{m,j}(\mathbf{k}),\psi_\ell\right)_w$$
$$\text{for } \ell=0,\dots,M-4,$$

for all $\mathbf{k}\in S_N\cap\{k_3\geq 0\}$. It can be easily shown that the following properties hold for the basis functions:

$$\begin{aligned}(\phi_\ell,\phi_i)_w&=0,\quad \text{for } i\neq \ell,\ell\pm 2,\\ \left(\frac{d^2\phi_\ell}{dx_2^2},\phi_i\right)_w&=0,\quad \text{for } i\neq\ell \text{ (Legendre case) and}\\ &\qquad \text{for } \ell<i \text{ or } \ell+i \text{ odd (Chebyshev case),}\end{aligned} \tag{3.53}$$

and

$$\begin{aligned}(\psi_\ell,\psi_i)_w&=0,\quad \text{for } i\neq \ell,\ell\pm 2,\ell\pm 4,\\ \left(\frac{d^2\psi_\ell}{dx_2^2},\psi_i\right)_w&=0,\quad \text{for } i\neq \ell,\ell\pm 2,\\ \left(\frac{d^4\psi_\ell}{dx_2^4},\psi_i\right)_w&=0,\quad \text{for } i\neq\ell \text{ (Legendre case) and}\\ &\qquad \text{for } \ell<i \text{ or } \ell+i \text{ odd (Chebyshev case).}\end{aligned} \tag{3.54}$$

The values of the nonzero terms in (3.53)–(3.54) are explicitly given in Shen (1994, 1995). In the Legendre case, the full discretized system reduces to the resolution of tri- and pentadiagonal matrices, while sparse upper triangular matrices are obtained in the case of a Chebyshev expansion. However, due to the particular structure of the matrices, the linear systems can be resolved in $O(M)$ operations.

Moreover, Shen showed that, in the Chebyshev case, the matrix with entries $(d^4\psi_\ell/dx_2^4,\ \psi_i)_w$ has a condition number behaving as M^4, while the matrix issued from the tau method leads to a condition number of the order of M^8. The system obtained with the Galerkin method is then more stable than that obtained with the tau method.

Computational Complexity of the Galerkin Approximation

The computation of the approximated velocity field reduces to s evaluations of the right-hand sides $\hat{R}_\omega^{m,j}(\mathbf{k})$ and $\hat{R}_u^{m,j}(\mathbf{k})$, for all $\mathbf{k}\in S_N\cap\{k_3\geq 0\}$ and

$j = 1, \ldots, s$, and to the resolution of $2sN(N/2+1)$ complex linear systems per time step. When the Chebyshev–Galerkin basis is used, the pseudospectral evaluation of the nonlinear terms requires $9s$ calls to three-dimensional Fourier–Chebyshev transforms, which corresponds to $9sN^2[c_4(M+1)\log_2 N + c_5 M \log_2 M]$ operations. As in the homogeneous case (Section 3.1), an explicit low-storage Runge–Kutta method has been used. The resolution of the linear systems can be achieved in $sc_6 N(N+2)M$ operations (see Shen (1995)).

Recent works, namely Alpert and Rokhlin (1991) and Shen (1996), show that even in the case of the Legendre–Galerkin basis, the nonlinear terms can be computed efficiently using fast transforms. Indeed, the Legendre coefficients can be easily computed in $O(M)$ operations knowing the Legendre–Galerkin ones; then the Chebyshev coefficients corresponding to the approximated velocity field can be obtained from the Legendre coefficients (see the references cited above). This transform (CL) from the Legendre to the Chebyshev coefficients can be done in $c_7 M$ operations with Alpert and Rokhlin's algorithm or in $M^2/4$ operations according to Shen (1996). The constant c_7 is quite large, so that the transform proposed by Shen is more efficient up to $M = 256$ (see Shen (1996)). Therefore, by solving the system (3.52) using the Legendre–Galerkin basis, the Chebyshev coefficients of the unknowns are carried during the computation, so that fast Chebyshev and Fourier transforms are used for the evaluation of the nonlinear term. We only apply the CL transform to the right-hand sides $\hat{R}_\omega^{m,j}(\mathbf{k})$ and $\hat{R}_u^{m,j}(\mathbf{k})$, for all $\mathbf{k} \in S_N \cap \{k_3 \geq 0\}$ and $j = 1, \ldots, s$. Once the linear systems have been solved, the CL transform is applied to $\tilde{\omega}_{2,M}^{m,j}(\mathbf{k})$ and $\tilde{u}_{2,M}^{m,j}(\mathbf{k})$ in order to recover their Chebyshev coefficients.

Hence, the total number of operations per time step can be estimated as follows:

1. Pseudospectral computation of the nonlinear term: $9sN^2[c_4(M+1)\log_2 N + c_5 M \log_2 M]$.
2. Chebyshev–Legendre transforms of the right-hand side of (3.52): $sN(N+2)M^2/2$.
3. Resolution of the linear systems: $sc_6 N(N+2)M$.
4. Legendre–Chebyshev transforms of $\tilde{u}_2(\mathbf{k})$ and $\tilde{\omega}_2(\mathbf{k})$: $sN(N+2)M^2/2$.

The total number of operations is then of the order of

$$9sN^2[c_4(M+1)\log_2 N + c_5 M \log_2 M] + sN(N+2)M(M+c_6).$$

Note that for a $128 \times 129 \times 128$ simulation (i.e. $N = M = 128$), the CL transforms represent approximately 15% of the whole CPU time.

3.2.5. A Legendre–Galerkin Approximation of the NSE

As in Chapter 1, we denote by V the space of divergence-free vector fields with finite enstrophy and satisfying the boundary conditions (3.20). Then the weak form of the Navier–Stokes equations can be written

$$\frac{d}{dt}(\mathbf{u}, \varphi) + \nu\,((\mathbf{u}, \varphi)) + (\mathbf{H}(\mathbf{u}, \mathbf{u}), \varphi) + K_p \int_\Omega \varphi_1(\mathbf{x})\, d\mathbf{x} = 0, \tag{3.55}$$

for any $\varphi \in V$; we have set $\mathbf{H}(\mathbf{u}, \mathbf{u}) = \omega \times \mathbf{u}$.

Let us denote by P_{NM} the space of polynomials that are trigonometric and of degree less than N in the streamwise and spanwise directions and of degree less than M in the direction normal to the wall. Note that $P_{NM} = \text{span}\{e^{i\underline{\mathbf{k}}\cdot\mathbf{x}} L_\ell(x_2), \underline{\mathbf{k}} \in S_N, \ell = 0, \ldots, M\}$, where $L_\ell(x_2)$ is the Legendre polynomial of degree ℓ. We now introduce

$$V_{NM} = \left\{\varphi_{NM}(\mathbf{x}) \in (P_{NM})^n;\ \ \varphi_{NM} \text{ satisfies (3.20) and } \nabla \cdot \varphi_{NM} = 0\right\},$$

so that $V_{NM} \subset V$. Therefore, the Galerkin approximation of (3.19)–(3.20) consists in the following: Find $\mathbf{u}_{NM}(\mathbf{x}, t) \in V_{NM}$ such that

$$\frac{d}{dt}(\mathbf{u}_{NM}, \varphi_{NM}) + \nu\left((\mathbf{u}_{NM}, \varphi_{NM})\right) \tag{3.56}$$
$$+ \left(\mathbf{H}(\mathbf{u}_{NM}, \mathbf{u}_{NM}), \varphi_{NM}\right) + K_p \int_\Omega \varphi_{NM}(\mathbf{x}) \cdot \mathbf{e}_1\, d\mathbf{x} = 0$$

for any $\varphi_{NM} \in V_{NM}$.

The main problem when using a Galerkin approximation is whether or not an efficient basis of V_{NM} is available. Using the recent Legendre–Galerkin basis proposed by Shen (1994) to solve second- and fourth-order differential equations and the same technique used by Moser, Moin, and Leonard (1983) to obtain free divergence functions, we derive hereafter a new efficient basis of V_{NM}.

Due to the periodic boundary conditions in the Ox_1x_3 plane, any vector field in V_{NM} can be expanded as follows:

$$\varphi_{NM}(\mathbf{x}) = \sum_{\mathbf{k} \in S_N} \hat{\varphi}_M(\mathbf{k}, x_2)\, e^{i\underline{\mathbf{k}}\cdot\mathbf{x}},$$

where $\hat{\varphi}_M(\mathbf{k}, x_2)$ is a complex-valued polynomial of degree less than M in x_2 which vanishes at the wall of the channel i.e. $x_2 = \pm h$. We recall that, according to the notation introduced in Section 3.2.2, the incompressibility constraint becomes

$$D_{\mathbf{k}} \cdot \hat{\varphi}_M(\mathbf{k}, x_2) = 0 \qquad \text{for all } \mathbf{k} \in S_N \cap \{k_3 \geq 0\}.$$

Hence, the Galerkin approximation (3.56) can be reformulated in this form: Find $\mathbf{u}_{NM}(\mathbf{x}, t) \in V_{NM}$ such that

$$\frac{d}{dt}\left(\hat{\mathbf{u}}_M(\mathbf{k}), \overline{\hat{\varphi}}_M(\mathbf{k})\right) + \nu\left(\left(\hat{\mathbf{u}}_M(\mathbf{k}), \overline{\hat{\varphi}}_M(\mathbf{k})\right)\right) + \nu \underline{k}^2\left(\hat{\mathbf{u}}_M(\mathbf{k}), \overline{\hat{\varphi}}_M(\mathbf{k})\right) \tag{3.57}$$
$$+ \left(\hat{\mathbf{H}}_{\mathbf{k}}(\mathbf{u}_{NM}, \mathbf{u}_{NM}), \overline{\hat{\varphi}}_{NM}(\mathbf{k})\right) + K_p \delta_{\mathbf{k},\mathbf{0}} \int_{-h}^{+h} \hat{\varphi}_{1M}(\mathbf{0})\, dx_2 = 0$$

for any $\mathbf{k} \in S_N \cap \{k_3 \geq 0\}$ and any $\varphi_{NM} \in V_{NM}$.
The scalar products $(\cdot, \cdot)$ and $((\cdot, \cdot))$ are similar to those introduced in Chapter 1 and used in (3.56), except that the integration extends here only over the normal direction of the channel, that is, on $(-h, h)$. For the sake of simplicity, we keep the same notation as previously. The overbar in (3.57) is used as usual to denote complex conjugates.

Helmholtz-like Decomposition

We first state the following result:

Lemma 3.1. *Any vector field $\varphi_{NM} \in (P_{NM})^n$ satisfying the boundary conditions (3.20) can be decomposed into*

$$\begin{aligned}\varphi_{NM}(\mathbf{x}) &= \varphi^1_{NM}(\mathbf{x}) + \varphi^2_{NM}(\mathbf{x}) \\ &= \sum_{\mathbf{k} \in S_N} \hat{\varphi}^1_M(\mathbf{k}, x_2)\, e^{i\underline{\mathbf{k}}\cdot\mathbf{x}} + \sum_{\mathbf{k} \in S_N} \hat{\varphi}^2_M(\mathbf{k}, x_2)\, e^{i\underline{\mathbf{k}}\cdot\mathbf{x}},\end{aligned} \tag{3.58}$$

where $D_{\mathbf{k}} \cdot \hat{\varphi}^2_M(\mathbf{k}, x_2) = 0$ for any $\mathbf{k} \in S_N \cap \{k_3 \geq 0\}$.

Proof. It suffices to set

$$\begin{aligned}\hat{\varphi}^1_M(\mathbf{k}, x_2) &= \frac{\underline{\mathbf{k}}}{\underline{k}^2}\left[\underline{\mathbf{k}} \cdot \hat{\varphi}_M(\mathbf{k}, x_2)\right] + \hat{\varphi}_{2M}(\mathbf{k}, x_2)\, \mathbf{e}_2, \\ \hat{\varphi}^2_M(\mathbf{k}, x_2) &= \frac{\underline{\mathbf{k}}^\perp}{\underline{k}^2}[\underline{\mathbf{k}}^\perp \cdot \hat{\varphi}_M(\mathbf{k}, x_2)]\end{aligned} \tag{3.59}$$

for any $\mathbf{k} \in S_N \cap \{k_3 \geq 0\} \backslash \{\mathbf{0}\}$ and

$$\hat{\varphi}^1_M(\mathbf{0}, x_2) = \hat{\varphi}_{2M}(\mathbf{0}, x_2)\, \mathbf{e}_2 \quad \text{and} \quad \hat{\varphi}^2_M(\mathbf{0}, x_2) = \begin{pmatrix} \hat{\varphi}_{1M}(\mathbf{0}, x_2) \\ 0 \\ \hat{\varphi}_{3M}(\mathbf{0}, x_2) \end{pmatrix}. \tag{3.60}$$

□

As a consequence of the decomposition (3.58)–(3.60), we have that, for any vector fields $\varphi_{NM}(\mathbf{x})$ and $\psi_{NM}(\mathbf{x})$ in $(P_{NM})^n$, $\hat{\varphi}^i_M(\mathbf{k}, x_2)$ and $\hat{\psi}^j_M(\mathbf{k}, x_2)$ for $i \neq j$ are orthogonal vectors in $\mathbb{C}^n$. Hence, the polynomial space V_{NM} can be written as

$$V_{NM} = V^1_{NM} \oplus V^2_{NM}.$$

By inserting the decomposition (3.58)–(3.60) into the Galerkin approximation (3.57) we obtain the following system: Find $\mathbf{u}^i_{NM}(\mathbf{x}, t) \in V^i_{NM}$ such that

$$\begin{aligned} &\frac{d}{dt}\left(\hat{\mathbf{u}}^i_M(\mathbf{k}), \overline{\hat{\varphi}}^i_M(\mathbf{k})\right) + \nu\left(\left(\hat{\mathbf{u}}^i_M(\mathbf{k}), \overline{\hat{\varphi}}^i_M(\mathbf{k})\right)\right) \\ &\quad + \nu \underline{k}^2 \left(\hat{\mathbf{u}}^i_M(\mathbf{k}), \overline{\hat{\varphi}}^i_M(\mathbf{k})\right) + \left(\hat{\mathbf{H}}^i_{\mathbf{k}}(\mathbf{u}_{NM}, \mathbf{u}_{NM}), \overline{\hat{\varphi}}^i_M(\mathbf{k})\right) \\ &\quad + K_p \delta_{\mathbf{k},\mathbf{0}}\, \delta_{i,2} \int_{-h}^{+h} \hat{\varphi}^2_{1M}(\mathbf{0})\, dx_2 = 0, \end{aligned} \tag{3.61}$$

for $\mathbf{k} \in S_N \cap \{k_3 \geq 0\}$, $\varphi^i_{NM} \in V^i_{NM}$ and $i = 1, 2$.

We can now state the following result.

Lemma 3.2. *Let us set*

$$\begin{aligned} \hat{\phi}_j(\mathbf{k}, x_2) &= \frac{\mathbf{k}^\perp}{\underline{k}^2}\, \phi_j(x_2), && j = 0, \ldots, M-2, \\ \hat{\psi}_j(\mathbf{k}, x_2) &= \mathbf{i}\, \frac{\mathbf{k}}{\underline{k}^2} \frac{\partial \psi_j}{\partial x_2}(x_2) + \psi_j(x_2)\mathbf{e}_2, && j = 0, \ldots, M-4, \end{aligned} \tag{3.62}$$

for any $\mathbf{k} \in S_N \cap \{k_3 \geq 0\} \backslash \{\mathbf{0}\}$, *and*

$$\begin{aligned} \hat{\phi}_j(\mathbf{0}, x_2) &= \left(\phi_j(x_2), 0, \phi_j(x_2)\right)^T, && j = 0, \ldots, M-2, \\ \hat{\psi}_j(\mathbf{0}, x_2) &= \mathbf{0}, && j = 0, \ldots, M-4, \end{aligned} \tag{3.63}$$

where $\phi_j(x_2)$ and $\psi_j(x_2)$ constitute the Legendre–Galerkin bases defined in the previous section. Then

$$V_{NM}^1 = \text{span}\left\{\hat{\psi}_0(\mathbf{k}, x_2), \ldots, \hat{\psi}_{M-4}(\mathbf{k}, x_2);\ \mathbf{k} \in S_N \cap \{k_3 \geq 0\} \backslash \{\mathbf{0}\}\right\}$$

and

$$V_{NM}^2 = \text{span}\left\{\hat{\phi}_0(\mathbf{k}, x_2), \ldots, \hat{\phi}_{M-2}(\mathbf{k}, x_2);\ \mathbf{k} \in S_N \cap \{k_3 \geq 0\}\right\}.$$

Proof. Let $\varphi_{NM}(\mathbf{x}) \in V_{NM}$. Then $\underline{\mathbf{k}}^\perp \cdot \hat{\varphi}_M(\mathbf{k}, x_2)$ is a complex-valued polynomial of order M vanishing at $x_2 = \pm h$. Due to Shen (1994) and according to Lemma 3.1, we have that

$$\hat{\varphi}_{jM}^2(\mathbf{k}, x_2) \in \text{span}\{\phi_0, \ldots, \phi_{M-2}\},$$
$$j = 1, \ldots, n, \quad \mathbf{k} \in S_N \cap \{k_3 \geq 0\} \backslash \{\mathbf{0}\}.$$

Moreover, since the incompressibility condition holds, we have $\hat{\varphi}_{2M}(\mathbf{0}, x_2) = 0$. Hence, the result holds for V_{MN}^2.

Since $\hat{\varphi}_{2M}(\mathbf{k}, \pm h) = (\partial\hat{\varphi}_{2M}/\partial x_2)(\mathbf{k}, \pm h) = 0$, we have

$$\hat{\varphi}_{2M}(\mathbf{k}, x_2) \in \text{span}\{\psi_0, \ldots, \psi_{M-4}\}.$$

The incompressibility condition for $\hat{\varphi}_M(\mathbf{k}, x_2)$ becomes

$$\underline{\mathbf{k}} \cdot \hat{\varphi}_M(\mathbf{k}, x_2) = \mathbf{i}\frac{\partial\hat{\varphi}_{2M}}{\partial x_2}(\mathbf{k}, x_2).$$

Then, the result follows from the definition of $\hat{\varphi}_M^1(\mathbf{k}, x_2)$ (see Lemma 3.1) on choosing $\hat{\psi}_j(\mathbf{k}, x_2)$ as in (3.62)–(3.63). □

Recalling the definitions of the polynomials ψ_j, ϕ_j as well as the derivation rule for the Legendre polynomials (see e.g. Canuto et al. (1988)), we can easily show that

$$\frac{\partial\psi_j}{\partial x_2} = -\phi_{j+1} \qquad \text{for} \quad j = 0, \ldots, M-4,$$

so that

$$\text{span}\{\phi_1, \ldots, \phi_{M-3}\} = \text{span}\left\{\frac{\partial\psi_0}{\partial x_2}, \ldots, \frac{\partial\psi_{M-4}}{\partial x_2}\right\}.$$

Therefore, for any $\varphi_{NM} \in V_{NM}$ we have $\underline{\mathbf{k}} \cdot \hat{\varphi}_M(\mathbf{k}, x_2) \in \text{span}\,\{\phi_1, \ldots, \phi_{M-3}\}$ for $\mathbf{k} \in S_N \cap \{k_3 \geq 0\}$.

The Galerkin approximation (3.61) above can then be reformulated as follows: Find $\mathbf{u}^i_{NM}(\mathbf{x}, t) \in V^i_{NM}$, $i = 1, 2$, such that

$$\begin{aligned}
&\frac{\partial}{\partial t}\left(\hat{\mathbf{u}}^1_M(\mathbf{k}), \bar{\hat{\psi}}_j(\mathbf{k})\right) + \nu\left(\left(\hat{\mathbf{u}}^1_M(\mathbf{k}), \bar{\hat{\psi}}_j(\mathbf{k})\right)\right) \\
&\quad + \nu\underline{k}^2\left(\hat{\mathbf{u}}^1_M(\mathbf{k}), \bar{\hat{\psi}}_j(\mathbf{k})\right) + \left(\hat{\mathbf{H}}^1_{\mathbf{k}}(\mathbf{u}_{NM}, \mathbf{u}_{NM}), \bar{\hat{\psi}}_j(\mathbf{k})\right) = 0 \\
&\qquad\qquad\qquad \text{for } j = 0, \ldots, M-4, \\
&\frac{\partial}{\partial t}\left(\hat{\mathbf{u}}^2_M(\mathbf{k}), \bar{\hat{\phi}}_j(\mathbf{k})\right) + \nu\left(\left(\hat{\mathbf{u}}^2_M(\mathbf{k}), \bar{\hat{\phi}}_j(\mathbf{k})\right)\right) + \nu\underline{k}^2\left(\hat{\mathbf{u}}^2_M(\mathbf{k}), \bar{\hat{\phi}}_j(\mathbf{k})\right) \\
&\quad + \left(\hat{\mathbf{H}}^2_{\mathbf{k}}(\mathbf{u}_{NM}, \mathbf{u}_{NM}), \bar{\hat{\phi}}_j(\mathbf{k})\right) + K_p\delta_{j,0}\delta_{\mathbf{k},\mathbf{0}} = 0, \\
&\qquad\qquad\qquad \text{for } j = 0, \ldots, M-2,
\end{aligned} \tag{3.64}$$

for any $\mathbf{k} \in S_N \cap \{k_3 \geq 0\}$.

According to Lemma 3.2, the approximated velocity field $\mathbf{u}_{NM} \in V_{NM}$ can be expanded as follows:

$$\begin{aligned}
\mathbf{u}_{NM}(\mathbf{x}) &= \sum_{\mathbf{k}\in S_N}\sum_{j=0}^{M-4} \hat{u}^1(\mathbf{k}, j)\, \hat{\psi}_j(\mathbf{k}, x_2)\, e^{i\underline{\mathbf{k}}\cdot\mathbf{x}} \\
&\quad + \sum_{\mathbf{k}\in S_N}\sum_{j=0}^{M-2} \hat{u}^2(\mathbf{k}, j)\, \hat{\phi}_j(\mathbf{k}, x_2)\, e^{i\underline{\mathbf{k}}\cdot\mathbf{x}},
\end{aligned}$$

where the coefficients $\hat{u}^i(\mathbf{k}, j)$, $i = 1, 2$, are complex numbers. Therefore, we have

$$\begin{aligned}
\hat{\mathbf{u}}^1_M(\mathbf{k}, x_2) &= \sum_{j=0}^{M-4} \hat{u}^1(\mathbf{k}, j)\, \hat{\psi}_j(\mathbf{k}, x_2) \quad \text{and} \\
\hat{\mathbf{u}}^2_M(\mathbf{k}, x_2) &= \sum_{j=0}^{M-2} \hat{u}^2(\mathbf{k}, j)\, \hat{\phi}_j(\mathbf{k}, x_2).
\end{aligned}$$

By substituting the above expansions into (3.64) as well as recalling the definitions of the basis vector functions, we derive the following system for the unknowns $\hat{u}^i(\mathbf{k}, j),\ i = 1, 2$:

$$\begin{aligned}
&\sum_{\ell=0}^{M-4}\left(m^1_{j\ell}\,\frac{\partial \hat{u}^1}{\partial t}(\mathbf{k},\ell) + a^1_{j\ell}\,\hat{u}^1(\mathbf{k},\ell)\right)\\
&\quad = \mathbf{i}\left(\underline{\mathbf{k}}\cdot\hat{\mathbf{H}}_{\mathbf{k}}(\mathbf{u}_{NM},\mathbf{u}_{NM}), \frac{\partial\psi_j}{\partial x_2}\right) - \underline{k}^2\left(\hat{H}_{2,\mathbf{k}}(\mathbf{u}_{NM},\mathbf{u}_{NM}),\psi_j\right)\\
&\qquad\qquad\qquad\qquad \text{for } j = 0,\ldots,M-4,\\
&\sum_{\ell=0}^{M-2}\left(m^2_{j\ell}\,\frac{\partial \hat{u}^2}{\partial t}(\mathbf{k},\ell) + a^2_{j\ell}\,\hat{u}^2(\mathbf{k},\ell)\right) = -\left(\underline{\mathbf{k}}^{\perp}\cdot\hat{\mathbf{H}}_{\mathbf{k}}(\mathbf{u}_{NM},\mathbf{u}_{NM}),\phi_j\right)\\
&\quad - \delta_{\mathbf{k},\mathbf{0}}\left[K_p\delta_{j,0} + \sum_{\ell=0}^{1}\left(\hat{H}_{2\ell+1,\mathbf{0}}(\mathbf{u}_{NM},\mathbf{u}_{NM}),\phi_j\right)\right]\\
&\qquad\qquad\qquad\qquad \text{for } j = 0,\ldots,M-2,
\end{aligned} \tag{3.65}$$

where

$$\begin{aligned}
m^1_{j\ell} &= \underline{k}^2\left(\psi_\ell,\psi_j\right) + \left(\frac{\partial\psi_\ell}{\partial x_2},\frac{\partial\psi_j}{\partial x_2}\right),\\
a^1_{j\ell} &= 2\nu\underline{k}^2\left(\frac{\partial\psi_\ell}{\partial x_2},\frac{\partial\psi_j}{\partial x_2}\right) + \nu\underline{k}^4\left(\psi_\ell,\psi_j\right) + \nu\left(\frac{\partial^2\psi_\ell}{\partial x_2^2},\frac{\partial^2\psi_j}{\partial x_2^2}\right)\\
&\qquad\qquad \text{for } j,\ell = 0,\ldots,M-4,\\
m^2_{j\ell} &= \left(\phi_\ell,\phi_j\right),\\
a^2_{j\ell} &= \nu\underline{k}^2\left(\phi_\ell,\phi_j\right) + \nu\left(\frac{\partial\phi_\ell}{\partial x_2},\frac{\partial\phi_j}{\partial x_2}\right) \qquad \text{for } j,\ell = 0,\ldots,M-2,
\end{aligned}$$

for any $\mathbf{k}\in S_N\cap\{k_3\geq 0\}$. After approximation of the time derivative in the system above and according to (3.53) and (3.54), the first equation in (3.65) leads to a pentadiagonal matrix while the second one leads to a tridiagonal matrix. Both matrices can be efficiently solved in $O(M)$ operations. Note that while the approach is completely different than in Section 3.2.4, the discrete system obtained is equivalent to (3.52).

A similar approach can be used to derive a divergence-free Galerkin basis in the case of flows in an infinite pipe. This basis, as well as numerical simulations for the channel and pipe flow problems, will be reported elsewhere.

4

Direct Numerical Simulation versus Turbulence Modeling

The direct numerical simulation (DNS) of turbulent flows consists in resolving all the scales of motion that are physically relevant, that is, from the large energy-containing scales, usually generated and sustained by the mean flow or possibly an external force, to the dissipative scales, which are quickly damped by the viscous effects. This statement implies very strong restrictions on the feasibility of DNS.

In the first section of this chapter, we derive several estimates on the number of unknowns, the time step, and the memory used in terms of the Reynolds number for both homogeneous and channel turbulent flows. In the second section, we briefly introduce the reader to the modeling approach. Since there is a very large literature on this topic, the review proposed here is not exhaustive and only a few major modeling theories are summarized. The reader is referred to the cited work and the references therein.

4.1. The Practical Limits of DNS

4.1.1. Homogeneous Turbulent Flows

According to the analysis in Section 2.2.4, the energy-containing scales in isotropic homogeneous turbulence are characterized by the integral scale L (see (2.103)). Hence, the size L_1 of the box $\Omega = (0, L_1)^n$ has to be larger than L. In such a case (see Example 2.1), the spectral tensors have a discrete spectrum. Let ℓ be any length scale; we define the associated wavenumber as follows:

$$k_\ell = \left[\frac{L_1}{2\pi\ell}\right],$$

where $[\cdot]$ denotes the integer part. The size of the grid, $\Delta x = L_1/N$, which defines the smallest scale that one is able to compute, has to be smaller than, or at least of the order of, the dissipative scale η. A restriction on the number of degrees of freedom follows:

$$N = \frac{L_1}{\Delta x} > \frac{L_1}{\eta}. \tag{4.1}$$

In practice, as was mentioned in Section 2.2.5, $\eta k_N \simeq 1.5$ (i.e. $N \simeq 3k_d$) is the minimum resolution required in order to accurately compute the velocity gradients (see Jiménez (1994)).

In Chapter 2 (see (2.110) and (2.119)), the ratio L/η was estimated as

$$\frac{L}{\eta} = \begin{cases} \mathrm{Re}_L^{1/2} & \text{in the two-dimensional case,} \\ c_1^{1/4}\mathrm{Re}_L^{3/4} & \text{in the three-dimensional case,} \end{cases} \tag{4.2}$$

where c_1 is a dimensionless constant introduced in (2.106). Hence, on assuming that $N \geq 3(k_d + 1)$, we deduce that the number of degrees of freedom needed for DNS has to satisfy

$$N^n \geq \beta^n \left(\frac{3L_1}{2\pi L} \right)^n \mathrm{Re}_L^{n^2/4}, \tag{4.3}$$

where $\beta = c_1^{(n-2)/4}$. Recalling the estimates derived in Section 3.1 on the memory size as well as the number of operations required per time step, we can derive estimates in terms of the Reynolds number:

$$\begin{aligned} \text{Memory size} &\geq (4n-3)\left(\frac{3\beta L_1}{2\pi L} \right)^n \mathrm{Re}_L^{n^2/4}, \\ \text{Number of operations} &\geq (4n-3)sc_2\left(\frac{3\beta L_1}{2\pi L} \right)^n \\ &\quad \times \mathrm{Re}_L^{n^2/4} \log_2\left(\frac{3\beta L_1}{2\pi L} \mathrm{Re}_L^{n/4} \right). \end{aligned} \tag{4.4}$$

The parameter s in (4.4) represents the number of sub-time-steps performed when a one-step explicit scheme is used (see (3.16)).

In addition to this length-scale restriction, additional constraints on the time step have to be considered. The first one is related to the numerical stability of the method used to discretize the equations. When treating the nonlinearity with an explicit scheme, the time step Δt has to satisfy a Courant–Friedrichs–Levy (CFL) condition of the type

$$\frac{2\pi}{L_1} k_N \, \Delta t \, U_\infty < \alpha, \tag{4.5}$$

where $U_\infty = \sup_{\mathbf{x}\in\Omega} |\mathbf{u}(\mathbf{x})|$ and α is a constant that is close to unity and depends on the numerical scheme used; for instance, $\alpha \simeq \sqrt{3}$ for an explicit RK3 scheme. An upper bound on the time step immediately follows:

$$\Delta t < \frac{2\alpha}{3\beta} \frac{L}{U_\infty} \mathrm{Re}_L^{-n/4}, \tag{4.6}$$

so that $\Delta t \simeq \mathrm{Re}_L^{-n/4}$. From (4.2) and (4.6) we deduce that

$$\Delta t < \frac{2\alpha}{3} \frac{\eta}{U_\infty},$$

where η/U_∞ is a characteristic time scale of the convection of the smale scales by the large ones.

Moreover, as the scales of the order of the Kolmogorov length scales η are computed, we require Δt to be at least of the order of the characteristic time scale τ_d associated with η, which can be defined as (see Orszag (1973))

$$\begin{aligned} \tau_d &= \left(\nu \frac{4\pi^2}{L_1^2} k_d^2 \right)^{-1} \qquad (4.7) \\ &= \begin{cases} \gamma^{-1/3} = \frac{L}{u} & \text{(two-dimensional case)}, \\ \left(\frac{\nu}{\epsilon}\right)^{1/2} = c_1^{-1/2} \frac{L}{u} \mathrm{Re}_L^{-1/2} & \text{(three-dimensional case)}. \end{cases} \end{aligned}$$

Note that the conditions (4.7) are weaker than the condition (4.6). In the two-dimensional case the characteristic time scale of the smallest length scale is of the order of the eddy-turnover time $\tau_e = L/u$; in fact, in the whole inertial range, the characteristic time scale is independent of the wavenumber k (see Orszag (1973)).

In order to reach a statistically steady state, when the turbulence is artificially maintained with an external volume force, the Navier–Stokes equations have to be integrated over several eddy-turnover times τ_e, that is, over a time interval $[0, n_{\tau_e}\tau_e]$ with $n_{\tau_e} > 1$.[1] The number of time iterations performed is

$$\frac{n_{\tau_e}\tau_e}{\Delta t} > n_{\tau_e} \frac{3\beta}{2\alpha} \frac{U_\infty}{u} \mathrm{Re}_L^{n/4} \sim n_{\tau_e} \mathrm{Re}_L^{n/4}.$$

The total number of operations required is then of the order of

$$(4n-3)\, n_{\tau_e} s c_2 \left(\frac{3\beta L_1}{2\pi L} \right)^n \mathrm{Re}_L^{n(n+1)/4} \log_2 \left(\frac{3\beta L_1}{2\pi L} \mathrm{Re}_L^{n/4} \right).$$

[1] In fact, n_{τ_e} is much larger than one; see Jiménez et al. (1993) and Dubois and Jauberteau (1998).

Hence, the number of operations per time unit, τ_e, as well as the global memory size of the codes, is drastically restrictive, so that DNS can only be achieved at low Reynolds numbers. One of the highest resolutions reached on a massively parallel computer is 512^3, and the computations were done on a computer with a large central memory ($\geq$20 gigabytes; see Table 3.1). On a Cray YMP-C98 with 4-gigabyte memory, we have been able to run a 300^3 simulation over a few time units. Such a resolution corresponds to a microscale Reynolds number Re_λ of the order of 170.

4.1.2. Turbulent Channel Flows

As in the homogeneous case, a restriction on the mesh size arises from the Kolmogorov length scale η defined by

$$\eta = \left(\frac{\nu^3}{\epsilon}\right)^{1/4}, \tag{4.8}$$

where ϵ is the mean dissipation rate per volume unit,

$$\epsilon = \nu \frac{1}{|\Omega|} \int_\Omega |\nabla \mathbf{u}|^2 \, d\mathbf{x} = \nu \langle |\nabla \mathbf{u}|^2 \rangle_\Omega. \tag{4.9}$$

In this particular case, the energy equation (1.38) can be rewritten as

$$\frac{dK}{dt} + \epsilon + K_p \ \langle u_1(\mathbf{x}, t) \rangle_\Omega = 0, \tag{4.10}$$

where $K = \frac{1}{2}\langle |\mathbf{u}|^2 \rangle_\Omega$. The mean velocity appearing in (4.10) is usually called the bulk velocity and is denoted by U_m. Note that U_m is related to the flow rate Q (see Chapter 1) by $Q = 2hL_3U_m$. Therefore, when a statistically steady state is reached in the sense that most statistical quantities are fluctuating with a small amplitude of a few percent from their mean values, we deduce that

$$\epsilon \simeq -K_p U_m.$$

The Kolmogorov (or dissipation) length scale can then be rewritten as

$$\eta = 2h \, C_f^{-1/4} \, \mathrm{Re}_m^{-3/4}, \tag{4.11}$$

where

$$\mathrm{Re}_m = \frac{2U_m h}{\nu} \quad \text{and} \quad C_f = \frac{\tau_w}{\frac{1}{2}U_m^2};$$

C_f is called the friction velocity coefficient. We recall (see Section 2.3) that the wall shear stress satisfies $\tau_w = \tau(-h) = -K_p h = u_*^2$, where u_*^2 is the friction velocity. As the mesh size in each spatial direction has to be of the order of η, the above relation provides an estimate of the number of points needed in each direction, namely,

$$\max\left(\frac{L_1}{N_1}, h\left(1 - \cos\frac{\pi}{M}\right), \frac{L_3}{N_3}\right) \leq \eta. \tag{4.12}$$

A constraint on the time step is induced by the above length-scale restriction. The nonlinear term (see Section 3.2) is treated by an explicit scheme, so that a CFL condition has to be satisfied by the time step; this CFL condition has the following form:

$$\frac{\Delta t\ U_\infty}{\min_{i=1,n} \Delta_i} < \alpha, \tag{4.13}$$

where $\{\Delta_i\}_{i=1,n}$ are the mesh sizes in each spatial direction of the channel. The strongest restriction comes from the direction normal to the wall. Indeed, Chebyshev–Gauss–Lobatto points are used in spectral methods (see Kim, Moin, and Moser (1987)), and a hyperbolic-tangent repartition is used in finite-difference methods (see Deardorff (1970)). In the former case, the smallest mesh size is given by the distance between the wall and the first point away from it, namely, $\min_j \Delta_2(x_{2,j}) = h[1 - \cos(\pi/M)]$, which can be bounded as follows:

$$\frac{h}{2}\left(\frac{\pi}{M}\right)^2 \geq \min_j \Delta_2(x_{2,j}) \geq h\,\frac{\sqrt{3}}{4}\left(\frac{\pi}{M}\right)^2 \qquad \text{for} \quad M \geq 6, \tag{4.14}$$

where $M+1$ points in the direction normal to the wall are retained. In that case, (4.13) yields

$$\frac{2}{h\pi^2}\ \Delta t\ M^2 U_\infty < \alpha.$$

Note that the above restriction on the time step is stronger than (4.6). A distribution of points closer and closer to each other near the wall, that is, for $x_2^+ < 5$ (see Section 2.3), is necessary in order to take into account the presence of boundary layers. This induces a severe restriction on the size of the time step.

Recalling that $\min_j \Delta_2(x_{2,j}) \leq \eta$ and the estimates (4.14) and (4.11), we obtain

$$M \geq \frac{3^{1/4}\pi}{2\sqrt{2}} C_f^{1/8} \mathrm{Re}_m^{3/8}. \tag{4.15}$$

We define, as a time unit, $\tau_e = h/u_*$, so that $\tau_e = \sqrt{-h/K_p}$. In order to reach a statistically steady state the numerical simulation has to be performed over about ten time units. Then, using (4.11), (4.12), and (4.15), a lower bound for the total number of operations (see Section 3.2.4), can be obtained, namely,

$$10\sqrt{\frac{-h}{K_p}}\, 9s \frac{3^{1/4}\pi L_1 L_3}{8\sqrt{2}h^2} C_f^{5/8}\, \mathrm{Re}_m^{15/8} \left[c_3 \log_2 \left(\frac{L_1}{2h} C_f^{1/4} \mathrm{Re}_m^{3/4} \right) \right.$$
$$\left. + c_4 \log_2 \left(\frac{3^{1/4}\pi}{2\sqrt{2}} C_f^{1/8} \mathrm{Re}_m^{3/8} \right) + c_5 \log_2 \left(\frac{L_3}{2h} C_f^{1/4} \mathrm{Re}_m^{3/4} \right) \right].$$

Finally, another condition has to be satisfied, in the case of the channel flow problem, which concerns the length of the channel in the streamwise direction in two dimensions and in the streamwise and spanwise directions in three dimensions. In these directions, the flow is assumed to be homogeneous, and periodic boundary conditions are often applied. These conditions are somewhat unrealistic, and their validity can be checked by analyzing the two-point correlation functions. Indeed, we can reasonably assume that the boundary conditions have no influence on the development of turbulence if the velocity at the center of the channel is fully decorrelated with the velocity at the entrance of the channel (see Kim, Moin, and Moser (1987) for details). When this condition is violated, the lengths L_1 and L_3 in three dimensions, or L_1 in two dimensions, have to be increased.

4.2. A Different Approach: Turbulence Modeling

As we have shown in the preceding section, DNS is limited to simple flows and to very low Reynolds numbers. Hence, in most engineering applications or studies of complex flows, DNS cannot be used. Different types of modeling have been developed in order to overcome this intrinsic difficulty. Here, we intend to give a brief overview of the basic and general ideas of the wide field of turbulence modeling. Special attention will be paid to the large-eddy simulation (LES) approach, which is in a sense closest to the dynamic multilevel methods (DML) proposed in the next chapters.

4.2.1. The Reynolds-Averaged Equations and the Closure Problem

All types of turbulence modeling are based on some averaging of the Navier–Stokes equations. In his pioneering work, Reynolds (1895) introduced the following decomposition of the flow field:

$$\mathbf{u} = \overline{\mathbf{u}} + \mathbf{u}', \tag{4.16}$$

where $\overline{\mathbf{u}}$ is the averaged velocity with respect to an ensemble or statistical average (see Section 2.1) and $\mathbf{u}'$ is its fluctuating counterpart.

Furthermore, it is assumed that the following Reynolds conditions hold for any flow variables φ and ψ:

$$\begin{aligned} \overline{\varphi + \psi} &= \overline{\varphi} + \overline{\psi}, \\ \overline{\alpha\varphi} &= \alpha\overline{\varphi} \qquad \text{for any constant } \alpha, \\ \overline{\alpha} &= \alpha \qquad \text{for any constant } \alpha, \\ \overline{\left(\frac{\partial \varphi}{\partial s}\right)} &= \frac{\partial \overline{\varphi}}{\partial s} \qquad \text{where } s \text{ is } x_1, x_2, x_3, \text{ or } t, \\ \overline{\overline{\varphi}\psi} &= \overline{\varphi}\,\overline{\psi}. \end{aligned} \tag{4.17}$$

By substituting the decomposition (4.16) into the momentum equation (1.6), and performing an ensemble average, we obtain the mean velocity equation:

$$\frac{\partial \overline{u}_i}{\partial t} + \nabla \cdot (\overline{u}_i \overline{\mathbf{u}}) = -\frac{\partial \overline{p}}{\partial x_i} - \sum_{j=1}^{n} \frac{\partial \tau_{ij}}{\partial x_j} + \nu\, \Delta \overline{u}_i, \qquad \text{for} \quad i = 1, \ldots, n. \tag{4.18}$$

Note that the form $\nabla \cdot (\overline{u}_i \overline{\mathbf{u}})$ of the nonlinear term is equivalent to $(\overline{\mathbf{u}} \cdot \nabla)\overline{u}_i$ (see (1.9)). From the conditions (4.17), the Reynolds stress tensor $\tau_{ij} = \overline{u_i u_j} - \overline{u}_i \overline{u}_j$ reduces to

$$\tau_{ij} = \overline{u'_i u'_j}. \tag{4.19}$$

Indeed, the Reynolds conditions imply that all cross terms of the form $\overline{u'_i\, \overline{u}_j}$ vanish. The mean continuity equation is simply derived by averaging (1.5) and using (4.17):

$$\nabla \cdot \overline{\mathbf{u}} = 0. \tag{4.20}$$

The system formed by (4.18) and (4.20) is not closed until a model is provided for determining the Reynolds stress tensor (4.19). A wide range of models have been proposed and developed. Reynolds-stress closures can be achieved in many different manners, but all models are based on the concept of eddy viscosity introduced by Boussinesq (1877), which expresses the stress tensor as

$$\tau_{ij} = -\nu_T \overline{S}_{ij},$$

where

$$S_{ij} = \frac{1}{2}\left(\frac{\partial u_i}{\partial x_j} + \frac{\partial u_j}{\partial x_i}\right)$$

is the strain-rate tensor, $\mathbf{S} = \frac{1}{2}(\nabla\mathbf{u} + \nabla\mathbf{u}^t)$. The eddy viscosity ν_T can be determined by using the concept of mixing length (Prandtl):

$$\nu_T \sim \frac{\ell_0^2}{\tau_0},$$

where ℓ_0 is the turbulent length scale and τ_0 is the turbulent time scale. The prescription of these two parameters is needed in such models, and they greatly depend on the flow under consideration. This very simple model is referred to as the zero-equation model. One- and two-equation models have been proposed in order to reduce the number of physical parameters needed. Indeed, in the one-equation (K–ℓ) model, the eddy viscosity is given as a function of the turbulent kinetic energy K and a length scale ℓ; K is then obtained from a modeled version of its exact transport equation. In the two-equation (K–ϵ) models, the turbulent viscosity is given as a function of K and the turbulent energy dissipation rate ϵ; the transport equations for K and ϵ are modeled. Such models cannot be integrated to a solid boundary, so that wall or damping functions are used.

All these models then provide an expression for the Reynolds stress tensor. However, a transport equation for τ_{ij} can be derived in which third-order moments and the pressure-strain moment appear and need to be modeled; such approach leads to second-order closure models. A detailed review of Reynolds-stress closures in turbulence can be found in Speziale (1990).

Hence, Reynolds stress models provide the computation of the first moments such as the mean velocity $\bar{\mathbf{u}}$ and the mean pressure, and of second moments such as the turbulent kinetic energy and the energy dissipation rate.

4.2.2. The Large-Eddy Simulation

Another approach, slightly different than the previous one, has been introduced by Smagorinsky (1963). It consists in computing the large scales of the motion containing most of the energy (80% of the total energy). Instead of computing mean flow quantities as in Reynolds-stress closures, LES deals with a wider range of scales, and it provides a more detailed description of the flow.

The decomposition (4.16) is retained, where the large-scale component of any flow variable φ is defined according to Leonard (1974) by

$$\overline{\varphi}(\mathbf{x}, t) = \int_\Omega G(\mathbf{x} - \mathbf{x}', \Delta_f)\, \varphi(\mathbf{x}', t)\, d\mathbf{x}', \tag{4.21}$$

where $G(\mathbf{x} - \mathbf{x}', \Delta_f)$ is a filter function and Δ_f is the filter width. The fluctuating part of the velocity,

$$\mathbf{u}' = \mathbf{u} - \bar{\mathbf{u}},$$

is called the subgrid-scale (SGS) velocity.

Filter functions are generally infinitely differentiable functions with compact support; several examples are given in Chapter 6. The net effect of this filter function is to remove or at least to reduce the scales of length smaller than Δ_f. In practice, the most efficient filter seems to be the Gaussian filter (see Bardina et al. (1983) and the references therein). In (4.21) the filter function is assumed to depend only on the separation $\mathbf{x} - \mathbf{x}'$, and it is then independent of the location $\mathbf{x}$. With such an assumption, the filter defined by (4.21) commutes with spatial and temporal derivatives and it satisfies the fourth condition in (4.17).

By applying the filter (4.21) to the momentum and continuity equations, we obtain

$$\begin{aligned} &\frac{\partial \overline{u}_i}{\partial t} + \nabla \cdot (\overline{u}_i \overline{\mathbf{u}}) = -\frac{\partial \overline{p}}{\partial x_i} + \nu \, \Delta \overline{u}_i - \sum_{j=1}^{n} \frac{\partial \tau_{ij}}{\partial x_j}, \qquad \text{for } i = 1, \ldots, n, \\ &\nabla \cdot \overline{\mathbf{u}} = 0, \end{aligned} \tag{4.22}$$

where the subgrid-scale tensor $\boldsymbol{\tau}$ is defined as $\tau_{ij} = \overline{u_i u_j} - \overline{u}_i \overline{u}_j$. By modeling τ_{ij} the trace term is usually removed from the SGS tensor and added to the pressure (see Bardina et al. (1983) and Voke and Collins (1983)), so that τ_{ij} and $\overline{p}$ are replaced, in (4.22), by

$$\mathcal{T}_{ij} = \tau_{ij} - \frac{1}{3} \sum_{l=1}^{n} \tau_{ll} \, \delta_{ij} \quad \text{and} \quad \overline{P} = \overline{p} + \frac{1}{3} \sum_{l=1}^{n} \tau_{ll}. \tag{4.23}$$

Recalling the decomposition (4.16), the subgrid-scale tensor can be rewritten as

$$\begin{aligned} \tau_{ij} &= \left(\overline{\overline{u}_i \overline{u}_j} - \overline{u}_i \overline{u}_j \right) + \left(\overline{u'_i \overline{u}_j + \overline{u}_i u'_j} \right) + \overline{u'_i u'_j} \\ &= L_{ij} + C_{ij} + R_{ij}, \end{aligned} \tag{4.24}$$

where $L_{ij} = \overline{\overline{u}_i \overline{u}_j} - \overline{u}_i \overline{u}_j$ is the Leonard stress tensor, $C_{ij} = \overline{u'_i \overline{u}_j} + \overline{\overline{u}_i \, u'_j}$ is the cross-stress tensor, and $R_{ij} = \overline{u'_i u'_j}$ is the Reynolds stress tensor. The Leonard tensor is an explicit term and can be directly approximated, while the cross and Reynolds stress tensors need to be modeled. A first approach introduced by Smagorinsky consists in modeling the sum $C_{ij} + R_{ij}$ by making an eddy-viscosity assumption:

$$C_{ij} + R_{ij} \simeq -2\nu_T \overline{S}_{ij}, \tag{4.25}$$

where $\nu_T = C_S \, \Delta_f^2 \overline{S}$ and $\overline{S} = (\sum_{i,j=1}^{n} \overline{S}_{ij}^2)^{1/2}$.

The Smagorinsky model is able to maintain the correct mean energy balance of the large scales. It has been used in different situations such as homogeneous isotropic or shear flows and channel flows (see Voke and Collins (1983) for a review) and lead to fairly good results. However, it has been shown (see Bardina et al. (1983) and the references therein) that the right-hand side of (4.25) correlates very poorly with the exact SGS tensor obtained from DNS. In Bardina et al. (1983), the authors proposed a scale-similarity model

$$C_{ij} + R_{ij} \simeq C_r \left(\overline{u}_i\overline{u}_j - \overline{\overline{u}}_i\overline{\overline{u}}_j\right),$$

which gives a better representation of the SGS tensor. This model is not an eddy-viscosity model and it does not dissipate energy, so that Bardina et al. (1983) proposed to use the following linear combination model (LCM):

$$C_{ij} + R_{ij} \simeq -2\nu_T \overline{S}_{ij} + C_r \left(\overline{u}_i\overline{u}_j - \overline{\overline{u}}_i\overline{\overline{u}}_j\right). \tag{4.26}$$

Some improvements over the Smagorinsky model have been noticed when the LCM is used (see Bardina et al. (1983), Speziale (1985), and Erlebacher et al. (1992)). Speziale (1985) has shown that, for the model to satisfy Galilean invariance, the constant C_r must be set equal to unity. In such a case, the total SGS tensor is modeled as follows:

$$\tau_{ij} = L_{ij} + C_{ij} + R_{ij} \simeq -2\nu_T \overline{S}_{ij} + \left(\overline{\overline{u}_i\overline{u}_j} - \overline{\overline{u}}_i\overline{\overline{u}}_j\right). \tag{4.27}$$

Finally, the modeled filtered Navier–Stokes equations are written

$$\begin{aligned} \frac{\partial \overline{u}_i}{\partial t} + \nabla \cdot (\overline{u}_i\, \overline{\mathbf{u}}) &= -\frac{\partial \overline{P}}{\partial x_i} + \nu \Delta \overline{u}_i \\ &\quad - \sum_{j=1}^{n} \frac{\partial}{\partial x_j} \left(-2\nu_T \overline{S}_{ij} + M_{ij}\right) \end{aligned} \tag{4.28}$$

for $i = 1, \ldots, n$, where

$$M_{ij} = \begin{cases} L_{ij} - \frac{1}{3}\sum_{l=1}^{n} L_{ll}\delta_{ij} & \text{for the Smagorinsky model,} \\ \left(\overline{\overline{u}_i\overline{u}_j} - \overline{\overline{u}}_i\overline{\overline{u}}_j\right) - \frac{1}{3}\sum_{l=1}^{n}\left(\overline{\overline{u}_l^2} - \overline{\overline{u}}_l^2\right)\delta_{ij} & \text{for the LCM model.} \end{cases}$$

Hence, as in (4.23), M_{ij} is a trace-free tensor.

As was noted above, the resolvable velocity $\overline{\mathbf{u}}$ mainly contains scales of length smaller than the filter width, so that (4.28) has to be resolved on a grid with mesh size $\Delta_c \leq \Delta_f$. In Erlebacher et al. (1992), the authors suggest that, when a Gaussian filter is used in a pseudospectral resolution of (4.28), $\Delta_c = \frac{1}{2}\Delta_f$ is the best choice.

Another deficiency of the Smagorinsky model is that the constant C_S depends on the grid size used in the approximation of the filtered equations (4.22), on the numerical method, and on the flow. Germano et al. (1991) introduced a dynamic model providing an algebraic expression for the Smagorinsky constant C_S; this relation is both space- and time-dependent. The basic ideas are summarized below. A *test filter* function $\tilde{G}$ of filter width $\tilde{\Delta}_f > \Delta_f$ is applied to the large-eddy field. By filtering the equations (4.22) with respect to the test filter, a test filter SGS tensor is introduced:

$$T_{ij} = \widetilde{\overline{u_i u_j}} - \tilde{\bar{u}}_i \tilde{\bar{u}}_j.$$

Therefore, the SGS tensors T_{ij} and τ_{ij} need to be modeled. Germano et al. (1991) have derived the algebraic relation

$$T_{ij} = \tilde{\tau}_{ij} + \mathcal{L}_{ij},$$

where $\mathcal{L}_{ij} = \widetilde{\bar{u}_i \bar{u}_j} - \tilde{\bar{u}}_i \tilde{\bar{u}}_j$ is equivalent to the Leonard stress tensor and is computable from the resolvable scales $\bar{\mathbf{u}}$. By modeling both τ_{ij} and T_{ij} with a Smagorinsky model, the following equation can be derived:

$$\mathcal{L}_{ij} - \frac{1}{3} \sum_{l=1}^{n} \mathcal{L}_{ll} \delta_{ij} = -2\, \tilde{\Delta}_f^2\, C_S\, \tilde{\bar{S}}\, \tilde{\bar{S}}_{ij} + 2\, \Delta_f^2\, \widetilde{s_{ij}}, \tag{4.29}$$

where $s_{ij} = C_S\, \bar{S}\, \bar{S}_{ij}$. Several methods have been proposed to solve (4.29) in order to compute C_S; see Germano et al. (1991), Lilly (1992), and Ghosal et al. (1995) for instance. They have been applied in different physical situations, such as homogeneous and wall-bounded flows; see the references above. Some mathematical difficulties related to the filtering process are encountered when generalizing this approach to fully nonhomogeneous flows. For these reasons, LES models are mainly restricted to simple geometries.

4.2.3. Spectral and Statistical Models

In order to conclude this section, we mention several other turbulence theories (or models) which have been investigated.

Two-Point Closures

Two-point closures are based on a modeling of the two-point correlation tensor or a spectral representation such as the energy spectrum function (or tensor). Such models provide more details than Reynolds-stress models. Indeed, many (global) characteristic quantities of turbulent flows (see Chapter 2) can be obtained directly from the energy spectrum function. However, these models are limited to simple flows for which a spectral analysis can be conducted. Several theoretical as well as operational difficulties are encountered in more complex flows such as wall-bounded ones. However, two-point closures such as the EDQNM model (see Orszag (1970), Lesieur (1990), and the references therein) have been successfully used in the analysis of homogeneous turbulent flows. They have been recently extended to the case of compressible flows.

Renormalization-Group Methods

Renormalization-Group (RNG) methods have been applied to turbulence by Foster, Nelson, and Stephen (1977) and by Rose (1977). These two approaches are slightly different. In Foster, Nelson, and Stephen (1977) a zero-mean Gaussian random forcing term is added to the Navier–Stokes equations. This random force is determined by its correlation function, which obeys a power law depending on a parameter ε. An expansion about ε is then made, and bands of high wavenumbers are systematically removed, introducing an eddy viscosity into the resolvable scale equation; this method is referred to as the ε-RNG method (see also Yakhot and Orszag (1986)). In Rose (1977), the ε-expansion is not applied and the high wavenumbers are also iteratively removed from the renormalized Navier–Stokes equations (see Zhou, Vahala, and Hossain (1988)). In this recursive RNG method, triple nonlinearities are retained and modeled, while their effects were neglected in ε-RNG. Both methods lead to a renormalized equation for the large (resolvable) scales in which the subgrid-scale effects are taken into account through the presence of an eddy viscosity. LES based on the RNG methods can then be performed.

Note that by successively removing larger and larger scales, RNG procedures give rise to Reynolds stress models. In Yakhot and Orszag (1986), a $K-\epsilon$ model based on ε-RNG methods has been derived.

PDF Models

PDF models based on the joint probability density functions (see Pope (1985)) have been proposed and are mainly applied to combustion problems.

5

Long-Time Behavior. Attractors and Their Approximation

In this chapter we address a few questions on the long-time behavior of the solutions of the Navier–Stokes equations. This problem, which has a mathematical interest on its own from the point of view of dynamical systems theory, is also connected with the study of turbulence in relation with the dynamical system approach to turbulence.

It is well known that there are two mathematical approaches to turbulence. The conventional approach is based on a statistical study of the flow: a probability measure, invariant under the flow, is associated with such a flow, and statistical properties of the flow are recovered from this measure. The measure itself is determined by time or space averaging, assuming ergodicity (see e.g. Batchelor (1971), Kolmogorov (1941a, 1941b, 1941c), Kraichnan (1967)); this point of view was used in Chapter 2.

The second, more recent approach to turbulence is the dynamical system approach; here the permanent regime of a flow is mathematically represented by a compact attractor in the phase space and one tries to recover information on the flow or to approximate it by studying the attractor and its approximations. Of course, these two approaches to turbulence are not exclusive of each other, as we will see below; in particular, the probability measure can be recovered from the attractor and it is concentrated on the attractor. This second approach underlies our construction of the multilevel algorithms presented in this volume.

Our purpose on this chapter is to recall a few results on the attractors for the Navier–Stokes equations and related concepts, following essentially Foias and Temam (1979), Constantin, Foias and Temam (1985), and Foias, Sell, and Temam (1985, 1988). Numerical approximations of the Navier–Stokes equations using related ideas are studied in the other chapters.

5.1. Attractors: Existence and Dimension

Attractors

We start with space dimension $n = 2$, and consider the Navier–Stokes equations with a driving force $\mathbf{f}$ as in (1.6); we assume that $\mathbf{f} \in H$ is independent of time: $\mathbf{f}(\mathbf{x}, t) = \mathbf{f}(\mathbf{x})$. We denote by $\mathbf{u}(t)$ the solution of (1.41), (1.42) given by Theorem 1.1, and we denote by $S(t)$, $t \geq 0$, the mapping

$$S(t) : \mathbf{u}_0 = \mathbf{u}|_{t=0} \rightarrow \mathbf{u}(t).$$

According to Theorem 1.1, $S(t)$ is a continuous operator in H ($\mathbf{f}$ is fixed).

The global attractor for the semigroup $\{S(t)\}_{t\geq 0}$ is a compact set $\mathcal{A}$ of the phase space H ($\mathcal{A} \subset H$) such that

$$S(t)\mathcal{A} = \mathcal{A} \quad \forall t \geq 0, \tag{5.1a}$$

$$\mathcal{A} \text{ attracts all bounded sets of } H, \tag{5.1b}$$

that is, $\forall \mathcal{B} \subset H$ *bounded,* $\forall \varepsilon > 0$, there exists $T_1 = T_1(\varepsilon, \mathcal{B})$ such that for $t \geq T_1(\varepsilon, \mathcal{B})$, $S(t)\mathcal{B}$ is included in an ε-neighborhood of $\mathcal{A}$.

It is easy to see that the global attractor, if it exists, is unique. The existence of such an attractor for the Navier–Stokes equations was proven in Foias and Temam (1979):

$$\begin{aligned} &\text{There exists a global attractor } \mathcal{A} \\ &\text{for the two-dimensional Navier–Stokes equations.} \\ &\text{It is a compact set in the phase space } H;\ \text{its Hausdorff} \\ &\text{and fractal dimensions are finite.} \end{aligned} \tag{5.2}$$

We will come back hereafter on the significance of the statement on the dimension of the attractor.

From the physical viewpoint, if for example the fluid is very viscous (ν is large), then the flow is laminar: there is a unique stationary solution to (1.41), and for any $\mathbf{u}_0$, the solution $\mathbf{u} = \mathbf{u}(t)$ of (1.41), (1.42), converges to the stationary solution as $t \rightarrow \infty$; in this case $\mathcal{A}$ is reduced to one point, the stationary solution. When ν is smaller, everything else being unchanged, or alternatively when the Reynolds number increases, the flow and the attractor become more complex: as $t \rightarrow \infty$, the flow may converge to a stationary solution different than the previous one, and the attractor consists of the stationary solutions and the orbits that connect them (called homoclinic and heteroclinic curves). Then, for ν smaller, more complex phenomena may occur with the appearance of time-periodic solutions (similar to the von Kármán streets), followed by quasiperiodic solutions and even more complex attractors in the case of fully turbulent flows. Of course these (qualitative) results on the nature of the attractor have not yet been proven for the Navier–Stokes equations; they are conjectured on the basis of experimental data in turbulence and by comparison with simpler differential equations.

For space dimension 3, the only difficulty in proving a result similar to (5.2) lies in the fact that the existence of global (in time) regular solutions in this case is not yet known, and therefore the semigroup $\{S(t)\}_{t\geq 0}$ cannot be defined. However (see Theorem 1.4), the operator $S(t)$ can be defined in V for $0 < t < T_*$, and, based on that, partial results can be proved, for invariant sets, that is, sets satisfying only (5.1a); see the references cited above, Temam (1997), and the references hereafter.

Dimension of Attractors and Invariant Sets

The results establishing the finite dimension of the attractor or functional invariant sets (Foias and Temam (1979)) have been subsequently improved, and physically relevant estimates on the dimension of these sets have been derived.

It was shown, for example, that

$$(n = 2) \qquad \dim \mathcal{A} \leq c \left(\frac{\ell_0}{\ell_\eta}\right)^2, \tag{5.3}$$

$$(n = 3) \qquad \dim \mathcal{A} \leq c \left(\frac{\ell_0}{\ell_d}^3\right). \tag{5.4}$$

In (5.3) ($n = 2$), $\mathcal{A}$ is the global attractor, ℓ_0 is a macroscopic length (e.g. the radius of Ω), and ℓ_η is the Kraichnan dissipation length properly defined on the attractor. In (5.4) ($n = 3$), $\mathcal{A}$ is an invariant set bounded in V (enstrophy norm), and ℓ_d is the Kolmogorov dissipation length properly defined on the attractor.

We now have a connection between the conventional approach to turbulence and the dynamical system approach; indeed, (5.3), (5.4) coincide with the estimates of the number of degrees of freedom of a turbulent flow in space dimension 2 (Kraichnan (1967)) and in space dimension 3 (Kolmogorov (1941a, 1941b, 1941c)). More details and the proofs of these results can be found in Constantin et al. (1985), Temam (1986). Further results concerning the space-periodic case appear in Constantin, Foias, and Temam (1985) (CFT); lower bounds on the dimension of attractors for Navier–Stokes equations were derived by Babin and Vishik (1983) and by Liu (1993). For the flow in an elongated two-dimensional channel driven by a constant pressure gradient, Ziane (1997) improved the results of Constantin et al. (1985) and derived an upper bound on the dimension of the attractor that matches the lower bound of Babin and Vishik (1983), thus showing that the CFT upper bound is optimal in this case. See also a related result for shear flows in Doering and Wang (1998).

5.2. Inertial Manifolds and Approximate Inertial Manifolds

Inertial Manifolds

After it was shown that the two-dimensional Navier–Stokes equations (and other related equations) have finite-dimensional behavior as $t \to \infty$, in agreement with the results of conventional theory of turbulence, the following questions were raised and partly answered:

1. Can we embed the attractor in a smooth finite-dimensional manifold?
2. Can we approximate the attractor by a smooth finite-dimensional manifold?
3. Can we approximate the dynamics on the attractor by the dynamics on an approximate finite-dimensional manifold?

The attempt to answer question 1 leads to the concept of inertial manifolds, which we now recall. The answer to question 2 leads to the concept of approximate inertial manifold discussed below. Question 3 has been only very partially answered; it will not be addressed here.

Definition 5.1. *An* inertial manifold *(IM) for the semigroup* $\{S(t)\}_{t\geq 0}$ *is a smooth finite-dimensional manifold* $\mathcal{M}$ *such that*

(i) $S(t)\mathcal{M} \subset \mathcal{M}, \quad \forall t \geq 0,$
(ii) $\mathcal{M}$ *attracts all orbits at an exponential rate*

$$\text{dist}(S(t)\mathbf{u}_0, \mathcal{M}) \leq \kappa_1 \exp(-\kappa_2 t),$$

where $\kappa_1, \kappa_2 \geq 0$ *depend boundedly on* $\mathbf{u}_0$.[1]

The inertial manifold, if it exists, is not unique; also, it is clear that when an inertial manifold $\mathcal{M}$ and the global attractor $\mathcal{A}$ both exist, then $\mathcal{A} \subset \mathcal{M}$.

Inertial manifolds were introduced in Foias, Sell, and Temam (1985, 1988).[2] It was shown that IMs exist for certain classes of equations similar to the Navier–Stokes equations (dissipative evolution equations); it is not known

[1] Usually, κ_2 is found independent of $\mathbf{u}_0$.
[2] See also the references therein and (for example) Temam (1997) for further bibliographical comments.

whether IMs exist for certain dissipative evolution equations that possess a finite-dimensional attractor, and in particular the two-dimensional Navier–Stokes equations. However, in the references cited above, it was shown that the Navier–Stokes equations with enhanced viscosity do possess an inertial manifold in space dimension 2 and 3 (an additional viscosity term $\varepsilon\Delta^2\mathbf{u}$ is added to the equation).

Inertial manifolds have been obtained as the graph of a function Φ:

$$\mathbf{z} = \Phi(\mathbf{y}), \tag{5.5}$$

where

$$\mathbf{u} = \mathbf{y} + \mathbf{z} \tag{5.6}$$

is, roughly speaking, a decomposition of $\mathbf{u}$ into its large-scale and small-scale components. In the simplest cases, $\mathbf{u}$ is expanded in a series of spectral functions of the form

$$\mathbf{u}(\mathbf{x}, t) = \sum_{j=1}^{\infty} \hat{u}_j(t)\mathbf{w}_j(\mathbf{x}), \tag{5.7}$$

and, for a suitable m,

$$\mathbf{y} = \sum_{j=1}^{m} \hat{u}_j\mathbf{w}_j, \qquad \mathbf{z} = \sum_{j=m+1}^{\infty} \hat{u}_j\mathbf{w}_j. \tag{5.8}$$

From the physical viewpoint we have

$$\mathbf{z}(t) = \Phi(\mathbf{y}(t)) \tag{5.9}$$

on the IM, that is, the small scales are slaved by the large ones. This is equivalent to a family of relations

$$\hat{u}_j(t) = \Phi_j(\hat{u}_1(t), \ldots, \hat{u}_m(t)), \qquad j \geq m+1, \tag{5.10}$$

and any flow tends to a state where (5.10) holds.

It is likely that, for turbulent flows, if an IM exists, then the cutoff wavenumber corresponding to m lies far inside the dissipation range. However, we speculate, on the basis of the numerical simulations, that approximate forms of (5.9), (5.10) exist not too far in the dissipation range or may be in the inertial range: this is the concept of approximate inertial manifold.

Approximate Inertial Manifolds

We recall the definition.

Definition 5.2. *An* approximate inertial manifold *(AIM) of order η is a smooth finite-dimensional manifold $\mathcal{M}_\eta$ such that*

(i) $\mathcal{A}$ is included in a neighborhood of $\mathcal{M}_\eta$ of order η,
(ii) all orbits enter such a neighborhood in finite time.

AIMs have been constructed for many dissipative evolution equations and in particular for the Navier–Stokes equations in space dimension 2. Based on a decomposition of $\mathbf{u}$ similar to (5.6), we obtain the equation of the AIM in the form (5.5); with an expansion of the form (5.7), (5.8), equations (5.9) and (5.10) become approximate relations, for example,

$$\operatorname{dist}(\mathbf{z}(t), \Phi(\mathbf{y}(t))) \leq \varepsilon. \tag{5.11}$$

Approximate inertial manifolds for the two-dimensional Navier–Stokes equations were constructed, for example, in Foias, Manley, and Temam (1987, 1988), where they are based on the fact that $\mathbf{z}$ is small in comparison with $\mathbf{y}$. More AIMs for the Navier–Stokes equations are constructed in Temam (1989) and Titi (1990). Another approach appears in Foias et al. (1988). A whole sequence of approximate inertial manifolds appear in Temam (1989), based on an asymptotic expansion corresponding to small $\mathbf{z}$. We refer the reader to Chapters 8 and 9 for the utilization of AIMs for the development of numerical methods namely as a *nonlinear filter*.

All the AIMs above are of polynomial order, that is, η is of the form $\beta m^{-\gamma}$ where $\beta,\gamma > 0$ and m is the dimension of the AIM (the dimension of $\mathbf{y}$). Approximate inertial manifolds of exponential order, that is, with $\eta = \alpha \exp(-\beta m^\gamma)$, $\alpha, \beta, \gamma > 0$, are constructed in Debussche and Temam (1994). Let us mention also that a related concept appears in meteorology, where it is used for short-term weather prediction: the concept of *slow manifolds*, similar, but not identical, to that of inertial manifolds. Also, the Machenhauer algorithm is based on a concept similar to that of approximate inertial manifold. For more details, see for example Debussche and Temam (1991), Temam (1990a), and the references therein.

6

Separation of Scales in Turbulence

As has been shown in Section 4.1, direct numerical simulation (DNS) of turbulent flows is limited to simple flows and to low or moderate Reynolds numbers. These restrictions come from the fact that flows at high Reynolds numbers contain a large number of scales with different sizes and with different characteristic times. So, in order to solve problems of physical and industrial interest, the small structures, implying length- and time-scale restrictions on the computational parameters, and their interaction with the large (energy-containing) eddies must be modeled.

Different approaches to turbulence modeling, such as large-eddy simulation (LES) (see Section 4.2.2), have been proposed. In all LES models, only the large scales are computed and the actions of the small (subgrid) scale structures are modeled by introducing an eddy viscosity, so that the viscosity in the large (resolved) scale equation is enhanced. In the dynamic multilevel (DML) methods, as in LES, the separation of scales is achieved with the help of a filter, so that the velocity field $\mathbf{u}$ is decomposed as $\mathbf{u} = \overline{\mathbf{u}} + \mathbf{u}'$.

In this chapter, we first investigate the problem of the filtering operation used in the scale decomposition. Then we consider the separation of scales for two particular cases: the homogeneous case (periodic boundary conditions) and the nonhomogeneous case (channel flow).

6.1. Scale Decomposition

6.1.1. Filter Functions

We consider the approximated velocity field $\mathbf{u}_N(\mathbf{x}, t)$ on the fine grid, that is, the velocity field obtained from DNS (see Chapter 3). Hence the smallest scales contained in $\mathbf{u}_N$ are of the order of $\Delta(N)$, which is the mesh size of the fine grid.

In order to remove in $\mathbf{u}_N$ the scales of length smaller than $\Delta(N_1)$ $(N_1 < N)$, or at least to reduce their energy, a filtering operator is applied to $\mathbf{u}_N$, leading to the definition of a filtered field $\overline{\mathbf{u}}_{N_1}$. The remaining (subgrid-scale) velocity component is denoted

$$\mathbf{u}'^{N}_{N_1} = \mathbf{u}_N - \overline{\mathbf{u}}_{N_1}. \tag{6.1}$$

The DML methods are based on such decompositions of the velocity field. From the above considerations, $\overline{\mathbf{u}}_{N_1}$ represents the large scales of $\mathbf{u}_N$, while $\mathbf{u}'^{N}_{N_1}$ represents its small scales.

The filtering operation defining $\overline{\mathbf{u}}_{N_1}$ can take different forms. The filtered function is obtained by a convolution product of $\mathbf{u}_N$ with a filter function G:

$$\overline{\mathbf{u}}_{N_1}(\mathbf{x}, t) = \int_{\Omega} \mathbf{u}_N(\mathbf{y}, t)\, G(\mathbf{x}, \mathbf{y}, \Delta_f)\, d\mathbf{y}, \tag{6.2}$$

where Δ_f is the width of the filter, which is of the order of $\Delta(N_1)$. In the case of periodic boundary conditions, the filtering operation defined by (6.2) commutes with spatial derivatives, provided the filter function G satisfies the following properties:

1. G depends only on the difference $\mathbf{x} - \mathbf{y}$, that is, $G(\mathbf{x}, \mathbf{y}, \Delta_f) = G(\mathbf{x} - \mathbf{y}, \Delta_f)$,
2. the filter width Δ_f is independent of the location $\mathbf{x}$, and
3. G is a periodic function.

Moreover, when periodic boundary conditions are considered, as in the homogeneous case, a spectral representation of G can be easily derived and (6.2) can be rewritten in the spectral space as

$$\hat{\overline{\mathbf{u}}}(\mathbf{k}, t) = |\Omega|\, \hat{G}(\mathbf{k}, \Delta_f)\, \hat{\mathbf{u}}(\mathbf{k}, t) \qquad \text{for all } k = |\mathbf{k}| \leq k_N. \tag{6.3}$$

Given a sequence of decreasing positive coefficients $\sigma(\mathbf{k}, \Delta_f)$ satisfying

$$\sigma(0, \Delta_f) = \frac{1}{|\Omega|} \quad \text{and} \quad \sigma(-\mathbf{k}, \Delta_f) = \sigma(\mathbf{k}, \Delta_f),$$

the function

$$G_N(\zeta, \Delta_f) = \sum_{\mathbf{k} \in S_N} \sigma(\mathbf{k}, \Delta_f)\, e^{\mathbf{i}\mathbf{k}_{L_1} \cdot \zeta}$$

corresponds, according to (6.3), to the spectral filtering operator

$$\overline{\mathbf{u}}_{N_1}(\mathbf{x}, t) = \sum_{\mathbf{k} \in S_N} |\Omega|\, \sigma(\mathbf{k}, \Delta_f)\, \hat{\mathbf{u}}(\mathbf{k}, t)\, e^{\mathbf{i}\mathbf{k}_{L_1} \cdot \mathbf{x}}.$$

We have noted that $\mathbf{k}_{L_1} = 2\mathbf{k}\pi/L_1$.

One of the most commonly used filter functions is the Gaussian filter given by

$$G_N(\zeta, \Delta_f) = \sum_{\mathbf{k}\in S_N} \sigma(\mathbf{k}, \Delta_f)\, e^{\mathrm{i}\mathbf{k}_{L_1}\cdot\zeta},$$

with

$$\sigma(\mathbf{k}, \Delta_f) = \frac{1}{|\Omega|}\, e^{-k^2\Delta_f^2}.$$

Note that with this filter, $\mathbf{u}_N$ and $\bar{\mathbf{u}}_{N_1}$ have the same support in the spectral space, that is, the spectral coefficients of the filtered field do not vanish for $k > k_{N_1}$; however, their amplitudes are decreased, so that the resulting field is smoother than $\mathbf{u}_N$. Note also that for small values of k (i.e. for the large scales), the energy spectrum of $\bar{\mathbf{u}}_{N_1}$ is close to that of $\mathbf{u}_N$. A detailed discussion of the effects of the Gaussian filter can be found in Zhou, Hossain, and Vahala (1989). Also, other examples of filter functions are given for instance in Leslie and Quarini (1979), Voke and Collins (1983), and Zhou, Hossain, and Vahala (1989).

Another filter often used in the case of finite difference methods is the top-hat filter, which consists in performing a volume average over a box of given size surrounding the point $\mathbf{x}$, namely,

$$G(\mathbf{x}, \mathbf{y}, \Delta_f) = \frac{1}{\Delta_f^n} \times \begin{cases} 1 & \text{if } |x_i - y_i| < \frac{1}{2}\,\Delta_f,\ i = 1, \ldots, n, \\ 0 & \text{otherwise,} \end{cases} \tag{6.4}$$

where n is the space dimension ($n = 2$ or 3). Such filter is used in Deardorff (1970), where the three-dimensional channel flow problem is studied.

6.1.2. Projective Filters Based on Galerkin Approximations

Let $(\phi_l)_{l\in\mathbb{N}}$ be a Galerkin basis, and let $\mathbf{u}_N(x,t)$ be the following approximated velocity field:

$$\mathbf{u}_N(x,t) = \sum_{l=0}^{N} \tilde{\mathbf{u}}(l,t)\,\phi_l(x). \tag{6.5}$$

Clearly $\mathbf{u}_N \in \mathrm{Span}\{\phi_l,\ l = 0, \ldots, N\}^n$. We can define the following filter:

$$\bar{\mathbf{u}}_{N_1}(x,t) = P_{N_1}\mathbf{u}_N(x,t) = \mathbf{u}_{N_1}(x,t) = \sum_{l=0}^{N_1} \tilde{\mathbf{u}}(l,t)\,\phi_l(x), \tag{6.6}$$

where P_{N_1} is the projection operator onto the space $\text{Span}\{\phi_l,\ l = 0, \ldots, N_1\}$; $\mathbf{u}_N$ and $\overline{\mathbf{u}}_{N_1}$ do not have the same support in the spectral space, but the spectrum of $\overline{\mathbf{u}}_{N_1}$ corresponds to the low part ($l \leq N_1$) of the spectrum of $\mathbf{u}_N$. Moreover, this filter is projective, that is,

$$\left(P_{N_1} \circ P_{N_1}\right) \mathbf{u}_N(x, t) = P_{N_1} \overline{\mathbf{u}}_{N_1}(x, t) = \overline{\mathbf{u}}_{N_1}(x, t). \tag{6.7}$$

If the Galerkin basis $(\phi_l)_{l \in \mathbf{N}}$ is orthogonal, the projector P_{N_1} is orthogonal.

Let us now consider the particular case of the Fourier expansion in several spatial directions x_i, $i = 1, \ldots, n$, $n = 2$ or 3:

$$\mathbf{u}_N(\mathbf{x}, t) = \sum_{\mathbf{k} \in S_N} \hat{\mathbf{u}}(\mathbf{k}, t)\, e^{\mathbf{i}\mathbf{k}_{L_1} \cdot \mathbf{x}}.$$

We have

$$\overline{\mathbf{u}}_{N_1}(\mathbf{x}, t) = P_{N_1} \mathbf{u}_N(\mathbf{x}, t) = \sum_{\mathbf{k} \in S_{N_1}} \hat{\mathbf{u}}(\mathbf{k}, t)\, e^{\mathbf{i}\mathbf{k}_{L_1} \cdot \mathbf{x}}, \tag{6.8}$$

with $S_{N_1} = \{\mathbf{k} \in \mathbb{Z}^n,\ \mathbf{k} \neq \mathbf{0}\ / |k_j| \leq N_1/2 - 1,\ j = 1, \ldots, n\}$. We note that (6.8) can be rewritten in the form (6.2) where

$$\hat{G}(\mathbf{k}, \Delta_f) = \hat{G}_{N_1}(\mathbf{k}) = \begin{cases} \frac{1}{|\Omega|} & \text{if } \mathbf{k} \in S_{N_1}, \\ 0 & \text{if } \mathbf{k} \in S_N \backslash S_{N_1} \end{cases} \tag{6.9}$$

is the sharp-cutoff filter function. Indeed, replacing in (6.8) the Fourier coefficients $\hat{\mathbf{u}}(\mathbf{k}, t)$ by their expression, we obtain

$$\begin{aligned} \overline{\mathbf{u}}_{N_1}(\mathbf{x}, t) &= \sum_{\mathbf{k} \in S_{N_1}} \left[\frac{1}{|\Omega|} \int_\Omega \mathbf{u}_N(\mathbf{y}, t) e^{-\mathbf{i}\mathbf{k}_{L_1} \cdot \mathbf{y}}\, d\mathbf{y} \right] e^{\mathbf{i}\mathbf{k}_{L_1} \cdot \mathbf{x}} \\ &= \frac{1}{|\Omega|} \int_\Omega \left[\sum_{\mathbf{k} \in S_{N_1}} e^{\mathbf{i}\mathbf{k}_{L_1} \cdot (\mathbf{x} - \mathbf{y})} \right] \mathbf{u}_N(\mathbf{y}, t)\, d\mathbf{y}, \end{aligned}$$

with $\Omega = \prod_{i=1}^n (0, L_1)$ (see Section 3.1). Equivalently

$$\overline{\mathbf{u}}_{N_1}(\mathbf{x}, t) = \int_\Omega \mathbf{u}_N(\mathbf{y}, t)\, G_{N_1}(\mathbf{x} - \mathbf{y})\, d\mathbf{y}, \tag{6.10}$$

with

$$G_{N_1}(\zeta) = \frac{1}{L_1^n} \sum_{\mathbf{k} \in S_{N_1}} e^{\mathbf{i}\mathbf{k}_{L_1} \cdot \zeta} = \prod_{i=1}^n D_{N_1}(\zeta_i),$$

where $D_{N_1}(\zeta_i)$ is the Dirichlet kernel (see Canuto et al. (1988))

$$D_{N_1}(\zeta_i) = \frac{1}{L_1}\begin{cases} \frac{\sin[(N_1+1)\pi\zeta_i/L_1]}{\sin(\pi\zeta_i/L_1)} & \text{if } \zeta_i \neq L_1 j,\ j \in \mathbb{Z}, \\ N_1+1 & \text{if } \zeta_i = L_1 j,\ j \in \mathbb{Z}. \end{cases}$$

Now we consider the particular case of the polynomial approximation using polynomial functions, which are orthogonal for the scalar product in $L^2_w(\Omega)$:

$$\mathbf{u}_N(x,t) = \sum_{l=0}^{N} \tilde{\mathbf{u}}(l,t)\,\phi_l(x)$$

for $x \in \Omega = (-h, +h)$. By definition we have

$$\overline{\mathbf{u}}_{N_1}(x,t) = \sum_{l=0}^{N_1} \tilde{\mathbf{u}}(l,t)\,\phi_l(x). \tag{6.11}$$

We can also rewrite (6.11) in the form (6.2). Indeed, if we take the filter function

$$G_{N_1}(x,y) = \sum_{l=0}^{N_1} \frac{1}{c_l}\,\phi_l(x)\,\phi_l(y)\,w(y)$$

with $c_l = \int_\Omega \phi_l^2(x)\,w(x)\,dx$, we find

$$\overline{\mathbf{u}}_{N_1}(x,t) = \int_\Omega \mathbf{u}_N(y,t)\,G_{N_1}(x,y)\,dy. \tag{6.12}$$

In the case of the Legendre (the Chebyshev) polynomials, we have $w(x) = 1$ ($w(x) = (1-x^2)^{-1/2}$) $\forall x \in \Omega$.

6.2. The Separation of Scales

Now, in order to separate the small and large scales for DML methods applied to the homogeneous and nonhomogeneous turbulent flows, we consider the application of Section 6.1 to the fully periodic case and to wall-bounded flows.

6.2.1. The Fully Periodic Case

In DML methods, several successive levels are used, and so it is advantageous to use a projective filter. Indeed, if $N_2 < N_1$, we notice that

$$P_{N_2}(P_{N_1}\mathbf{u}_N(\mathbf{x},t)) = P_{N_2}\overline{\mathbf{u}}_{N_1}(\mathbf{x},t) = \overline{\mathbf{u}}_{N_2}(\mathbf{x},t).$$

In the case of the periodic boundary conditions, we use the projection operator P_{N_1} previously defined in (6.8). This filter provides a complete separation between the small and large scales. For a given cutoff level $N_1 < N$, the large scales are associated with small wavenumbers $\mathbf{k}$, $|\mathbf{k}| \leq k_{N_1} = N_1/2$, and the small scales are associated with large wavenumbers $\mathbf{k}$, $|\mathbf{k}| > k_{N_1}$.

Note that when the filter operation P_N is directly applied to the velocity field $\mathbf{u}(\mathbf{x}, t)$ and when the cutoff wavenumber $k_N = N/2$ is large enough ($k_N > k_d$, with k_d the Kolmogorov wavenumber defined in Section 2.2.4), the fluctuating velocity

$$\mathbf{u}'(\mathbf{x}, t) = \mathbf{u}(\mathbf{x}, t) - P_N\mathbf{u}(\mathbf{x}, t) \tag{6.13}$$

is sufficiently small to be neglected, due to the decrease of the Fourier coefficients of $\mathbf{u}$ (see e.g. Canuto et al. (1988), Foias and Temam (1989)). So we have

$$\mathbf{u}(\mathbf{x}, t) \simeq P_N\mathbf{u}(\mathbf{x}, t) = \mathbf{u}_N(\mathbf{x}, t). \tag{6.14}$$

According to Section 3.1, the approximated velocity field $\mathbf{u}_N(\mathbf{x}, t)$ satisfies the following evolution equation:

$$\frac{\partial \mathbf{u}_N}{\partial t} - \nu\, \Delta \mathbf{u}_N + P_N B(\mathbf{u}_N, \mathbf{u}_N) = \mathbf{g}_N. \tag{6.15}$$

As previously, we consider a cutoff value $k_{N_1} < k_d < k_N$, and we apply to $\mathbf{u}_N$ the filter function G_{N_1} introduced in (6.10). We obtain the large-scale component of the velocity, which we denote by

$$\mathbf{y}_{N_1}(\mathbf{x}, t) = P_{N_1}\mathbf{u}_N(\mathbf{x}, t) = \int_\Omega \mathbf{u}_N(\mathbf{y}, t)\, G_{N_1}(\mathbf{x} - \mathbf{y})\, d\mathbf{y}. \tag{6.16}$$

The remaining subgrid-scale velocity, representing the small-scale component, is denoted by

$$\mathbf{z}_{N_1}^N(\mathbf{x}, t) = Q_{N_1}^N \mathbf{u}_N(\mathbf{x}, t) = \mathbf{u}_N(\mathbf{x}, t) - \mathbf{y}_{N_1}(\mathbf{x}, t). \tag{6.17}$$

Since the filter is projective, we have $P_{N_1}(\mathbf{z}_{N_1}^N) = 0$. If $N_1 < N$, by projecting (6.15) with respect to P_{N_1}, we obtain the large-scale equation

$$\frac{\partial \mathbf{y}_{N_1}}{\partial t} - \nu\, \Delta \mathbf{y}_{N_1} + P_{N_1} B(\mathbf{y}_{N_1}, \mathbf{y}_{N_1}) = P_{N_1}\mathbf{g}_N - P_{N_1} B_{\text{int}}\left(\mathbf{y}_{N_1}, \mathbf{z}_{N_1}^N\right), \tag{6.18}$$

where $P_{N_1} B_{\text{int}}(\mathbf{y}_{N_1}, \mathbf{z}_{N_1}^N)$ is the nonlinear interaction term. Note that

$$P_{N_1} B_{\text{int}}\left(\mathbf{y}_{N_1}, \mathbf{z}_{N_1}^N\right) = P[\nabla \cdot (\mathbf{C} + \mathbf{R})]$$

where $\mathbf{C} = (C_{ij})$ is the cross-stress tensor, $\mathbf{R} = (R_{ij})$ is the Reynolds stress tensor introduced in Section 4.2.2, and P is the projection operator introduced in Chapter 1. By identification, we can easily show that

$$P_{N_1} B_{\text{int}}\left(\mathbf{y}_{N_1}, \mathbf{z}_{N_1}^N\right) = P_{N_1} B\left(\mathbf{y}_{N_1}, \mathbf{z}_{N_1}^N\right) + P_{N_1} B\left(\mathbf{z}_{N_1}^N, \mathbf{y}_{N_1}\right) + P_{N_1} B\left(\mathbf{z}_{N_1}^N, \mathbf{z}_{N_1}^N\right). \tag{6.19}$$

By projecting (6.15) with respect to the operator $Q_{N_1}^N$, we obtain the subgrid-scale equation

$$\frac{\partial \mathbf{z}_{N_1}^N}{\partial t} - \nu\, \Delta \mathbf{z}_{N_1}^N + Q_{N_1}^N B\left(\mathbf{y}_{N_1} + \mathbf{z}_{N_1}^N, \mathbf{y}_{N_1} + \mathbf{z}_{N_1}^N\right) = Q_{N_1}^N \mathbf{g}_N. \tag{6.20}$$

For deriving the equations (6.18) and (6.20), the commutation property of the projection operators P_{N_1} and $Q_{N_1}^N$ with the spatial derivatives has been used.

In summary, the filter operator P_{N_1} defined in (6.8) for the homogeneous case enjoys the following properties:

1. P_{N_1} is projective,
2. it commutes with partial derivatives,
3. it preserves the boundary conditions on $\mathbf{y}_{N_1}$ and $\mathbf{z}_{N_1}^N$, and
4. it provides a complete separation between small and large scales.

6.2.2. Wall-Bounded Flows

Separation of Scales in the Physical Space

Let us consider the top-hat filter defined in (6.4). With such a filter, $\overline{\mathbf{u}}_{N_1}$ cannot be defined at the boundary, and artificial boundary conditions must be imposed. A natural way to impose these boundary conditions is to enforce $\overline{\mathbf{u}}_{N_1}(x_1, \pm h, x_3, t) = 0$, where $x_2 = \pm h$ corresponds to the walls (see Section 3.2). However, there is no reason for $\overline{\mathbf{u}}_{N_1}(\mathbf{x}, t)$ defined by (6.2) and (6.4) to tend to 0 when $\mathbf{x}$ tends to the boundary. Hence, discontinuities at the wall appear on the filtered function. When spectral discretizations such as Chebyshev or Legendre decompositions are used, these discontinuities at the boundary generate oscillations in the interior of the domain. Consequently, the commutation property is not satisfied any more. Furthermore, these oscillations increase in amplitude when the width of the filter is decreased; this phenomenon is known as the Gibbs phenomenon (see Gottlieb and Orszag (1977), Canuto et al. (1988)).

In order to overcome the difficulty due to the boundaries, we can apply a filter similar to (6.4) but with variable filter widths, so that Δ_f depends on the location $\mathbf{x}$, for instance

$$G(\mathbf{x}, \mathbf{y}, h(\mathbf{x})) = \left(\frac{1}{h(\mathbf{x})}\right)^n \begin{cases} 1 & \text{if } |x_i - y_i| < h(\mathbf{x})/2,\ i = 1, \ldots, n, \\ 0 & \text{otherwise,} \end{cases} \tag{6.21}$$

where $h(\mathbf{x}) = \Delta_f(\mathbf{x})$ decreases towards 0 when $\mathbf{x}$ tends to the boundary. In the channel flow problem, Moin and Kim (1982) applied a filter similar to (6.21) in the direction normal to the walls and a Gaussian filter in the homogeneous directions. Such a filter is well adapted to the physics of the problem; indeed, the turbulence length scales decrease near the wall in the wall-normal direction. So Moin and Kim considered a filter function of the form (6.21) depending only upon x_2, the direction normal to the wall:

$$G(x_2, y_2, h(x_2)) = \frac{1}{h(x_2)} \begin{cases} 1 & \text{if } |x_2 - y_2| < h(x_2)/2, \\ 0 & \text{otherwise.} \end{cases} \tag{6.22}$$

In such a case, (6.2) becomes

$$\overline{u}_i(\mathbf{x}, t) = \frac{1}{h(x_2)} \int_{x_2 - h(x_2)/2}^{x_2 + h(x_2)/2} u_i(\mathbf{y})\, dy_2 \qquad \text{for} \quad i = 1, \ldots, n. \tag{6.23}$$

By differentiating (6.23) with respect to x_2 we can easily show that

$$\begin{aligned} \frac{\partial \overline{u}_i}{\partial x_2}(\mathbf{x}, t) &= \overline{\frac{\partial u_i}{\partial x_2}}(\mathbf{x}, t) \\ &\quad + h(x_2)^{-1} \frac{dh}{dx_2}(x_2)[u_i(\mathbf{x}, t) - \overline{u}_i(\mathbf{x}, t) + o(h^2(x_2))]. \end{aligned} \tag{6.24}$$

Hence, if the filter width is not constant, $(dh/dx_2)(x_2)$ does not vanish, so that the commutation property is not valid anymore. Thus, the concept of the filter operation (6.2) commuting with all spatial derivatives cannot be generalized to boundary conditions other than the periodic ones.

This lack of commutativity between filtering and differentiation causes every spatial derivative operator in the Navier–Stokes equations to generate terms that cannot be expressed solely in terms of the filtered fields. Therefore, a closure problem is introduced in the Navier–Stokes equations, not only for the nonlinear terms, but for the linear terms as well. Ghosal and Moin (1995) proposed a definition of a filtering operation on a nonuniform grid. The commutation error can be reduced at any given order of accuracy by introducing higher spatial derivatives in the equation associated with the filtered field; thus additional (artificial) boundary conditions are required. In order to overcome this difficulty, a perturbative solution can be obtained by using an asymptotic expansion with respect to a small parameter (grid spacing); no additional boundary conditions need to be enforced for the filtered equations with this method. However, the complexity of the problem increases with the order of accuracy required, since a correction must be computed at each order of accuracy.

Now we consider the particular case of a grid in the physical space, based on collocation points (collocation methods; see Gottlieb, and Orszag (1977) and Canuto et al. (1988)). A separation into small and large scales has been proposed by Dettori, Gottlieb, and Temam (1995). In this separation, two coarse grids in the nonhomogeneous direction $(-h, +h)$ are used:

$$\left\{x_{2,j}^{M}\right\}_{j=0}^{M} = \left\{\xi_j^{M/2}\right\}_{j=0}^{M/2} \cup \left\{\eta_j^{M/2}\right\}_{j=0}^{M/2-1}, \tag{6.25}$$

where $\xi_j^{M/2} = h\cos(2j\pi/M)$ are the Chebyshev Gauss–Lobatto points, and $\eta_j^{M/2} = h\cos[(2j+1)\pi/M]$ are the Chebyshev Gauss points. Consider $I_{M/2}$ and $\tilde{I}_{M/2}$, the interpolation polynomials associated with these two coarse grids. Let $\mathbf{u}_M \in (Y_M)^n$, where Y_M is the set of polynomials of degree $\leq M$ and $n = 2$ or 3 (spatial dimension); a linear combination of the operators $I_{M/2}$ and $\tilde{I}_{M/2}$ gives the following decomposition of $\mathbf{u}_M$:

$$\mathbf{u}_M = \tilde{J}_{M/2}\mathbf{u}_M + \tilde{G}_M\mathbf{u}_M, \tag{6.26}$$

where $\tilde{J}_{M/2}\mathbf{u}_M \in (Y_{M/2-1})^n$ and $\tilde{G}_M\mathbf{u}_M \in (Y_{M/2-1}^{M})^n$ (orthogonal complement of $(Y_{M/2-1})^n$ in $(Y_M)^n$); $\tilde{J}_{M/2}$ and $\tilde{G}_M$ are defined as follows:

$$\tilde{J}_{M/2} = \tilde{I}_{M/2}\,\frac{I_{M/2} + \tilde{I}_{M/2}}{2}, \tag{6.27}$$

$$\tilde{G}_M = I_M - \tilde{J}_{M/2}. \tag{6.28}$$

In the decomposition (6.26), $\tilde{J}_{M/2}\mathbf{u}_M = \overline{\mathbf{u}}_{M/2}$ is associated with the large scales and $\tilde{G}_M\mathbf{u}_M = \mathbf{u}'^{M}_{M/2}$ with the small scales.

Another filter in the physical space can be defined as the interpolation on a coarse grid of a solution defined on a fine grid. Let $\{x_{2,j}^M\}_{j=0}^M = \{h\cos(j\pi/M)\}_{j=0}^M$ be the fine grid (Chebyshev Gauss–Lobatto points) defined on $(-h, +h)$. Let $M_1 < M$, and $\{x_{2,j}^{M_1}\}_{j=0}^{M_1} = \{h\cos(j\pi/M_1)\}_{j=0}^{M_1}$ be the coarse grid. The interpolating polynomial $\mathbf{u}_M(\mathbf{x}, t)$, of degree M, of the continuous function $\mathbf{u}(\mathbf{x}, t)$ satisfies

$$\begin{aligned}\mathbf{u}_M\left(x_1, x_{2,j}^M, x_3\right) &= (I_M\mathbf{u})\left(x_1, x_{2,j}^M, x_3\right) \\ &= \mathbf{u}\left(x_1, x_{2,j}^M, x_3\right), \qquad \text{for} \quad j = 0, \ldots, M,\end{aligned} \tag{6.29}$$

where I_M is the interpolating operator of degree M on the fine grid. Moreover, $\mathbf{u}_M$ can be rewritten as

$$\mathbf{u}_M\left(x_1, x_{2,j}^M, x_3\right) = \sum_{l=0}^{M} \tilde{\mathbf{u}}_M(x_1, l, x_3)\, T_l\left(x_{2,j}^M\right) \qquad \text{for} \quad j = 0, \ldots, M. \tag{6.30}$$

We define a numerical filter such that the filtered function $\bar{\mathbf{u}}_{M_1}$ is the interpolating polynomial of degree M_1 of $\mathbf{u}_M$ on the coarse grid $\{x_{2,j}^{M_1}\}$, that is,

$$\begin{aligned}\bar{\mathbf{u}}_{M_1}\left(x_1, x_{2,j}^{M_1}, x_3\right) &= I_{M_1}(\mathbf{u}_M)\left(x_1, x_{2,j}^{M_1}, x_3\right) \\ &= \mathbf{u}_M\left(x_1, x_{2,j}^{M_1}, x_3\right) \qquad \text{for} \quad j = 0, \ldots, M_1.\end{aligned}$$

Note that, as in the homogeneous case (cf. (6.14)), when M is sufficiently large, we have

$$\mathbf{u}_M(x_1, x_2, x_3) \simeq \mathbf{u}(x_1, x_2, x_3) \qquad \text{for} \quad x_2 \in (-h, +h).$$

I_{M_1} being the interpolating polynomial of degree M_1 on the coarse grid, we have

$$\bar{\mathbf{u}}_{M_1}(\mathbf{x}) = I_{M_1}\mathbf{u}_M(\mathbf{x}). \tag{6.31}$$

Defined as previously, this filter and the filtered solution enjoy the following properties:

1. I_{M_1} is a projective filter, that is, $I_{M_1}(\bar{\mathbf{u}}_{M_1}) = (I_{M_1} \circ I_{M_1})(\mathbf{u}_M) = \bar{\mathbf{u}}_{M_1}$,
2. $\bar{\mathbf{u}}_{M_1}(\pm h) = 0$,
3. $\bar{\mathbf{u}}_{M_1} \in (Y_{M_1})^n = \text{Span}\{T_l,\ l = 0, \ldots, M_1\}^n$.

We now define

$$\mathbf{u}'^M_{M_1} = \mathbf{u}_M - \bar{\mathbf{u}}_{M_1}. \tag{6.32}$$

We can easily verify that

$$\mathbf{u}'^M_{M_1}(\mathbf{x}) = \sum_{l=0}^{M} \tilde{\mathbf{u}}'^M_{M_1}(x_1, l, x_3)\, T_l(x_2),$$

with

$$\tilde{\mathbf{u}}'^M_{M_1}(x_1, l, x_3) = \begin{cases} \tilde{\mathbf{u}}_M(x_1, l, x_3) - \tilde{\bar{\mathbf{u}}}_{M_1}(x_1, l, x_3), & \text{for } l = 0, \ldots, M_1, \\ \tilde{\mathbf{u}}_M(x_1, l, x_3) & \text{for } l = M_1 + 1, \ldots, M. \end{cases}$$

Hence, with (6.31) and (6.32) we have defined a decomposition of the solution

$$\mathbf{u}_M = \overline{\mathbf{u}}_{M_1} + \mathbf{u}'^M_{M_1}, \tag{6.33}$$

where $\overline{\mathbf{u}}_{M_1} \in (Y_{M_1})^n$ and $\mathbf{u}'^M_{M_1} \in (Y_M)^n$.

The numerical evaluation of the filter (6.31) is quite efficient. Indeed, the evaluation of the spectral coefficients of $\overline{\mathbf{u}}_{M_1}$ from the spectral coefficients of $\mathbf{u}_M$ requires:

- the evaluation of $\overline{\mathbf{u}}_{M_1}(x_1, x_{2,j}^{M_1}, x_3) = \mathbf{u}_M(x_1, x_{2,j}^{M_1}, x_3)$, $j = 0, \dots, M_1$, which can be done with one fast Chebyshev transform (FCT), at $M_1 + 1$ modes;
- the transformation in the spectral space to obtain the coefficients $\tilde{\overline{\mathbf{u}}}_{M_1}(x_1, l, x_3)$, $l = 0, \dots, M_1$, which can be done with one FCT at $M_1 + 1$ modes.

Hence, the total number of operations needed for the computation of $\{\tilde{\overline{\mathbf{u}}}_{M_1}(x_1, l, x_3)\}$ for $l = 0, \dots, M_1$ is $O(M_1 \log_2 M_1)$.

In Figures 6.1 and 6.2, we have compared a given scalar function φ_M with its filtered one $\overline{\varphi}_{M_1}$, and with residual part $\varphi'^M_{M_1}$, in both the physical and spectral spaces. As we can see on Figure 6.1, $\overline{\varphi}_{M_1}$ is smoother than φ_M. Figure 6.2 shows

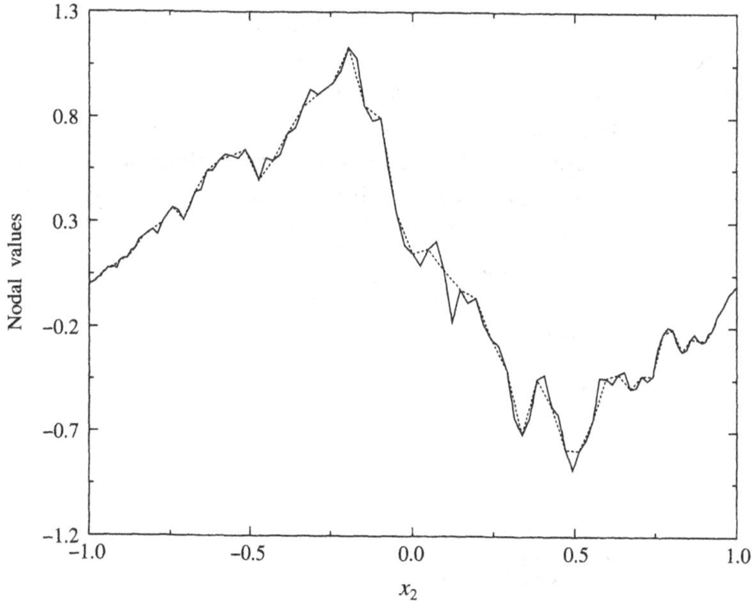

Figure 6.1. Comparison of the values of φ (solid line) and $\overline{\varphi}$ (dotted line) in the physical space.

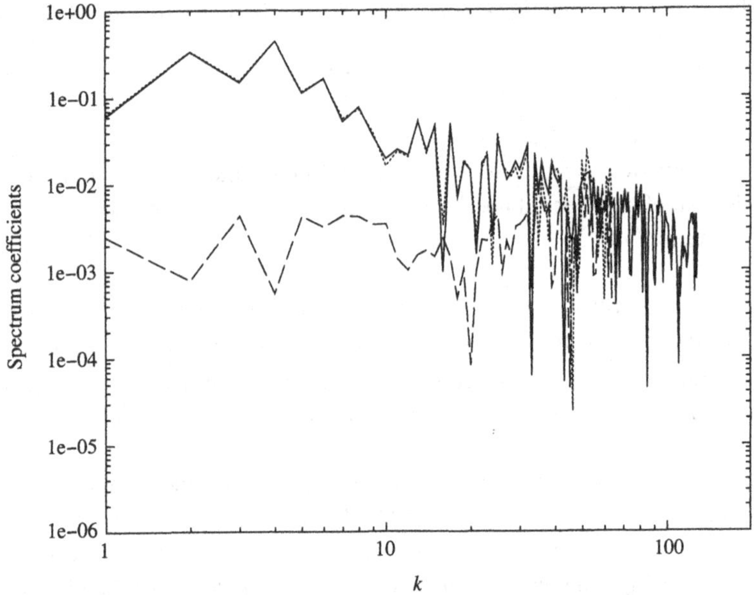

Figure 6.2. Comparison of the spectral coefficients associated to the Chebyshev polynomials of φ (solid line), $\overline{\varphi}$ (dotted line), and $\varphi' = \varphi - \overline{\varphi}$ (dashed line).

that the spectrum of $\overline{\varphi}_{M_1}$ is close to the low part of the spectrum of φ_M. Also the spectrum of $\varphi'^{M}_{M_1}$ is quite similar to the high part of the spectrum of φ_M.

Now, the problem of the commutation of the filter (6.31) with the derivative operators is considered. In Figure 6.3, the nodal values of $\overline{\partial \varphi_{M_1}/\partial x_2}$ and $\partial \overline{\varphi}_{M_1}/\partial x_2$ are represented on $(-h, +h)$. As we can see, both functions are of the same order; however, the difference between them is not negligible. In order to quantify the difference, we consider a scalar function $\varphi \in H^m_w(-h, +h)$ where w is the weight function associated with the Chebyshev polynomials (see Section 6.1.2). We denote by $\|\cdot\|_{m,w}$ the usual norm in $H^m_w(-h, +h)$. We can then derive the following estimate:

$$\begin{aligned}
&\left\| \left(\overline{\frac{\partial \varphi_{M_1}}{\partial x_2}}\right) - \frac{\partial \overline{\varphi}_{M_1}}{\partial x_2} \right\|_{0,w} \\
&\quad = \left\| I_{M_1}\left(\frac{\partial \varphi}{\partial x_2}\right) - \frac{\partial (I_{M_1}\varphi)}{\partial x_2} \right\|_{0,w} \\
&\quad \le \left\| I_{M_1}\left(\frac{\partial \varphi}{\partial x_2}\right) - \frac{\partial \varphi}{\partial x_2} \right\|_{0,w} + \left\| \frac{\partial \varphi}{\partial x_2} - \frac{\partial (I_{M_1}\varphi)}{\partial x_2} \right\|_{0,w} \\
&\quad \le c M_1^{1-m} \|\varphi\|_{m,w}.
\end{aligned}$$

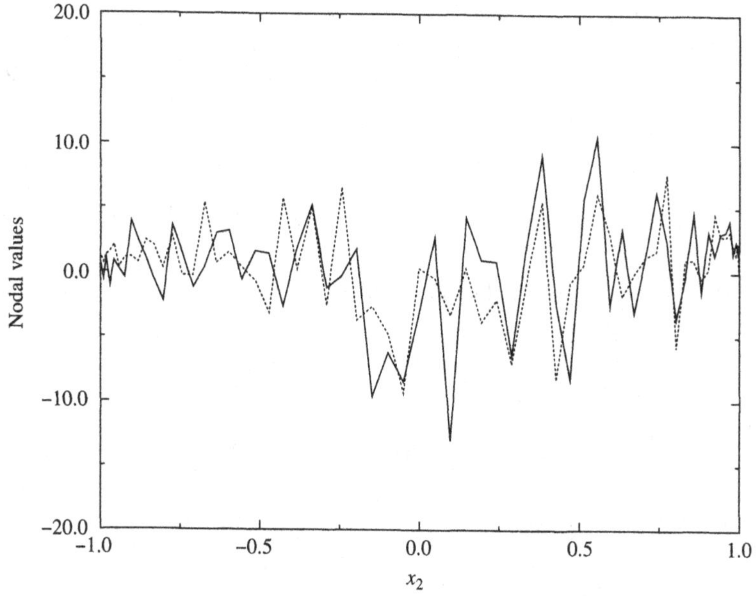

Figure 6.3. Comparison of $\overline{\partial\varphi/\partial x_2}$ (solid line) and $\partial\overline{\varphi}/\partial x_2$ (dotted line) in the physical space.

So the error on the commutation is of the order of the interpolation error on the coarse level M_1, that is, $\|I_{M_1}(\partial\varphi/\partial x_2) - \partial\varphi/\partial x_2\|_{0,w}$, since

$$\left\| I_{M_1}\left(\frac{\partial\varphi}{\partial x_2}\right) - \frac{\partial\varphi}{\partial x_2}\right\|_{0,w} \leq cM_1^{1-m}\|\varphi\|_{m,w} \qquad \text{for} \quad \varphi \in H_w^m(-h, +h)$$

(see Canuto et al. (1988)). We now consider a sequence of coarse levels $\{M_i\}_{i=1,p}$ satisfying $M_1 < M_2 < \cdots < M_i < \cdots < M_p < M$; then, M_p being the highest coarse level, we define $\overline{\mathbf{u}}_{M_p}$ and $\mathbf{u}'^{M}_{M_p}$ as (see (6.31) and (6.32))

$$\overline{\mathbf{u}}_{M_p} = I_{M_p}(\mathbf{u}_M) \quad \text{and} \quad \mathbf{u}'^{M}_{M_p} = \mathbf{u}_M - I_{M_p}(\mathbf{u}_M) = \mathbf{u}_M - \overline{\mathbf{u}}_{M_p}. \tag{6.34}$$

Now, for any level $M_i \in [M_1, M_p)$, we define $\overline{\mathbf{u}}_{M_i}$ as

$$\begin{aligned} \overline{\mathbf{u}}_{M_i} = I_{M_i}\left(\overline{\mathbf{u}}_{M_{i+1}}\right) \quad \text{and} \quad \mathbf{u}'^{M_{i+1}}_{M_i} &= \overline{\mathbf{u}}_{M_{i+1}} - I_{M_i}\left(\overline{\mathbf{u}}_{M_{i+1}}\right) \\ &= \overline{\mathbf{u}}_{M_{i+1}} - \overline{\mathbf{u}}_{M_i}. \end{aligned} \tag{6.35}$$

So the values of $\overline{\mathbf{u}}_{M_p}$ only are directly deduced from the values of $\mathbf{u}_M$. On the other levels $M_i < M_p$, the values of $\overline{u}_{M_i}$ are deduced from the values of $\overline{\mathbf{u}}_{M_{i+1}}$.

With the definitions (6.34) and (6.35) we have

$$\overline{\mathbf{u}}_{M_i} = I_{M_i}\left(\overline{\mathbf{u}}_{M_{i+1}}\right) = \left(I_{M_i} \circ I_{M_{i+1}}\right)\left(\mathbf{u}_{M_{i+2}}\right) \tag{6.36}$$

and

$$\overline{\mathbf{u}}_{M_{i+2}} = \overline{\mathbf{u}}_{M_{i+1}} + \mathbf{u}'^{M_{i+2}}_{M_{i+1}} = \overline{\mathbf{u}}_{M_i} + \left(\mathbf{u}'^{M_{i+1}}_{M_i} + \mathbf{u}'^{M_{i+2}}_{M_{i+1}}\right). \tag{6.37}$$

Repeating this process, we obtain a recurrent definition of $\overline{\mathbf{u}}_{M_i}$ and of $\mathbf{u}'^{M}_{M_i}$ for $i = 1, \ldots, p$. Indeed,

$$\begin{aligned} \mathbf{u}_M &= I_{M_p}(\mathbf{u}_M) + \mathbf{u}'^{M}_{M_p} \\ &= \left(I_{M_{p-1}} \circ I_{M_p}\right)(\mathbf{u}_M) + \left(\mathbf{u}'^{M_p}_{M_{p-1}} + \mathbf{u}'^{M}_{M_p}\right), \end{aligned}$$

which can be rewritten as

$$\mathbf{u}_M = J_{M_{p-1}}(\mathbf{u}_M) + \left(\mathbf{u}'^{M_p}_{M_{p-1}} + \mathbf{u}'^{M}_{M_p}\right),$$

with $J_{M_{p-1}} = I_{M_{p-1}} \circ I_{M_p}$. We finally obtain

$$\mathbf{u}_M = J_{M_1}(\mathbf{u}_M) + \sum_{j=1}^{p-1} \mathbf{u}'^{M_j+1}_{M_j} + \mathbf{u}'^{M}_{M_p},$$

which can be written as

$$\mathbf{u}_M = J_{M_1}(\mathbf{u}_M) + G^M_{M_1}(\mathbf{u}_M), \tag{6.38}$$

where

$$J_{M_1}(\mathbf{u}_M) = \left(I_{M_1} \circ I_{M_1+1} \circ \ldots \circ I_{M_{p-1}} \circ I_{M_p}\right)\mathbf{u}_M$$

and

$$G^M_{M_1}(\mathbf{u}_M) = \sum_{j=1}^{p-1} \mathbf{u}'^{M_j+1}_{M_j} + \mathbf{u}'^{M}_{M_p}.$$

Finally, we can write

$$\mathbf{u}_M = \overline{\mathbf{u}}_{M_1} + \mathbf{u}'^{M}_{M_1},$$

with $\overline{\mathbf{u}}_{M_1} = J_{M_1}(\mathbf{u}_M) \in (Y_{M_1})^n$ and $\mathbf{u}'^{M}_{M_1} = G^M_{M_1}(\mathbf{u}_M) \in (Y_M)^n$.

Separation of Scales in the Spectral Space

Let us now consider the decomposition in the spectral space, since we study the DML methodology in the particular case of the spectral methods. In spectral decompositions, the small (large) wavenumbers are associated with the large (small) scales. Firstly, we recall that the tau method enforces the boundary conditions on the highest modes. So the tau method precludes the separation of scales in the spectral space. However, if we consider the Galerkin bases developed by Shen for the Legendre polynomials (Shen (1994)) or for the Chebyshev polynomials (Shen (1995)), we can use projective filters based on these Galerkin bases (cf. Section 6.1.2). As we can see in Section 3.2.4, two Galerkin bases are used: $(\phi_l)_{l\in\mathbf{N}}$ and $(\psi_l)_{l\in\mathbf{N}}$ such that

$$V_M^n = \{\mathbf{v} \in (Y_M)^n,\ \mathbf{v}(\pm h) = 0\} = \mathrm{Span}\{\phi_0, \phi_1, \ldots, \phi_{M-2}\}^n \tag{6.39}$$

and

$$\begin{aligned} W_M^n &= \left\{\mathbf{v} \in (Y_M)^n,\ \mathbf{v}(\pm h) = 0,\ \frac{\partial \mathbf{v}}{\partial x_2}(\pm h) = 0\right\} \\ &= \mathrm{Span}\{\psi_0, \psi_1, \ldots, \psi_{M-4}\}^n. \end{aligned} \tag{6.40}$$

For $\mathbf{u}_M \in V_M^n$, we have

$$\mathbf{u}_M(x_2) = \sum_{l=0}^{M-2} \tilde{\mathbf{u}}(l)\ \phi_l(x_2). \tag{6.41}$$

As in Section 6.1.2, we define

$$\overline{\mathbf{u}}_{M_1}(x_2) = \mathbf{u}_{M_1}(x_2) = \sum_{l=0}^{M_1-2} \tilde{\mathbf{u}}(l)\ \phi_l(x_2), \tag{6.42}$$

and we obtain the decomposition of $\mathbf{u}_M$ into

$$\mathbf{u}_M = \overline{\mathbf{u}}_{M_1} + \mathbf{u}'^{M}_{M_1}, \tag{6.43}$$

with $\overline{\mathbf{u}}_{M_1} \in V_{M_1}^n$ and $\mathbf{u}'^{M}_{M_1} \in V_M^n \backslash V_{M_1}^n$. We can see readily that both $\overline{\mathbf{u}}_{M_1}$ and $\mathbf{u}'^{M}_{M_1}$ verify the boundary conditions at $\pm h$:

$$\overline{\mathbf{u}}_{M_1}(\pm h) = \mathbf{u}'^{M}_{M_1}(\pm h) = 0. \tag{6.44}$$

In the same manner, for $\mathbf{u}_M \in W_M^n$, we have

$$\mathbf{u}_M(x_2) = \sum_{l=0}^{M-4} \tilde{\mathbf{u}}(l)\ \psi_l(x_2), \tag{6.45}$$

and we obtain the following decomposition:

$$\mathbf{u}_M = \overline{\mathbf{u}}_{M_1} + \mathbf{u}'^{M}_{M_1}, \tag{6.46}$$

in which

$$\overline{\mathbf{u}}_{M_1}(x_2) = \mathbf{u}_{M_1}(x_2) = \sum_{l=0}^{M_1-4} \tilde{\mathbf{u}}(l)\ \psi_l(x_2) \in W^n_{M_1}$$

and

$$\mathbf{u}'^{M}_{M_1}(x_2) = \sum_{l=M_1-3}^{M-4} \tilde{\mathbf{u}}(l)\ \psi_l(x_2) \in W^n_M \backslash W^n_{M_1}.$$

For this decomposition, $\overline{\mathbf{u}}_{M_1}$ and $\mathbf{u}'^{M}_{M_1}$ verify the boundary conditions at $\pm h$:

$$\overline{\mathbf{u}}_{M_1}(\pm h) = \mathbf{u}'^{M}_{M_1}(\pm h) = 0, \tag{6.47}$$

$$\frac{\partial \overline{\mathbf{u}}_{M_1}}{\partial x_2}(\pm h) = \frac{\partial \mathbf{u}'^{M}_{M_1}}{\partial x_2}(\pm h) = 0.$$

The filters used to obtain $\overline{\mathbf{u}}_{M_1}$ are projective filters. However, they do not commute with partial derivatives.

As was said in Section 3.2, in order to compute the solution of (3.19), we compute solutions ω_2 and u_2 of (3.22) and (3.23) respectively. Then we deduce u_1 and u_3, knowing ω_2 and using the incompressibility constraint.

We first consider the decomposition in the periodic directions x_1 and x_3. The approximated velocity field $\mathbf{u}_N$ is a solution of the system (3.31). Let $N_1 < N$ be a given cutoff value; we decompose $\mathbf{u}_N$ in the following manner:

$$\mathbf{u}_N = \mathbf{y}_{N_1} + \mathbf{z}^N_{N_1}, \tag{6.48}$$

with

$$\mathbf{y}_{N_1}(\mathbf{x}, t) = \sum_{\mathbf{k} \in S_{N_1}} \tilde{\mathbf{u}}(\mathbf{k}, x_2, t)\ e^{\underline{\mathbf{i}}\mathbf{k}\cdot\mathbf{x}} = P_{N_1}\mathbf{u}_N(\mathbf{x}, t),$$

where P_{N_1} is the projection operator defined in (6.8), and

$$\mathbf{z}^N_{N_1}(\mathbf{x}, t) = \sum_{\mathbf{k} \in S_N \backslash S_{N_1}} \tilde{\mathbf{u}}(\mathbf{k}, x_2, t)\ e^{\underline{\mathbf{i}}\mathbf{k}\cdot\mathbf{x}} = Q^N_{N_1}\mathbf{u}_N(\mathbf{x}, t),$$

with $Q^N_{N_1} = I_N - P_{N_1}$. The decomposition (6.48) induces a similar decomposition on Δu_2 and ω_2, namely

$$\Delta u_{2,N} = \Delta y_{2,N_1} + \Delta z^N_{2,N_1}, \tag{6.49}$$

and

$$\omega_{2,N} = y_{\omega_{2,N_1}} + z^N_{\omega_{2,N_1}}. \tag{6.50}$$

By applying the projector P_{N_1} to (3.22) and (3.23), we obtain the large-scale equations

$$\begin{aligned}
&\frac{\partial y_{\omega_{2,N_1}}}{\partial t} - \nu\,\Delta y_{\omega_{2,N_1}} + \frac{\partial}{\partial x_3} P_{N_1} H_1(\mathbf{y}_{N_1}, \mathbf{y}_{N_1}) - \frac{\partial}{\partial x_1} P_{N_1} H_3(\mathbf{y}_{N_1}, \mathbf{y}_{N_1})\\
&\quad = -\frac{\partial}{\partial x_3} P_{N_1} H_{\text{int},1}\left(\mathbf{y}_{N_1}, \mathbf{z}^N_{N_1}\right) + \frac{\partial}{\partial x_1} P_{N_1} H_{\text{int},3}\left(\mathbf{y}_{N_1}, \mathbf{z}^N_{N_1}\right),\\
&\frac{\partial \Delta y_{2,N_1}}{\partial t} - \nu\,\Delta^2 y_{2,N_1} + \Delta P_{N_1} H_2(\mathbf{y}_{N_1}, \mathbf{y}_{N_1}) - \frac{\partial}{\partial x_2} P_{N_1}\,(\boldsymbol{\nabla}\cdot\mathbf{H})\,(\mathbf{y}_{N_1}, \mathbf{y}_{N_1})\\
&\quad = -\Delta P_{N_1} H_{\text{int},2}\left(\mathbf{y}_{N_1}, \mathbf{z}^N_{N_1}\right) + \frac{\partial}{\partial x_2} P_{N_1}\,(\boldsymbol{\nabla}\cdot\mathbf{H}_{\text{int}})\,\left(\mathbf{y}_{N_1}, \mathbf{z}^N_{N_1}\right),
\end{aligned} \tag{6.51}$$

where $\mathbf{H}_{\text{int}}$ is the nonlinear interaction term:

$$\mathbf{H}_{\text{int}}\left(\mathbf{y}_{N_1}, \mathbf{z}^N_{N_1}\right) = \mathbf{H}\left(\mathbf{y}_{N_1}, \mathbf{z}^N_{N_1}\right) + \mathbf{H}\left(\mathbf{z}^N_{N_1}, \mathbf{y}_{N_1}\right) + \mathbf{H}\left(\mathbf{z}^N_{N_1}, \mathbf{z}^N_{N_1}\right). \tag{6.52}$$

By applying the projector $Q^N_{N_1}$ to (3.22) and (3.23), we obtain the subgrid-scale equations

$$\begin{aligned}
&\frac{\partial z^N_{\omega_{2,N_1}}}{\partial t} - \nu\,\Delta z^N_{\omega_{2,N_1}} + \frac{\partial}{\partial x_3} Q^N_{N_1} H_1(\mathbf{y}_{N_1}, \mathbf{y}_{N_1}) - \frac{\partial}{\partial x_1} Q^N_{N_1} H_3(\mathbf{y}_{N_1}, \mathbf{y}_{N_1})\\
&\quad = -\frac{\partial}{\partial x_3} Q^N_{N_1} H_{\text{int},1}\left(\mathbf{y}_{N_1}, \mathbf{z}^N_{N_1}\right) + \frac{\partial}{\partial x_1} Q^N_{N_1} H_{\text{int},3}\left(\mathbf{y}_{N_1}, \mathbf{z}^N_{N_1}\right),\\
&\frac{\partial \Delta z^N_{2,N_1}}{\partial t} - \nu\,\Delta^2 z^N_{2,N_1} + \Delta Q^N_{N_1} H_2(\mathbf{y}_{N_1}, \mathbf{y}_{N_1}) - \frac{\partial}{\partial x_2} Q^N_{N_1}\,(\boldsymbol{\nabla}\cdot\mathbf{H})\,(\mathbf{y}_{N_1}, \mathbf{y}_{N_1})\\
&\quad = -\Delta Q^N_{N_1} H_{\text{int},2}\left(\mathbf{y}_{N_1}, \mathbf{z}^N_{N_1}\right) + \frac{\partial}{\partial x_2} Q^N_{N_1}\,(\boldsymbol{\nabla}\cdot\mathbf{H}_{\text{int}})\,\left(\mathbf{y}_{N_1}, \mathbf{z}^N_{N_1}\right).
\end{aligned} \tag{6.53}$$

Once y_{2,N_1} and y_{ω_2,N_1} have been computed, we deduce y_{1,N_1} and y_{3,N_1} using the incompressibility constraint.

Now we consider the decomposition in the direction x_2 normal to the walls. To solve (3.31), we look for a solution $\mathbf{u}_N$ and $\omega_{2,N}$ such that, for each $\mathbf{k} \in S_N$, we have $\tilde{u}_1(\mathbf{k}, x_2), \tilde{u}_3(\mathbf{k}, x_2), \tilde{\omega}_2(\mathbf{k}, x_2) \in V_M$ and $\tilde{u}_2(\mathbf{k}, x_2) \in W_M$. So we have

$$\tilde{u}_{2,M}(\mathbf{k}, x_2) = \sum_{l=0}^{M-4} \tilde{u}_2(\mathbf{k}, l)\,\psi_l(x_2), \tag{6.54}$$

$$\tilde{u}_{j,M}(\mathbf{k}, x_2) = \sum_{l=0}^{M-2} \tilde{u}_j(\mathbf{k}, l)\,\phi_l(x_2) \tag{6.55}$$

for $j = 1,\ 3$, and similarly for $\tilde{\omega}_2(\mathbf{k}, x_2)$. Let $M_1 < M$ be a given cutoff value; according to (6.46) we obtain the following decomposition:

$$\tilde{u}_{2,M}(\mathbf{k}, x_2) = \tilde{u}_{2,M_1}(\mathbf{k}, x_2) + \tilde{u}'^{M}_{2,M_1}(\mathbf{k}, x_2), \tag{6.56}$$

with $\tilde{u}_{2,M_1}(\mathbf{k}) \in W_{M_1}$ and $\tilde{u}'^{M}_{2,M_1}(\mathbf{k}) \in W_M \backslash W_{M_1}$. Similarly, we have

$$\tilde{u}_{j,M}(\mathbf{k}, x_2) = \tilde{u}_{j,M_1}(\mathbf{k}, x_2) + \tilde{u}'^{M}_{j,M_1}(\mathbf{k}, x_2) \tag{6.57}$$

for $j = 1,\ 3$ with $\tilde{u}_{j,M_1}(\mathbf{k}) \in V_{M_1}$ and $\tilde{u}'^{M}_{j,M_1}(\mathbf{k}) \in V_M \backslash V_{M_1}$. As for the periodic directions, the decompositions (6.56) and (6.57) induce a similar decomposition on Δu_2 and ω_2:

$$\Delta\tilde{u}_{2,M}(\mathbf{k}, x_2) = \Delta\tilde{u}_{2,M_1}(\mathbf{k}, x_2) + \Delta\tilde{u}'^{M}_{2,M_1}(\mathbf{k}, x_2), \tag{6.58}$$

and

$$\begin{aligned}\tilde{\omega}_{2,M}(\mathbf{k}, x_2) &= ik_3\tilde{u}_{1,M}(\mathbf{k}, x_2) - ik_1\tilde{u}_{3,M}(\mathbf{k}, x_2) \\ &= \tilde{\omega}_{2,M_1}(\mathbf{k}, x_2) + \tilde{\omega}'^{M}_{2,M_1}(\mathbf{k}, x_2).\end{aligned} \tag{6.59}$$

Consider then equations (3.31): by projecting the equation for $\tilde{\omega}_2$ onto the space V_{M_1} and the equation for $\tilde{u}_2$ onto the space W_{M_1}, we obtain the large-scale equations

$$\begin{aligned}&\left(\frac{\partial}{\partial t}D^2_{\mathbf{k}}\tilde{u}_{2,M_1}(\mathbf{k}), \psi_l\right) - \nu\left(D^4_{\mathbf{k}}\tilde{u}_{2,M_1}(\mathbf{k}), \psi_l\right) + \left(\hat{h}_{\mathbf{k}}(\mathbf{u}_{NM_1}, \mathbf{u}_{NM_1}), \psi_l\right) \\ &\quad = -\left(\frac{\partial}{\partial t}D^2_{\mathbf{k}}\tilde{u}'^{M}_{2,M_1}(\mathbf{k}), \psi_l\right) + \nu\left(D^4_{\mathbf{k}}\tilde{u}'^{M}_{2,M_1}(\mathbf{k}), \psi_l\right) \\ &\qquad - \left(\hat{h}_{\text{int},\mathbf{k}}\left(\mathbf{u}_{NM_1}, \mathbf{u}'^{M}_{NM_1}\right), \psi_l\right) \qquad \text{for } l = 0, \ldots, M_1 - 4, \\ &\left(\frac{\partial}{\partial t}\tilde{\omega}_{2,M_1}(\mathbf{k}), \phi_l\right) - \nu\left(D^2_{\mathbf{k}}\tilde{\omega}_{2,M_1}(\mathbf{k}), \phi_l\right) + \left(i\underline{\mathbf{k}}^{\perp} \cdot \hat{\mathbf{H}}_{\mathbf{k}}(\mathbf{u}_{NM_1}, \mathbf{u}_{NM_1}), \phi_l\right) \\ &\quad = -\left(\frac{\partial}{\partial t}\tilde{\omega}'^{M}_{2,M_1}(\mathbf{k}), \phi_l\right) + \nu\left(D^2_{\mathbf{k}}\tilde{\omega}'^{M}_{2,M_1}(\mathbf{k}), \phi_l\right) \\ &\qquad - \left(i\underline{\mathbf{k}}^{\perp} \cdot \hat{\mathbf{H}}_{\text{int},\mathbf{k}}(\mathbf{u}_{NM_1}, \mathbf{u}'^{M}_{NM_1}), \phi_l\right) \qquad \text{for } l = 0, \ldots, M_1 - 2,\end{aligned} \tag{6.60}$$

where $\hat{h}_{\text{int},\mathbf{k}}$ and $\hat{\mathbf{H}}_{\text{int},\mathbf{k}}$ are the nonlinear interaction terms

$$\begin{aligned}&\hat{h}_{\text{int},\mathbf{k}}\left(\mathbf{u}_{NM_1}, \mathbf{u}'^{M}_{NM_1}\right) \\ &\quad = \hat{h}_{\mathbf{k}}\left(\mathbf{u}_{NM_1}, \mathbf{u}'^{M}_{NM_1}\right) + \hat{h}_{\mathbf{k}}\left(\mathbf{u}'^{M}_{NM_1}, \mathbf{u}_{NM_1}\right) + \hat{h}_{\mathbf{k}}\left(\mathbf{u}'^{M}_{NM_1}, \mathbf{u}'^{M}_{NM_1}\right),\end{aligned} \tag{6.61}$$

and similarly for $\hat{\mathbf{H}}_{\text{int},\mathbf{k}}$. In the same manner, considering again equation (3.31) and projecting the equation for $\tilde{\omega}_2$ on the space $V_M \backslash V_{M_1}$ and the equation for $\tilde{u}_2$ on the space $W_M \backslash W_{M_1}$, we obtain the small-scale equations

$$\begin{aligned}
&\left(\frac{\partial}{\partial t} D_{\mathbf{k}}^2 \tilde{u}_{2,M_1}(\mathbf{k}), \psi_l\right) - \nu\left(D_{\mathbf{k}}^4 \tilde{u}_{2,M_1}(\mathbf{k}), \psi_l\right) + \left(\hat{h}_{\mathbf{k}}(\mathbf{u}_{NM_1}, \mathbf{u}_{NM_1}), \psi_l\right) \\
&= -\left(\frac{\partial}{\partial t} D_{\mathbf{k}}^2 \tilde{u}_{2,M_1}^{\prime M}(\mathbf{k}), \psi_l\right) + \nu\left(D_{\mathbf{k}}^4 \tilde{u}_{2,M_1}^{\prime M}(\mathbf{k}), \psi_l\right) \\
&\quad - \left(\hat{h}_{\text{int},\mathbf{k}}\left(\mathbf{u}_{NM_1}, \mathbf{u}_{NM_1}^{\prime M}\right), \psi_l\right) \qquad \text{for } l = M_1 - 3, \ldots, M - 4, \\
&\left(\frac{\partial}{\partial t} \tilde{\omega}_{2,M_1}(\mathbf{k}), \phi_l\right) - \nu\left(D_{\mathbf{k}}^2 \tilde{\omega}_{2,M_1}(\mathbf{k}), \phi_l\right) + \left(\mathbf{i}\underline{\mathbf{k}}^{\perp} \cdot \hat{\mathbf{H}}_{\mathbf{k}}(\mathbf{u}_{NM_1}, \mathbf{u}_{NM_1}), \phi_l\right) \\
&= -\left(\frac{\partial}{\partial t} \tilde{\omega}_{2,M_1}^{\prime M}(\mathbf{k}), \phi_l\right) + \nu\left(D_{\mathbf{k}}^2 \tilde{\omega}_{2,M_1}^{\prime M}(\mathbf{k}), \phi_l\right) \\
&\quad - \left(\mathbf{i}\underline{\mathbf{k}}^{\perp} \cdot \hat{\mathbf{H}}_{\text{int},\mathbf{k}}\left(\mathbf{u}_{NM_1}, \mathbf{u}_{NM_1}^{\prime M}\right), \phi_l\right) \qquad \text{for } l = M_1 - 1, \ldots, M - 2.
\end{aligned} \tag{6.62}$$

Having computed $\tilde{u}_{2,M_1}$ and $\tilde{\omega}_{2,M_1}$, and using the incompressibility constraint, we deduce $\tilde{u}_{1,M_1}$ and $\tilde{u}_{3,M_1}$ in the following manner:

$$\begin{aligned}
ik_1\tilde{u}_{1,l} + ik_3\tilde{u}_{3,l} &= -\tilde{u}_{2,l}^{(1)}, \\
ik_3\tilde{u}_{1,l} - ik_1\tilde{u}_{3,l} &= \tilde{\omega}_{2,l}
\end{aligned} \tag{6.63}$$

for $l = 0, \ldots, M_1 - 2$, where $\tilde{u}_{2,l}^{(1)}$ represent the coefficients of the spatial derivative $\partial\tilde{u}_{2,M}/\partial x_2(\mathbf{k}, x_2)$ in the Galerkin basis ϕ_l ($\partial\tilde{u}_{2,M}/\partial x_2(\mathbf{k}) \in V_M$).

In the dynamic multilevel methods described in Chapter 8, we will use the separation of the small- and large-scale equations to compute differently the small and large scales. In the fully periodic case, to compute the large (small) scales associated with the decomposition (6.16), (6.17), we use equation (6.18) (equation (6.20)). In the wall-bounded flows (channel flow problem), to compute the large (small) scales obtained with the decomposition (6.16), (6.17) in the two periodic directions, we use equations (6.51) (equations (6.53)). Finally, in the normal direction at the walls, we use equations (6.60) and (6.62) to compute, respectively, the large and the small scales associated with the decomposition (6.56), (6.57). The computation of the small components from the previous equations will be less accurate than the computation of the large ones, since the quantities associated with the small scales are small in comparison with that associated with the large scales (see Chapter 8).

7

Numerical Analysis of Multilevel Methods

Our object in this chapter is to present some elements of the numerical analysis of the multilevel methods that we are using in this book. In fact, instead of conducting our analysis on the Navier–Stokes equations themselves, we will present it on a very simple model, namely a pair of coupled ordinary differential equations. It happens that the numerical analysis, for this system, of the schemes that we study is very similar to that of the multilevel methods for the Navier–Stokes equations, and we avoid in this way the full functional analysis framework; more mathematically oriented readers are referred for the Navier–Stokes analysis to the following references: Burie and Marion (1997), Marion and Temam (1989, 1990), Temam (1990b, 1991, 1993, 1994).

In Section 7.1 we present our simple model and make a few pertinent remarks. In Section 7.2 we study numerical schemes with the same time step for both variables. Finally, in Section 7.3 we consider schemes using different time steps for the two unknowns.

7.1. A Simple Model

The simple differential system that we propose reads

$$\begin{pmatrix} \dot{y} \\ \dot{z} \end{pmatrix} + \begin{pmatrix} 1 & 1 \\ 1 & 1/\varepsilon \end{pmatrix} \begin{pmatrix} y \\ z \end{pmatrix} + \begin{pmatrix} yz \\ -y^2 \end{pmatrix} = \begin{pmatrix} f \\ g \end{pmatrix}, \qquad t > 0, \tag{7.1}$$

$$y(0) = y_0, \qquad z(0) = z_0. \tag{7.2}$$

Here y and z play the same role respectively as the low- and high-frequency components of the flow (i.e. of the velocity vector). The important feature of the matrix in (7.1) are (1) that its eigenvalues are $O(1)$ and $O(1/\varepsilon)$, $\varepsilon > 0$ small; (2) that the couplings between the y and z terms and between the z and y

terms are not important and in fact are not present in the related case of Fourier and spectral discretizations.

For this very simple system, and also the associated linear system obtained by dropping the nonlinear terms, the stability analysis is elementary. We will show how one can build appropriate numerical schemes based on different treatments and different time steps for y and z $(\Delta t_y \neq \Delta t_z)$. As we said, and although it may not be transparent at first glance, much of the analysis is very similar to what have been done elsewhere in infinite dimension for partial differential equations and in particular the Navier–Stokes equations (see the references cited above). Also, as we will see, the conclusions are sometimes counterintuitive.

Before starting the analysis, we conclude this section with a few simple remarks on the system (7.1). For simplicity f and g are constants independent of time and of ε, that is, $f, g = O(1)$. We use the analog of the energy equation in fluid mechanics, obtained by multiplying the first equation (7.1) by y, the second by z, and adding the resulting equations.[1] From this we find that for any y_0, z_0, the functions $y(t)$, and $z(t)$ are respectively $O(1)$ and $O(\varepsilon)$ for large time and hence $\dot{y} = O(1)$, $\dot{z} = O(1)$, so that $y/\dot{y} = O(1)$ and $z/\dot{z} = O(\varepsilon)$. If $y_0 = O(1)$ and $z_0 = O(\varepsilon)$, the same result is valid for all times. Here large time plays the role of times for which the statistical equilibrium has been reached, and small time plays the role of transient time in turbulence. Note that $y/\dot{y}$ and $z/\dot{z}$ have the dimensions of time and they play respectively the roles of the eddy-turnover time for the large and the small structures (see Section 2.2); the remark on their order of magnitude is in agreement with the conventional theory of turbulence (see e.g. Batchelor (1971), Orszag (1973, p. 279)).

Another feature of equation (7.1) is the following: at first glance, because of the factor $1/\varepsilon$, it seems to be a stiff differential equation. In fact the factor $1/\varepsilon$, corresponding to an increased viscosity, has a beneficial damping effect, and the stiffness of (7.1) may only appear in the initial transient if z_0 is not already small, $O(\varepsilon)$.

7.2. Two-Level Discretization Schemes

As indicated before, the schemes that we consider correspond to differentiated treatment for y and z; by this we mean that the time discretization (explicit/implicit) is not the same for y and z. By comparison with a partial differential equation, the introduction of the decomposition $u = y + z$ yields

[1] The stability analysis, as well, will be based on energy methods, that is, on the discrete analog of the energy equation.

a number of new schemes by combining explicit/implicit discretizations for y and z, for the linear and nonlinear terms.

To bring a little more generality to the computations, we replace the matrix

$$\begin{pmatrix} 1 & 1 \\ 1 & 1/\varepsilon \end{pmatrix} \text{ by } \begin{pmatrix} \nu & \nu \\ \nu & \nu/\varepsilon \end{pmatrix},$$

and we start by considering the linear equation. An ordinary explicit Euler scheme with time step Δt reads

$$\begin{aligned} &\frac{y^n - y^{n-1}}{\Delta t} + \nu y^{n-1} + \nu z^{n-1} = f, \\ &\frac{z^n - z^{n-1}}{\Delta t} + \nu y^{n-1} + \frac{\nu}{\varepsilon} z^{n-1} = g. \end{aligned} \tag{7.3}$$

It is standard that the stability (CFL-type) condition for (7.3) reads

$$\Delta t < \frac{2\varepsilon}{\nu}, \tag{7.4}$$

so that for e.g. $\nu = O(1)$ and ε small, Δt must be less than $O(\varepsilon)$.

If we treat y and z differently, we can consider the scheme

$$\begin{aligned} &\frac{y^n - y^{n-1}}{\Delta t} + \nu y^{n-1} + \nu z^{n-1} = f, \\ &\frac{z^n - z^{n-1}}{\Delta t} + \nu y^n + \frac{\nu}{\varepsilon} z^n = g. \end{aligned} \tag{7.5}$$

It is elementary to derive the stability analysis for (7.5) by considering the eigenvalues of the amplification matrix; we find (compare (7.4))

$$\Delta t < \frac{2 - 4\varepsilon}{\nu} \tag{7.6}$$

and

$$\Delta t < \frac{\nu}{\varepsilon}. \tag{7.7}$$

This stability result can be exactly recovered by the energy method used below for the more involved nonlinear case.

For the full nonlinear equation, we can consider similarly the scheme

$$\begin{aligned} &\frac{y^n - y^{n-1}}{\Delta t} + \nu y^{n-1} + \nu z^{n-1} + y^{n-1} z^{n-1} = f, \\ &\frac{z^n - z^{n-1}}{\Delta t} + \nu y^n + \frac{\nu}{\varepsilon} z^n - (y^n)^2 = g. \end{aligned} \tag{7.8}$$

This scheme is linearly implicit in z and fully explicit in y (y^n is known when we compute z^n in the second equation (7.8)).

We can conduct the stability analysis of this scheme. We properly combine the three equations obtained as follows: we multiply the first equation (7.8) by y^{n-1}, the second equation (7.8) by z^n, and the first equation (7.8) by $y^n - y^{n-1}$. We multiply the nonlinear terms by γ, where $\gamma = 1$ corresponds to the nonlinear equation, and $\gamma = 0$ to the linear equation.

We find, for every $\delta > 0$,

$$
\begin{aligned}
&|y^n|^2 - |y^{n-1}|^2 - |y^n - y^{n-1}|^2 + 2\nu\,\Delta t\,|y^{n-1}|^2 \\
&\quad + 2\nu\,\Delta t\, y^{n-1} z^{n-1} + 2\gamma\,\Delta t\,(y^{n-1})^2 z^{n-1} = 2\,\Delta t\, f y^{n-1},
\end{aligned}
$$

$$
\begin{aligned}
|y^n - y^{n-1}|^2 &= \Delta t^2\,|f - \nu y^{n-1} - \nu z^{n-1} - \gamma y^{n-1} z^{n-1}|^2 \\
&\le \Delta t^2\,\nu^2 (1+\delta)|y^{n-1}|^2 + \Delta t^2\,\left(1 + \tfrac{3}{\delta}\right) \\
&\quad \times \left(|f|^2 + \nu^2 |z^{n-1}|^2 + \gamma^2 |y^{n-1} z^{n-1}|^2\right).
\end{aligned}
$$

Here we have used Lemma 7.1 below. Combining these relations and using also

$$
2\nu\,\Delta t\, y^{n-1} z^{n-1} \ge -2\varepsilon\nu\,\Delta t\,|y^{n-1}|^2 - \frac{\nu\,\Delta t}{2\varepsilon}|z^{n-1}|^2,
$$

we arrive at

$$
\begin{aligned}
&|y^n|^2 - |y^{n-1}|^2 + \nu\,\Delta t\,[2 - 2\varepsilon - \nu\,\Delta t\,(1+\delta)]\,|y^{n-1}|^2 \\
&\quad + \frac{\nu\,\Delta t}{\varepsilon}\left[\frac{1}{2} + \varepsilon\nu\,\Delta t\left(1 + \frac{3}{\delta}\right)\right]|z^{n-1}|^2 \\
&\le 2\,\Delta t\, f y^{n-1} + \Delta t^2\left(1 + \frac{3}{\delta}\right)|f|^2 \\
&\quad - 2\gamma\,\Delta t\,(y^{n-1})^2 z^{n-1} + \Delta t^2\,\gamma^2 |y^{n-1} z^{n-1}|^2.
\end{aligned}
\tag{7.9$_n$}
$$

For the z-equation we obtain similarly

$$
\begin{aligned}
&|z^n|^2 - |z^{n-1}|^2 + |z^n - z^{n-1}|^2 + 2\nu\,\Delta t\, y^n z^n \\
&\quad + \frac{2\nu\,\Delta t}{\varepsilon}|z^n|^2 - 2\gamma\,\Delta t (y^n)^2 z^n = 2\,\Delta t\, g z^n, \\
&|z^n|^2 - |z^{n-1}|^2 + |z^n - z^{n-1}|^2 - 2\varepsilon\nu\,\Delta t\,|y^n|^2 + \frac{3}{2}\frac{\nu\,\Delta t}{\varepsilon}|z^n|^2 \\
&\le 2\,\Delta t\, g z^n + 2\gamma\,\Delta t\,(y^n)^2 z^n.
\end{aligned}
\tag{7.10$_n$}
$$

We now add $(7.9)_{n+1}$ and $(7.10)_n$, which yields

$$\begin{aligned}
&\left(|y^{n+1}|^2 + |z^n|^2\right) - \left(|y^n|^2 + |z^{n-1}|^2\right) + |z^n - z^{n-1}|^2 \\
&\quad + \nu\,\Delta t\,\left[2 - 4\varepsilon - \nu\,\Delta t\,(1+\delta)\right]|y^n|^2 \\
&\quad + \frac{\nu\,\Delta t}{\varepsilon}\left[1 - \varepsilon\nu\,\Delta t\left(1 + \frac{3}{\delta}\right)\right]|z^n|^2 \\
&\leq 2\,\Delta t\, f y^n + \Delta t^2\left(1+\frac{3}{\delta}\right)|f|^2 + 2\,\Delta t\, g z^n + \Delta t^2\,|y^n z^n|^2.
\end{aligned} \tag{7.11}$$

Now assume that

$$\Delta t < \frac{2 - 4\varepsilon}{\nu}; \tag{7.12}$$

then there exists $\delta > 0$ such that

$$2\kappa_1 = 2 - 4\varepsilon - \nu\,\Delta t\,(1+\delta) > 0. \tag{7.13}$$

Assume also that

$$\Delta t \leq \frac{1}{2\varepsilon\nu\left(1+\frac{3}{\delta}\right)}; \tag{7.14}$$

then

$$1 - \varepsilon\,\Delta t\left(1+\frac{3}{\delta}\right) \geq \frac{1}{2}.$$

Finally we write

$$\begin{aligned}
2\,\Delta t\, f y^n &\leq \kappa_1\,\Delta t\,|y^n|^2 + \frac{\Delta t}{\kappa_1}|f|^2, \\
2\,\Delta t\, g z^n &\leq \frac{\nu\,\Delta t}{4\varepsilon}|z^n|^2 + \frac{4\varepsilon\,\Delta t}{\nu}|g|^2,
\end{aligned}$$

and we find that

$$\begin{aligned}
&\left(|y^{n+1}| + |z^n|^2\right) - \left(|y^n|^2 + |z^{n-1}|^2\right) + |z^n - z^{n-1}|^2 \\
&\quad + \kappa_1\nu\,\Delta t\,|y^n|^2 + \frac{\nu\,\Delta t}{4\varepsilon}|z^n|^2 \leq \alpha\,\Delta t + \Delta t^2\,\gamma^2|y^n z^n|^2,
\end{aligned} \tag{7.15}$$

with

$$\alpha = \frac{1}{\kappa_1}|f|^2 + \Delta t\left(1+\frac{3}{\delta}\right)|f|^2 + \frac{4\varepsilon}{\nu}|g|^2. \tag{7.16}$$

In the linear case ($\gamma = 0$), stability follows from (7.15) under the assumptions (7.12) and (7.14).

In the nonlinear case, we set

$$\begin{aligned} L_1 &= |y^1|^2 + |z^0|^2, \\ L_j &= L_1 + \alpha j\,\Delta t, \qquad j \geq 1, \\ L_j \leq L_N = L &= L_1 + \alpha T \qquad \text{for} \quad 1 \leq j \leq N = T/\Delta t; \end{aligned} \tag{7.17}$$

we replace (7.14) by

$$\Delta t \leq \frac{1}{\varepsilon} \min \left(\frac{1}{2 + \left(1 + \frac{3}{\delta}\right)}, \frac{\nu}{8L^2\gamma^2} \right), \tag{7.18}$$

and we prove by induction on n that

$$|y^n|^2 + |z^{n-1}|^2 \leq L_{n-1}. \tag{7.19}$$

Indeed assuming (7.19), we infer from (7.14)–(7.18) that

$$\begin{aligned} &\left(|y^{n+1}| + |z^n|^2\right) - \left(|y^n|^2 + |z^{n-1}|^2\right) + |z^n - z^{n-1}|^2 \\ &+ \kappa_1 \nu\,\Delta t\,|y^n|^2 + \frac{\nu\,\Delta t}{8\varepsilon}|z^n|^2 \leq \alpha\,\Delta t, \end{aligned} \tag{7.20}$$

and (7.19) follows at order $n + 1$.

In summary, the scheme (7.8), which is explicit in y and implicit in z, is stable under the CFL conditions (7.12), and (7.18). For ε small, (7.18) is a consequence of (7.12), and therefore the stability condition is essentially that of the explicit Euler scheme for the large structure (first equation in (7.1) with $z = 0$),[2] namely

$$\Delta t < 2/\nu. \tag{7.21}$$

Remark 7.1. This presentation of the stability analysis based on energy methods is slightly different than the classical one; as seen before, it allows us to obtain, in the linear case, the optimal stability condition, exactly as with spectral methods based on the study of the eigenvalues of the amplification matrix. □

We conclude this section with the following technical lemma used before:

[2] That is, $(y^n - y^{n-1})/\Delta t + \nu y^{n-1} = f$ for solving $dy/dt + \nu y = f$.

Lemma 7.1. *For every $\delta > 0$ and every $a_1, \ldots, a_n \in \mathbb{R}$,*

$$\left|\sum_{i=1}^{n} a_i\right|^2 \leq (1+\delta)a_1^2 + \left(1 + \frac{n-1}{\delta}\right)\sum_{i=2}^{n} a_i^2. \tag{7.22}$$

Proof. By expanding, we need to show that

$$\delta a_1^2 + \frac{R}{\delta}\sum_{i=2}^{n} a_i^2 - 2\sum_{1\leq i<j\leq n} a_i a_j \geq 0, \tag{7.23}$$

for $R \geq n-1$. The determinant $\Delta_n(R)$ of the quadratic form above is shown, by induction, to be equal to

$$\Delta_n(R) = \left(\frac{R}{\delta} - 1\right)^{n-1}\delta - (1-\delta)(n-1)\left(\frac{R}{\delta} - 1\right)^{n-2},$$

$$\Delta_n(R) = \left(\frac{R}{\delta} - 1\right)^{n-2}[R - \delta - (1-\delta)(n-1)].$$

Hence $\Delta_n(n-1) \geq 0$. We prove similarly that $\Delta_j(n-1) \geq 0$, $j = 1, \ldots, n-1$, and we conclude that the quadratic form in the left-hand side of (7.23) is positive semidefinite for $R = n-1$. □

7.3. Multilevel Discretization Schemes

Pursuing the idea of treating y and z (the large and small structures) differently, we consider now schemes with $\Delta t_y \neq \Delta t_z$; see Burie and Marion (1997) and Lions, Temam and Wang (1996) for related schemes.

We set $\Delta t_y = \Delta t$, $\Delta t_z = q\,\Delta t$, where $q > 1$ is an integer. One of the many schemes one can think of is the following:

$$\begin{aligned}
&\frac{y^{n+1/q} - y^n}{\Delta t} + \nu y^n + \nu z^n + \gamma y^n z^n = f,\\
&\frac{y^{n+2/q} - y^{n+1/q}}{\Delta t} + \nu y^{n+1/q} + \nu z^n + \gamma y^n z^n = f,\\
&\qquad\vdots\\
&\frac{y^{n+1} - y^{n+(q-1)/q}}{\Delta t} + \nu y^{n+(q-1)/q} + \nu z^n + \gamma y^n z^n = f;
\end{aligned} \tag{7.24$_{n,j}$}$$

$$\frac{z^{n+1} - z^n}{q\,\Delta t} + \nu y^{n+1} + \frac{\nu}{\varepsilon} z^{n+1} - \frac{\gamma}{q}\sum_{j=1}^{q} |y^{n+(j-1)/q}|^2 = g. \tag{7.25}$$

Here $\gamma = 0$ in the linear case and $\gamma = 1$ for the nonlinear equation. Note that the scheme is always explicit in y (y^{n+1} is known when computing z^{n+1} in (7.25)); it is linearly implicit in z (in (7.25)).

In the linear case, the stability analysis conducted as before leads to the following stability conditions

$$\Delta t < \frac{2 - 2\varepsilon - 2q\varepsilon}{\nu} \tag{7.26}$$

and

$$\Delta t < \frac{1 - \varepsilon}{2\varepsilon\nu(1 + 2/\delta)} \tag{7.27}$$

for some suitable $\delta > 0$. Of course (7.27) follows from (7.26) for ε small, and we conclude that the (main) stability condition (7.26) is very close to (essentially the same as) the stability condition for the y-equation only (i.e. (7.1) when $z = 0$).

In the nonlinear case the stability analysis is more involved than in the two-level case. However, the method is the same and we just give here a sketch of the calculations.

Multiplying (7.25) by $2q\,\Delta t\, z^{n+1}$, we find

$$\begin{aligned} &|z^{n+1}|^2 - |z^n|^2 + |z^{n+1} - z^n|^2 + 2\nu q\,\Delta t\; y^{n+1} z^{n+1} \\ &\quad + 2\frac{\nu q\,\Delta t}{\varepsilon}|z^{n+1}|^2 - 2\gamma q\,\Delta t\; |y^{n+1}|^2 z^{n+1} \\ &= 2q\,\Delta t\; g z^{n+1} \le \frac{q\,\Delta t}{\nu}|g|^2 + \nu q\,\Delta t\; |z^{n+1}|^2. \end{aligned}$$

Hence

$$\begin{aligned} &|z^{n+1}|^2 - |z^n|^2 + |z^{n+1} - z^n|^2 - 2\varepsilon\nu q\,\Delta t\; |y^{n+1}|^2 \\ &\quad + \frac{\nu q\,\Delta t}{\varepsilon}\left(\frac{3}{2} - \varepsilon\right)|z^{n+1}|^2 - 2\gamma q\,\Delta t\; |y^{n+1}|^2 z^{n+1} \le \frac{q\,\Delta t}{\nu}|g|^2. \end{aligned} \tag{7.28$_{n+1}$}$$

Similarly, multiplying equation $(7.24)_{n,j}$ by $2\,\Delta t\; y^{n+(j-1)/q}$ and then by Δt $(y^{n+j/q} - y^{n+(j-1)/q})$, we find

$$\begin{aligned} &\left|y^{n+j/q}\right|^2 - \left|y^{n+(j-1)/q}\right|^2 - \left|y^{n+j/q} - y^{n+(j-1)/q}\right|^2 + 2\nu\,\Delta t\; \left|y^{n+(j-1)/q}\right|^2 \\ &\quad + 2\nu\,\Delta t\; y^{n+(j-1)/q} z^n + 2\gamma\,\Delta t\; y^n y^{n+(j-1)/q} z^n = 2\,\Delta t\; f y^{n+(j-1)/q}, \end{aligned}$$

$$
\begin{aligned}
\left|y^{n+j/q}-y^{n+(j-1)/q}\right|^2 &= \Delta t^2 \left|f-\nu y^{n+(j-1)/q}-\nu z^n-\gamma y^n z^n\right|^2 \\
&\le \Delta t^2 \Big\{ \nu^2(1+\delta)\left|y^{n+(j-1)/q}\right|^2 \\
&\quad + \left(1+\frac{3}{\delta}\right)(|f|^2+\nu^2|z^n|^2+\gamma^2|y^n z^n|^2) \Big\}
\end{aligned}
$$

(by Lemma 7.1). Also

$$
2\nu\,\Delta t\, y^{n+(j-1)/q} z^n \ge -\frac{\nu\,\Delta t}{2\varepsilon}|z^n|^2 - 2\varepsilon\nu\,\Delta t\,\left|y^{n+(j-1)/q}\right|^2 .
$$

Hence

$$
\begin{aligned}
&\left|y^{n+j/q}\right|^2 - \left|y^{n+(j-1)/q}\right|^2 \qquad (7.29)_{n,j}\\
&\quad +\nu\,\Delta t\,[2-2\varepsilon-\nu\,\Delta t\,(1+\delta)]\,\left|y^{n+(j-1)/q}\right|^2 \\
&\quad -\frac{\nu\,\Delta t}{\varepsilon}\left[\frac{1}{2}+\varepsilon\nu\,\Delta t\left(1+\frac{3}{\delta}\right)\right]|z^n|^2 + 2\gamma\,\Delta t\, y^n y^{n+(j-1)/q} z^n \\
&\le 2\,\Delta t\, f y^{n+(j-1)/q} + \Delta t^2\left(1+\frac{3}{\delta}\right)|f|^2 + \Delta t^2\gamma^2|y^n z^n|^2.
\end{aligned}
$$

We now add $(7.28)_n$ to equation $(7.29)_{n,j}$, $j=1,\ldots,q$. This yields

$$
\begin{aligned}
&\left(|y^{n+1}|^2+|z^n|^2\right) - \left(|y^n|^2+|z^{n-1}|^2\right) + |z^n-z^{n-1}|^2 \qquad (7.30)\\
&\quad + 2\kappa_1\nu\,\Delta t \sum_{j=1}^{q}\left|y^{n+(j-1)/q}\right|^2 + 2\kappa_1'\frac{q\nu\,\Delta t}{\varepsilon}|z^n|^2 \\
&\le I_1 + I_2 + \frac{q\,\Delta t}{\nu}|g|^2 + q\,\Delta t^2\left(1+\frac{3}{\delta}\right)|f|^2 + 2\,\Delta t\sum_{j=1}^{q} f y^{n+(j-1)/q},
\end{aligned}
$$

with

$$
\begin{aligned}
2\kappa_1 &= 2-4\varepsilon-\nu\,\Delta t\,(1+\delta),\\
2\kappa_1' &= 1-\varepsilon-\varepsilon\nu\,\Delta t\left(1+\frac{3}{\delta}\right),\\
I_1 &= \gamma^2\,\Delta t^2\, q|y^n z^n|^2,\\
I_2 &= 2\gamma\,\Delta t\, z^n y^n \sum_{j=1}^{q}\left(y^n - y^{n+(j-1)/q}\right).
\end{aligned}
$$

For $\gamma = 0$ (the linear case) we conclude as before. In the nonlinear cases

($\gamma = 1$), we show by induction that

$$\begin{aligned} \xi^j &= |y^j|^2 + |z^{j-1}|^2 \le L_{j-1} \le L_N = L, \\ L_j &= \xi^1 + \alpha j\,\Delta t, \qquad L_N = L = \xi^1 + \alpha T, \end{aligned} \tag{7.31}$$

with α given below. Assuming (7.31) for $j = 1, \dots, n$, we write

$$I_1 \le \gamma^2\,\Delta t^2\, qL|z^n|^2,$$

$$2\kappa_1' \frac{q\nu\,\Delta t}{\varepsilon}|z^n|^2 - I_1 \ge 2\kappa_2' \frac{q\nu\,\Delta t}{\varepsilon}|z^n|^2,$$

$$2\kappa_2' = 2\kappa_1' - \frac{\gamma^2\,\Delta t^2\,\varepsilon}{\nu}L = 1 - \varepsilon - \varepsilon\nu\,\Delta t\left(1 + \frac{3}{\delta}\right) - \frac{\gamma^2\,\Delta t^2\,\varepsilon}{\nu}L.$$

Also

$$\begin{aligned} I_2 &\le 2\gamma\,\Delta t\,L^{1/2}|z^n|\sum_{j=1}^{q}\left|y^{n+j/q} - y^{n+(j-1)/q}\right| \\ &\le 2\gamma\,\Delta t^2\,L^{1/2}|z^n|\sum_{j=1}^{q}\left|f - \nu y^{n+(j-1)/q} - \nu z^n - \gamma y^n z^n\right| \\ &\le 2\gamma\,\Delta t^2\,L^{1/2}|z^n|\left(q\,|f| + q\nu|z^n| + q\gamma|y^n||z^n| + \nu\sum_{j=1}^{q}\left|y^{n+(j-1)/q}\right|\right) \\ &\le 2q\gamma\,\Delta t^2\,L^{1/2}\,|f|\,|z^n| + 2q\gamma\nu\,\Delta t^2\,L^{1/2}|z^n|^2 + 2q\gamma^2\,\Delta t^2\,L|z^n|^2 \\ &\quad + 2\gamma\nu\,\Delta t^2\,L^{1/2}|z^n|\sum_{j=1}^{q}\left|y^{n+(j-1)/q}\right| \\ &\le q\gamma\,\Delta t^2\left(L|z^n|^2 + |f|^2\right) + 2q\gamma\nu\,\Delta t^2\,L^{1/2}|z^n|^2 + 2q\gamma^2\,\Delta t^2\,L|z^n|^2 \\ &\quad + \gamma\nu\,\Delta t^2\,L^{1/2}\left(q\frac{|z^n|^2}{\sqrt{\varepsilon}} + \sqrt{\varepsilon}\sum_{j=1}^{q}|y^{n+(j-1)/q}|^2\right). \end{aligned}$$

Thus

$$\begin{aligned} &\xi^{n+1} - \xi^n + |z^n - z^{n-1}|^2 + \nu\,\Delta t\left(2\kappa_1 - \gamma\sqrt{\varepsilon}\,\Delta t\,L^{1/2}\right)\sum_{j=1}^{q}\left|y^{n+(j-1)/q}\right|^2 \\ &+ \frac{q\nu\,\Delta t}{\varepsilon}\left(2\kappa_1' - \frac{\varepsilon\,\Delta t}{\nu}(\gamma^2 + \gamma)L - \gamma\,\Delta t\,\sqrt{\varepsilon}L^{1/2}\right. \\ &\left. - 2\gamma\varepsilon\,\Delta t\,L^{1/2} - 2\frac{\gamma^2\varepsilon\,\Delta t\,L}{\nu}\right)|z^n|^2 \end{aligned}$$

$$\leq \frac{q\,\Delta t}{\nu}|g|^2 + 2\,\Delta t \sum_{j=1}^{q} f y^{n+(j-1)/q}$$

$$+q\,\Delta t^2\left(1+\frac{3}{\delta}\right)|f|^2 + \gamma^2\,\Delta t^2\,|f|^2,$$

with

$$2\kappa_2' = 2\kappa_1' - \frac{\varepsilon\,\Delta t}{\nu}(\gamma^2+\gamma)L - \gamma\,\Delta t\,\sqrt{\varepsilon}L^{1/2}$$

$$-\,2\gamma\varepsilon\,\Delta t\,L^{1/2} - 2\frac{\gamma^2\varepsilon}{\nu}\,\Delta t\,L.$$

Now assume that $2\kappa_2' > 0$ and that

$$\Delta t < \frac{2-4\varepsilon}{\nu+\sqrt{\varepsilon}L^{1/2}}. \tag{7.32}$$

Then there exists $\delta > 0$ such that

$$2\kappa_2 = 2\kappa_1 - \gamma\,\Delta t\,\sqrt{\varepsilon}L^{1/2}$$
$$= 2-4\varepsilon-\nu\,\Delta t\,(1+\delta) - \gamma\,\Delta t\,\sqrt{\varepsilon}L^{1/2} > 0,$$

and we can write

$$2\,\Delta t\sum_{j=1}^{q} f y^{n+(j-1)/q} \leq \kappa_2\nu\,\Delta t\sum_{j=1}^{q}\left|y^{n+(j-1)/q}\right|^2 + \frac{q\,\Delta t}{\kappa_2\nu}|f|^2.$$

Thus we obtain

$$\xi^{n+1} - \xi^n + |z^n - z^{n-1}|^2 + \kappa_2\nu\,\Delta t\sum_{j=1}^{q}\left|y^{n+(j-1)/q}\right|^2 \tag{7.33}$$

$$+\,2\kappa_2'\frac{q\nu\,\Delta t}{\varepsilon}|z^n|^2 \leq \alpha\,\Delta t,$$

with ($\Delta t \leq T$)

$$\alpha = \frac{q}{\kappa_2\nu}|f|^2 + q\,\Delta t\left(1+\frac{3}{\delta}\right)|f|^2 + \gamma^2\Delta t|f|^2 + \frac{q}{\nu}|g|^2. \tag{7.34}$$

Therefore (7.31) follows for $j = n+1$, and the induction is complete.

It remains to make explicit the second CFL condition, $2\kappa_2' > 0$. It reads

$$\Delta t < \frac{1-\varepsilon}{\varepsilon\nu[1+(3/\delta)] + \frac{\varepsilon}{\nu}(\gamma^2+\gamma)L + \gamma(\varepsilon L)^{1/2} + 2\gamma\varepsilon L^{1/2} + (2\gamma^2\varepsilon L/\nu)}. \tag{7.35}$$

This condition follows from (7.32) when ε is sufficiently small. Remembering that Δt is the time step for y ($\Delta t = \Delta t_y$) and that $q\,\Delta t = \Delta t_z$ is the time step for z, we conclude, as in the two-level case, that the stability condition for this multilevel scheme is essentially that of the large structure, that is, the y's. The counterintuitive conclusion here is that the time mesh $\Delta t_z = q\,\Delta t$ for the high frequencies is larger than the time mesh $\Delta t_y = \Delta t$ for low frequencies (see Chapters 8, 9, 10).

We refer the reader to Chapters 8 and 9 for the actual implementation of such multilevel schemes for the Navier–Stokes equations. Note that, as explained there, we have the option, in the infinite-dimensional case, to replace or supplement (7.25) by the utilization of a *nonlinear filter* $z = \Phi(y)$ provided by an approximate inertial manifold; of course, this additional feature is not applicable to the simple ordinary differential equation (7.1).

8

Dynamic Multilevel Methodologies

In Chapter 6, we studied the separation of scales in turbulence. In dynamic multilevel (DML) methodologies, as in LES, a filter is used to decompose the velocity field $\mathbf{u} = \overline{\mathbf{u}} + \mathbf{u}'$. However, in DML methods, the closure is achieved by using a completely different methodology, which is based on physical as well as mathematical (numerical) arguments. According to the theoretical study of the Navier–Stokes equations (see Foias, Manley, and Temam (1988)) and under some hypotheses that will be described hereafter, we have that norm$(\mathbf{u}') \ll$ norm$(\overline{\mathbf{u}})$ (for appropriate norms such as the energy norm $|\cdot|_0$ or the enstrophy norm $\|\cdot\|$ (see Section 1.4)). Note that this result is consistent with the Kolmogorov theory (see Section 2.2.4). Moreover, following the latter theory, the characteristic (eddy-turnover) time of the small scales $\tau(\mathbf{u}')$ decreases as a function of the length scale, and hence it is much smaller than that of the large scales $\tau(\overline{\mathbf{u}})$, that is, we have $\tau(\mathbf{u}') \ll \tau(\overline{\mathbf{u}})$ (see Section 2.2.4). Hence, the velocity components $\overline{\mathbf{u}}$ and $\mathbf{u}'$ have different temporal and spatial behaviors. Using these properties, in DML methods, small and large scales of the flow are systematically treated by different numerical schemes. An example of such multilevel schemes on a simpler differential system was given in Chapter 7.

In this chapter, three-dimensional isotropic homogeneous and three-dimensional nonhomogeneous turbulent flows are considered. Several estimates comparing the small and large scales are derived. Based on this study, numerical schemes are developed and proposed; their effect is to speed up the computation of the subgrid scales. These schemes are based on an adaptive strategy and a quasistatic approximation of the small scales and of the interaction terms.

8.1. Behavior of the Small and Large Scales

8.1.1. The Homogeneous Case

Let $k_{N_1} < k_N$ be a cutoff level; we consider the decomposition $\mathbf{u}_N = \mathbf{y}_{N_1} + \mathbf{z}_{N_1}^N$ (see Section 6.2.1 for the definitions of $\mathbf{y}_{N_1}$ and $\mathbf{z}_{N_1}^N$). In two dimensions, Foias, Manley, and Temam (1988) have established estimates comparing the velocity components $\mathbf{y}_{N_1}$ and $\mathbf{z}_{N_1}^N$. They have shown that

$$|\mathbf{z}_{N_1}^N|_0 \le |\mathbf{y}_{N_1}|_0 \tag{8.1}$$

and

$$|\nabla \mathbf{z}_{N_1}^N|_0 \le |\nabla \mathbf{y}_{N_1}|_0 \tag{8.2}$$

for a sufficiently large cutoff level N_1; here $|\cdot|_0$ is the norm defined on the space $L^2(\Omega)^n$ (see Chapter 1). The inequalities (8.1), and (8.2) compare the kinetic energy and the enstrophy in the small scales with the kinetic energy and the enstrophy in the large scales.

Several two-dimensional numerical simulations have been presented by Jauberteau (1990), Dubois (1993), and Debussche, Dubois, and Temam (1995), for instance. They confirm the estimates (8.1), (8.2). In three dimensions, simulations have been performed in order to check the validity of the inequalities (8.1), (8.2). In that case, the numerical results presented by Dubois and Jauberteau (1998) have shown that

$$|\mathbf{z}_{N_1}^N|_0 \le |\mathbf{y}_{N_1}|_0$$

when $k_{N_1} > k_L$, with k_L the wavenumber associated with the integral scale L (see Section 2.2.4). Similarly,

$$|\nabla \mathbf{z}_{N_1}^N|_0 \le |\nabla \mathbf{y}_{N_1}|_0$$

for $k_{N_1} > k_\epsilon$, where k_ϵ is the wavenumber for which the enstrophy spectrum function reaches its maximum, satisfying $k_\lambda = [L_1/2\pi\lambda] \le k_\epsilon \le k_d$ where $[\cdot]$ denotes the integer part. We recall that L_1 is the spatial period of the flow (see Section 3.1) and that λ is the Taylor microscale (see Section 2.2.2).

Results similar to the previous ones can be directly recovered from the Kolmogorov and Kraichnan theories. According to Sections 2.2.4 and 4.1.1 we can assume that the energy of the flow is extended in the wavenumber range $[1, k_N]$ with $k_N \simeq k_d$. The scales corresponding to wavenumbers of order k_d are dissipative scales and contain most of the enstrophy. In DNS, the integral scale is generally close to unity. Indeed, in the case of forced turbulence, the forcing acts on small wavenumbers $k_F \simeq 1$ (see Vincent and Ménéguzzi (1991), Jiménez et al. (1993), Dubois and Jauberteau (1998)), so that $k_L \simeq 1$. Hence, the scales corresponding to wavenumbers $k \simeq 1$ contain most of the kinetic energy.

According to the Kolmogorov theory (Section 2.2.4), we can assume that the energy in three-dimensional flows is distributed across wavenumbers following the $-\frac{5}{3}$ law

$$E(k) = C_K \epsilon^{2/3} \left(\frac{2\pi}{L_1} k \right)^{-5/3} \quad \text{for} \quad 1 < k < k_d. \tag{8.3}$$

As a model for the far dissipation range, we use (2.114), that is,

$$E(k) = C_K \epsilon^{2/3} \left(\frac{2\pi}{L_1} k \right)^{-5/3} e^{-\alpha_1 (2\pi/L_1) k\eta} \quad \text{for} \quad k \in [1, k_N], \tag{8.4}$$

with $\alpha_1 = [2C_K \Gamma(\frac{4}{3})]^{3/4}$ $(\simeq 1.545 C_K^{3/4})$.

Let us now assume that the cutoff level $k_{N_1} = N_1/2 \in (1, k_N)$. Using the Parseval relation, the L^2 norm is related to the Fourier coefficients as follows:

$$|\varphi(\mathbf{x}, t)|_0^2 = |\Omega| \sum_{\mathbf{k} \in \mathbb{Z}^3} |\hat{\varphi}(\mathbf{k}, t)|^2 \tag{8.5}$$

for any flow variable φ. Recalling (3.11) and (6.16), we deduce

$$\left| \mathbf{y}_{N_1} \right|_0^2 = L_1^n \sum_{\mathbf{k} \in S_{N_1}} |\tilde{\mathbf{u}}(\mathbf{k})|^2. \tag{8.6}$$

Similarly, using (3.11) and (6.17), we obtain

$$\left| \mathbf{z}_{N_1}^N \right|_0^2 = L_1^n \sum_{\mathbf{k} \in S_N \setminus S_{N_1}} |\tilde{\mathbf{u}}(\mathbf{k})|^2. \tag{8.7}$$

The ratio $|\mathbf{z}_{N_1}^N|_0 / |\mathbf{y}_{N_1}|_0$ is then a decreasing function of k_{N_1}. Recalling the definition of the energy spectrum function (2.46) and observing that $\bigcup_{k=1}^{k_{N_1}-1} S_{k,1/2} \subset S_{N_1}$ we derive a lower bound for $|\mathbf{y}_{N_1}|_0^2$, namely

$$\left| \mathbf{y}_{N_1} \right|_0^2 \geq 2L_1^n \sum_{k=1}^{k_{N_1}-1} E(k). \tag{8.8}$$

Using the fact that $E(k)$ is a decreasing function of k and recalling the asymptotic law (8.4), we find

$$\left(\frac{2\pi}{L_1}\right)^{5/3} \frac{|\mathbf{y}_{N_1}|_0^2}{2L_1^n C_K \epsilon^{2/3}} \geq \int_1^{k_{N_1}} x^{-5/3} e^{-\alpha_1(2\pi/L_1)k_{N_1}\eta}\, dx.$$

Now, by noticing that $e^{-\alpha_1(2\pi/L_1)x\eta}$ is a positive decreasing function of x, the following inequalities can be obtained:

$$\left(\frac{2\pi}{L_1}\right)^{5/3} \frac{|\mathbf{y}_{N_1}|_0^2}{2L_1^n C_K \epsilon^{2/3}} \geq \frac{3}{2}\left[1 - k_{N_1}^{-2/3}\right] e^{-\alpha_1(2\pi)/(L_1)k_{N_1}\eta}. \tag{8.9}$$

Similarly, observing that $S_N \backslash S_{N_1} \subset \bigcup_{k=k_{N_1}}^{k_N} S_{k,1/2}$, an upper bound can be derived for $|\mathbf{z}_{N_1}^N|_0$, namely

$$\begin{aligned} &\left(\frac{2\pi}{L_1}\right)^{5/3} \frac{|\mathbf{z}_{N_1}^N|_0^2}{2L_1^n C_K \epsilon^{2/3}} \\ &\leq \frac{3}{2}\left[(k_{N_1} - 1)^{-2/3} - k_N^{-2/3}\right] e^{-\alpha_1(2\pi/L_1)k_{N_1}\eta}. \end{aligned} \tag{8.10}$$

From (8.9) and (8.10), an upper bound on the ratio $|\mathbf{z}_{N_1}^N|_0/|\mathbf{y}_{N_1}|_0$ immediately follows:

$$\frac{|\mathbf{z}_{N_1}^N|_0}{|\mathbf{y}_{N_1}|_0} \leq \left(\frac{(k_{N_1} - 1)^{-2/3} - k_N^{-2/3}}{1 - k_{N_1}^{-2/3}}\right)^{1/2}. \tag{8.11}$$

Therefore $|\mathbf{z}_{N_1}^N|_0 \leq |\mathbf{y}_{N_1}|_0$ as soon as $k_{N_1} \geq 2^{3/2}(1 + k_N^{-2/3})^{-3/2}$. Since $k_N > k_d$, upon using (4.2) we see that the previous estimate becomes

$$k_{N_1} \geq 2^{3/2}\left[1 + \left(\frac{2\pi L}{c_1^{1/4} L_1}\right)^{2/3} \mathrm{Re}_L^{-1/2}\right]^{-3/2},$$

which is asymptotically of the order of $2\sqrt{2}$, since the quantity in square brackets is close to 1 for large Reynolds numbers.

We now consider the enstrophy-related norm $\|\cdot\|$ on the space V (see Section 1.4). Using the Parseval relation (8.5) and recalling the definition of the enstrophy spectrum function (2.48), we find

$$\|\mathbf{y}_{N_1}\|^2 = L_1^n \sum_{\mathbf{k}\in S_{N_1}} \frac{4\pi^2}{L_1^2} |\mathbf{k}|^2 |\tilde{\mathbf{u}}(\mathbf{k}, t)|^2 \geq L_1^n \sum_{k=1}^{k_{N_1}-1} \epsilon(k). \tag{8.12}$$

Similarly

$$\left\|\mathbf{z}_{N_1}^N\right\|^2 = L_1^n \sum_{\mathbf{k}\in S_N\setminus S_{N_1}} \frac{4\pi^2}{L_1^2} |\mathbf{k}|^2 |\tilde{\mathbf{u}}(\mathbf{k},t)|^2 \le L_1^n \sum_{k=k_{N_1}}^{k_N} \epsilon(k). \tag{8.13}$$

By observing that $\epsilon(k)$ behaves at large wavenumbers like $2\,[(2\pi/L_1)k]^2\, E(k)$, we use the same model as in (8.4) for the dissipation range:

$$\epsilon(k) = C_K' \epsilon^{2/3} \left(\frac{2\pi}{L_1}k\right)^{1/3} e^{-\alpha_1(2\pi/L_1)k\eta} \qquad \text{for} \quad k \in [1, k_N]. \tag{8.14}$$

Note that $\epsilon(k)$ reaches its maximum value near the wavenumber $k_\epsilon = [1/3\beta]$, where $[\cdot]$ denotes the integer part and where we have set for convenience $\beta = \alpha_1\, 2\eta\pi/L_1$. Substituting the law (8.14) into (8.12), we obtain

$$\left\|\mathbf{y}_{N_1}\right\|^2 \ge C_K' L_1^n \epsilon^{2/3} \left(\frac{2\pi}{L_1}\right)^{1/3} \sum_{k=1}^{k_{N_1}-1} k^{1/3} e^{-\beta k}. \tag{8.15}$$

We now assume that $k_{N_1} > k_\epsilon$, and we split the sum in the right-hand side of (8.15) as follows:

$$\sum_{k=1}^{k_{N_1}-1} k^{1/3} e^{-\beta k} = \sum_{k=1}^{k_\epsilon} k^{1/3} e^{-\beta k} + \sum_{k=k_\epsilon+1}^{k_{N_1}-1} k^{1/3} e^{-\beta k}.$$

Observing that the behavior of $k^{1/3}e^{-\beta k}$ is dominated by $k^{1/3}$ in the range $[1, k_\epsilon]$ and by $e^{-\beta k}$ in the range $[k_\epsilon + 1, k_{N_1} - 1]$, we derive the following lower bound:

$$\sum_{k=1}^{k_{N_1}-1} k^{1/3} e^{-\beta k} \ge \frac{3}{4} e^{-\beta k_\epsilon} \left(k_\epsilon^{4/3} + \frac{1}{3}\right) + \frac{(k_\epsilon + 1)^{1/3}}{\beta} \left(e^{-\beta(k_\epsilon+1)} - e^{-\beta k_{N_1}}\right).$$

We finally obtain a lower bound for $\|\mathbf{y}_N\|^2$, namely

$$\begin{aligned} \left\|\mathbf{y}_{N_1}\right\|^2 \ge{}& C_K' L_1^n \epsilon^{2/3} \left(\frac{2\pi}{L_1}\right)^{1/3} e^{-\beta k_\epsilon} \\ &\times \left[\frac{3}{4}\left(k_\epsilon^{4/3} + \frac{1}{3}\right) + \frac{(k_\epsilon + 1)^{1/3}}{\beta}\left(e^{-\beta} - e^{-\beta(k_{N_1} - k_\epsilon)}\right)\right]. \end{aligned} \tag{8.16}$$

Similarly, we derive the following upper bound for $\|\mathbf{z}_{N_1}^N\|^2$:

$$\left\|\mathbf{z}_{N_1}^N\right\|^2 \le C_K' L_1^n \epsilon^{2/3} \left(\frac{2\pi}{L_1}\right)^{1/3} \frac{k_N^{1/3}}{\beta} \left(e^{-\beta(k_{N_1}-1)} - e^{-\beta k_N}\right). \tag{8.17}$$

From (8.16) and (8.17), we deduce an estimate of the ratio $\|\mathbf{z}_{N_1}^N\|/\|\mathbf{y}_{N_1}\|$:

$$\frac{\|\mathbf{z}_{N_1}^N\|}{\|\mathbf{y}_{N_1}\|} \leq \frac{k_N^{1/6}}{\sqrt{\beta}} e^{(\beta/2)k_\epsilon} \frac{\left(e^{-\beta(k_{N_1}-1)} - e^{-\beta k_N}\right)^{1/2}}{\left(\frac{3}{4}\left(k_\epsilon^{4/3} + \frac{1}{3}\right) + \frac{(k_\epsilon+1)^{1/3}}{\beta}\left(e^{-\beta} - e^{-\beta(k_{N_1}-k_\epsilon)}\right)\right)^{1/2}}. \tag{8.18}$$

Hence, $\|\mathbf{z}_{N_1}^N\| \leq \|\mathbf{y}_{N_1}\|$ as soon as

$$k_{N_1} \geq k_\epsilon \left\{1 + 3\log\left[1 + \left(\frac{k_\epsilon + 1}{k_N}\right)^{1/3}\right]\right\}$$
$$- 3(k_\epsilon + 1)\log\left[\left(\frac{k_\epsilon + 1}{k_N}\right)^{1/3} + e^{-\beta(k_N - k_\epsilon)} + \frac{3\beta}{4}\left(\frac{k_\epsilon^{4/3}}{k_N^{1/3}} + \frac{1}{3k_N^{1/3}}\right)\right].$$

Recalling the definitions of k_ϵ, k_d, and β, we can show that the lower bound for k_{N_1} asymptotically (i.e. when k_N goes to infinity) behaves like

$$k_\epsilon[1 + 3\varphi(r)],$$

where $r = k_N/k_d$ (>1) and

$$\varphi(r) = \log\left(1 + \frac{1}{(3\alpha_1)^{1/3} r}\right) - \log\left(\frac{(4 + r^{2/3})}{4r(3\alpha)^{1/3}} + e^{(1-3\alpha_1 r)/3}\right).$$

Moreover, it can be easily shown that

$$3.2\, k_\epsilon \geq k_\epsilon[1 + 3\varphi(r)] \geq 4.8\, k_\epsilon \qquad \text{for} \quad r \in [1, 2].$$

Recalling that k_ϵ is of the order of $k_d/[3\alpha_1]$, we find $k_\epsilon[1 + 3\varphi(r)] \simeq 0.5k_d$.

Similar estimates can be derived in the two-dimensional case, where the energy spectrum function satisfies (2.120):

$$E(k) \simeq C_K \gamma^{2/3}\left(\frac{2\pi}{L_1}k\right)^{-3} \qquad \text{for} \quad k \in [1, k_d]. \tag{8.19}$$

The enstrophy spectrum function is then given by

$$\epsilon(k) \simeq C'_K \gamma^{2/3}\left(\frac{2\pi}{L_1}k\right)^{-1} \qquad \text{for} \quad k \in [1, k_d]. \tag{8.20}$$

Note that both the energy and enstrophy spectrum functions are decreasing in the inertial range. Hence, the kinetic energy as well as the enstrophy is dominated by the large scales, that is, $k \simeq k_L$.

Let us choose a cutoff wavenumber $k_{N_1} \in [1, k_N]$, and let us set $k_N = k_d$. Then, proceeding as before, we obtain

$$|\mathbf{y}_{N_1}|_0^2 \geq C_K L_1^n \gamma^{2/3} \left(\frac{2\pi}{L_1}\right)^{-3} (1 - k_{N_1}^{-2}) \tag{8.21}$$

and

$$|\mathbf{z}_{N_1}^N|_0^2 \leq C_K L_1^n \gamma^{2/3} \left(\frac{2\pi}{L_1}\right)^{-3} k_{N_1}^{-2} \left[1 - \left(\frac{k_{N_1}}{(k_N+1)}\right)^2\right]. \tag{8.22}$$

The ratio $|\mathbf{z}_{N_1}^N|_0 / |\mathbf{y}_{N_1}|_0$ is then bounded as follows:

$$\frac{|\mathbf{z}_{N_1}^N|_0}{|\mathbf{y}_{N_1}|_0} \leq \frac{1}{k_{N_1}} \left(\frac{1 - \left(\frac{k_{N_1}}{(k_N+1)}\right)^2}{1 - \frac{1}{k_{N_1}^2}}\right)^{1/2}. \tag{8.23}$$

From (8.23), we deduce that $|\mathbf{z}_{N_1}^N|_0 \leq |\mathbf{y}_{N_1}|_0$ if $k_{N_1} \geq \sqrt{2}[1 + 1/(k_N+1)^2]^{-1/2}$, which is asymptotically of the order of $\sqrt{2}$. Similarly, we see that

$$\|\mathbf{y}_{N_1}\|^2 \geq C_K' L_1^n \gamma^{2/3} \left(\frac{2\pi}{L_1}\right)^{-1} \log k_{N_1}, \tag{8.24}$$

and

$$\|\mathbf{z}_{N_1}^N\|^2 \leq C_K' L_1^n \gamma^{2/3} \left(\frac{2\pi}{L_1}\right)^{-1} \log\left(\frac{k_N+1}{k_{N_1}}\right). \tag{8.25}$$

The condition $\|\mathbf{z}_{N_1}^N\| \leq \|\mathbf{y}_{N_1}\|$ is then satisfied if $k_{N_1} \geq \sqrt{1 + k_N}$.

Time and Space Behavior of the Small and Large Scales and of the Nonlinear Interaction Term

Now we analyze the behavior, in space and time, of the small and large scales using the DNS results obtained.

From the Parseval relation (8.5) and from the definitions of the projection operators P_{N_1} and $Q_{N_1}^N$ (see Section 6.2.1), we deduce that the ratio $|\mathbf{z}_{N_1}^N|_0/|\mathbf{y}_{N_1}|_0$ is a decreasing function of N_1. Moreover, $|\mathbf{z}_{N_1}^N|_0/|\mathbf{y}_{N_1}|_0$ is less than unity for any cutoff wavenumber k_{N_1}, and it behaves like $k_{N_1}^{-1}$ for $k_{N_1} \leq k_d/3$ (see

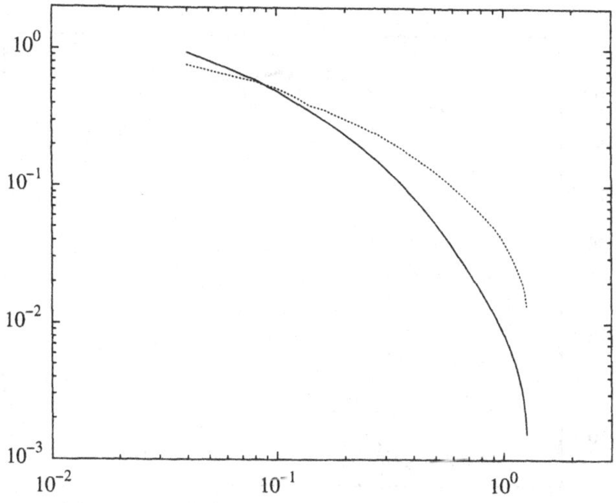

Figure 8.1. $|\mathbf{z}_{N_1}^N|_0/|\mathbf{y}_{N_1}|_0$ (solid line) and $|P_{N_1}B_{\text{int}}(\mathbf{y}_{N_1}, \mathbf{z}_{N_1}^N)|_0/|P_{N_1}B(\mathbf{y}_{N_1}, \mathbf{y}_{N_1})|_0$ (dotted line) as functions of the normalized wavenumber ηk_{N_1}.

Figure 8.1).[1] Hence, for high cutoff levels, the small scales carry a small percentage of the whole energy. Indeed, for the 128^3 simulation corresponding to Figure 8.1, $|\mathbf{z}_{N_1}^N|_0$ for $k_{N_1} = k_d/3$ represents less that 12% of the total energy of the flow.

By recalling the large-scale equation (6.18) we note that $\mathbf{z}_{N_1}^N$ acts on the evolution of $\mathbf{y}_{N_1}$ via the nonlinear interaction term $P_{N_1}B_{\text{int}}(\mathbf{y}_{N_1}, \mathbf{z}_{N_1}^N)$. From Figure 8.1, we deduce, as previously, that the ratio $|P_{N_1}B_{\text{int}}(\mathbf{y}_{N_1}, \mathbf{z}_{N_1}^N)|_0/|P_{N_1}B(\mathbf{y}_{N_1}, \mathbf{y}_{N_1})|_0$ is a decreasing function of N_1 and is less than the unity for any cutoff wavenumber. Moreover, it behaves like $k_{N_1}^{-0.8}$ for $k_{N_1} \leq k_d/3$. The nonlinear interaction term then represents a small correction to the large-scale evolution. The quantity analyzed above are global. A more detailed analysis of the nonlinear interaction can be achieved by considering the following spectrum functions:

$$
\begin{aligned}
\text{NL}_{\text{int}}(k, k_{N_1}) &= \frac{1}{2} \sum_{\mathbf{k} \in S_{k,1/2}} \left| \hat{B}_{\text{int},\mathbf{k}}\left(\mathbf{y}_{N_1}, \mathbf{z}_{N_1}^N\right) \right|^2, \\
\text{NL}(k, k_{N_1}) &= \frac{1}{2} \sum_{\mathbf{k} \in S_{k,1/2}} \left| \hat{B}_{\mathbf{k}}\left(\mathbf{y}_{N_1}, \mathbf{y}_{N_1}\right) \right|^2 \quad \text{for} \quad 1 \leq k \leq k_N.
\end{aligned}
\tag{8.26}
$$

[1] Figures 8.1 to 8.4 correspond to a 128^3-mode direct numerical simulation.

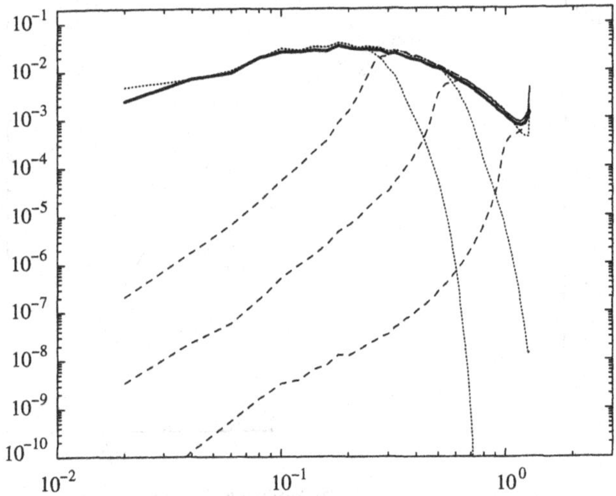

Figure 8.2. $NL(k, k_{N_1})$ (dotted line) and $NL_{int}(k, k_{N_1})$ (dashed line) for $\eta k_{N_1} = 0.245, 0.49, 0.98$, and $T(k)$ (solid line) as functions of the normalized wavenumber $\eta\, k$.

We also define the spectrum function

$$T(k) = \frac{1}{2} \sum_{\mathbf{k} \in S_{k,1/2}} |\tilde{\dot{\mathbf{u}}}(\mathbf{k})|^2, \tag{8.27}$$

where $\dot{\mathbf{u}}$ denotes the time derivative $\partial \mathbf{u}/\partial t$. In Figure 8.2 we note that

$$\mathrm{NL}_{\mathrm{int}}\big(k, k_{N_1}\big) < \mathrm{NL}\big(k, k_{N_1}\big) \simeq T(k) \qquad \text{for} \quad k < k_{N_1}. \tag{8.28}$$

For a fixed k_{N_1}, $\mathrm{NL}_{\mathrm{int}}(k, k_{N_1})$ is an increasing function of the wavenumber k, while $\mathrm{NL}(k, k_{N_1})$ remains approximatively constant; moreover, $\mathrm{NL}_{\mathrm{int}}(k_{N_1}, k_{N_1}) = \mathrm{NL}(k_{N_1}, k_{N_1})$. Note that for $k \le k_{N_1}/2$ the nonlinear interaction represents less than 0.5% of the time derivative of the large-scale component. Hence, most of the nonlinear interactions occur near the cutoff wavenumber. Consequently, the nonlinear interaction term $P_{N_1} B_{\mathrm{int}}(\mathbf{y}_{N_1}, \mathbf{z}_{N_1}^N)$, and thus the small scales, cannot be neglected in the simulation. The net effect of this term is to maintain the $k^{-5/3}$ decrease of the energy spectrum and to preserve the turbulence statistics of the flow (see Dubois, Jauberteau, and Zhou (1997) for instance).

Nevertheless, the time variations of the small scales as well as of the nonlinear interaction term appear to be small. Indeed, from Figure 8.3 we deduce that

$$\Delta t \left|\dot{\mathbf{z}}_{N_1}^N\right|_0 < \left|\mathbf{z}_{N_1}^N\right|_0, \tag{8.29}$$

$$\Delta t \left|P_{N_1} \dot{B}_{\mathrm{int}}\big(\mathbf{y}_{N_1}, \mathbf{z}_{N_1}^N\big)\right|_0 < \left|P_{N_1} B_{\mathrm{int}}\big(\mathbf{y}_{N_1}, \mathbf{z}_{N_1}^N\big)\right|_0 \qquad \text{for} \quad 1 \le k_{N_1} \le k_N.$$

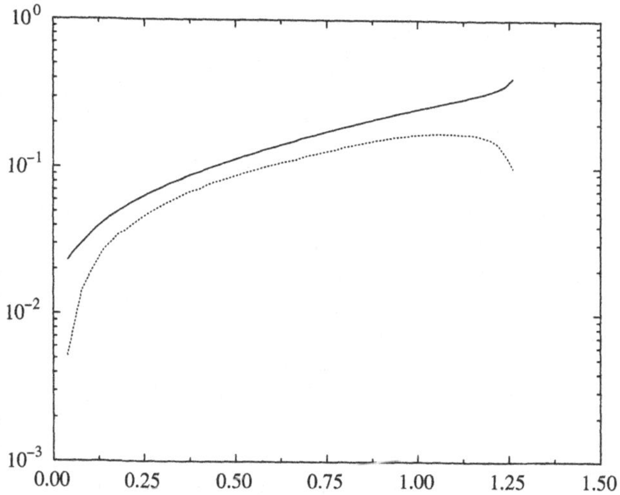

Figure 8.3. $(k_N U_\infty)^{-1}|\dot{\mathbf{z}}^N_{N_1}|_0/|\mathbf{z}^N_{N_1}|_0$ (solid line) and $(k_N U_\infty)^{-1}|P_{N_1}\dot{B}_{\text{int}}(\mathbf{y}N_1, \mathbf{z}^N_{N_1})|_0/|P_{N_1}B_{\text{int}}(\mathbf{y}N_1, \mathbf{z}^N_{N_1})|_0$ (dotted line) as functions of ηk_{N_1}.

As previously, instead of comparing global quantities, we can analyze the spectrum functions of these quantities. We then define

$$\text{TNL}_{\text{int}}(k, k_{N_1}) = \frac{1}{2}\sum_{\mathbf{k}\in S_{k,1/2}} \left|\frac{\partial}{\partial t}\hat{B}_{\text{int},\mathbf{k}}\left(\mathbf{y}_{N_1}, \mathbf{z}^N_{N_1}\right)\right|^2,$$

$$\text{TNL}(k, k_{N_1}) = \frac{1}{2}\sum_{\mathbf{k}\in S_{k,1/2}} \left|\frac{\partial}{\partial t}\hat{B}_{\mathbf{k}}\left(\mathbf{y}_{N_1}, \mathbf{y}_{N_1}\right)\right|^2 \quad \text{for} \quad 1\leq k\leq k_N. \tag{8.30}$$

As for $\text{NL}_{\text{int}}(k, k_{N_1})$ and $\text{NL}(k, k_{N_1})$, in Figure 8.4 we note that

$$\text{TNL}_{\text{int}}(k, k_{N_1}) < \text{TNL}(k, k_{N_1}) \qquad \text{for} \quad k < k_{N_1}. \tag{8.31}$$

For a fixed k_{N_1}, $\text{TNL}_{\text{int}}(k, k_{N_1})$ is an increasing function of the wavenumber k while $\text{TNL}(k, k_{N_1})$ remains approximatively constant; moreover, we have $\text{TNL}_{\text{int}}(k_{N_1}, k_{N_1}) = \text{TNL}(k_{N_1}, k_{N_1})$.

In summary, the small scales as well as the interaction term have space and time behaviors completely different than those of the large ones. According to (8.28) and (8.31), the small scales represent a small correction to the dynamics of the large-energy-containing scales. A fundamental question arises at this point: can we numerically treat the former quantities in a simplified way? Firstly, equations (6.18) and (6.20) have different properties. Indeed, the spectral discretization of the dissipative operator $\Delta\mathbf{u}_N$ gives

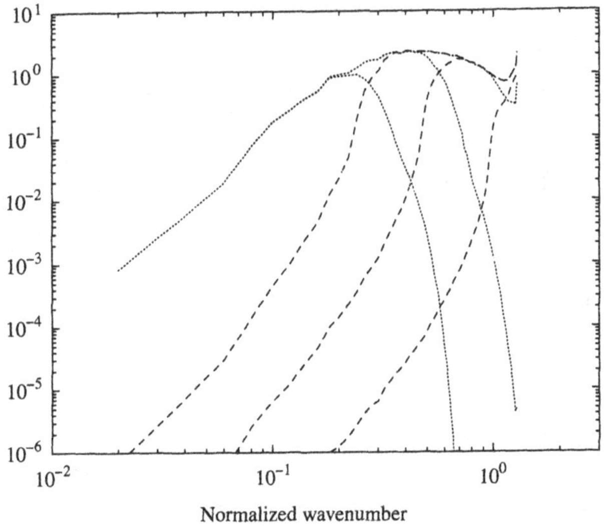

Figure 8.4. TNL(k, k_{N_1}) (dotted line) and TNL$_{\text{int}}(k, k_{N_1})$ (dashed line) for $\eta k_{N_1} = 0.245,\ 0.49,\ 0.98$, as functions of the normalized wavenumber ηk.

$k^2\tilde{\mathbf{u}}(\mathbf{k})$. Following the universal Kolmogorov theory, there is no amplification of the energy injection rate in the small scales via the nonlinearity. Hence, since $\mathbf{y}_{N_1}$ is associated with the small wavenumbers of the velocity field $\mathbf{u}_N$, and $\mathbf{z}_{N_1}^N$ is associated with the high wavenumbers of $\mathbf{u}_N$, the term $\Delta\mathbf{z}_{N_1}^N$ is more significant in (6.20) than the term $\Delta\mathbf{y}_{N_1}$ in (6.18). As a direct consequence, the small scales via (6.20) are less sensitive to small disturbances than the large ones via (6.18). We can then expect to be able to compute the small scales with less-accurate discretization schemes. Since the time variations of the small scales and of the nonlinear interaction appear to be small (see (8.29) and (8.31)), we neglect the time variations of the nonlinear interaction term and of the small scales over small time intervals. In this dynamic multilevel (DML) method (see Section 8.2.1), the small scales as well as the interaction term are kept constant over small time intervals (quasistatic approximation). Note that such numerical treatment reduce the computational efforts required by DNS.

The spectrum function $T(k)$ corresponding to the time derivative of the velocity is increasing for $k \leq k_d/4$ and decreasing for larger wavenumbers (see Figure 8.2). It is then necessary to control the size of the allowable cutoff level used in the scale separation. In Chapter 9, the error introduced via a quasistatic approximation of the small scales will be estimated and used in order to estimate these cutoff levels.

To conclude this section, we want to link the above study with the phenomenological theory of turbulence. It is well known that the small scales reach a statistically steady state faster than the large ones (see Batchelor (1971), Orszag (1973)). Indeed, the characteristic time scale $\tau(k)$ associated with each wavenumber k is a decreasing function of the wavenumber k (see Section 2.2.4). Hence, the smallest physical one is associated with the Kolmogorov scale η and can be defined (see Section 2.2.4) as

$$\tau_d = \left(\nu \frac{4\pi^2}{L_1^2} k_d^2 \right)^{-1}.$$

By assuming that $k_N \geq 3(k_d + 1)/2$, and recalling the CFL condition (4.5), (4.7), (2.110) and the definition of $k_d = [L_1/(2\pi\eta)]$, we deduce that

$$\tau_d \geq \frac{3}{2\alpha c_1^{1/4}} \frac{U_\infty}{u} \mathrm{Re}_L^{1/4} \Delta t, \tag{8.32}$$

where c_1 is a dimensionless constant of the order of unity (see Section 2.2.4). So $\tau_d/\Delta t$ behaves like $\mathrm{Re}_L^{1/4}$. Hence, applying a quasistatic approximation to the small scales is not in contradiction with the fact that they have the smallest characteristic time scale.

8.1.2. The Nonhomogeneous Case

In this section, we analyze the behavior of the small and large scales in the case of the channel flow problem. We consider the decomposition (6.48) of the velocity field into small and large scales, in the two periodic directions x_1 (streamwise direction) and x_3 (spanwise direction).

Firstly, we consider the ratio $\langle |z_{i,N_1}^N|^2 \rangle^{1/2} / \langle |y_{i,N_1}|^2 \rangle^{1/2}$, where $i = 1, 2, 3$ and $\langle \cdot \rangle$ is the averaging operator defined in Section 3.2.1. As for the periodic flow problem, from Parseval identity (8.5) and from the definitions of $\mathbf{y}_{N_1}$ and $\mathbf{z}_{N_1}^N$, we have that this ratio is a decreasing function of N_1. In Figure 8.5[2] we have represented $\langle |\mathbf{z}_{i,N_1}^N|^2 \rangle^{1/2} / \langle |\mathbf{y}_{i,N_1}|^2 \rangle^{1/2}$ as a function of x_2, for each component of the velocity. We notice that this ratio increases near the walls, especially for u_2. It becomes less than unity as soon as the cutoff level N_1 is greater than 20 for u_2. As for u_1 and u_3, the ratio is less than one for all the chosen cutoff levels. Moreover, the small scales associated with the Fourier development of u_2 carry a greater percentage of the kinetic energy than the small scales associated with u_3 and than those associated with u_1.

[2] Figures 8.5 to 8.11 correspond to the $128 \times 129 \times 128$ direct simulation described in detail in Chapter 10.

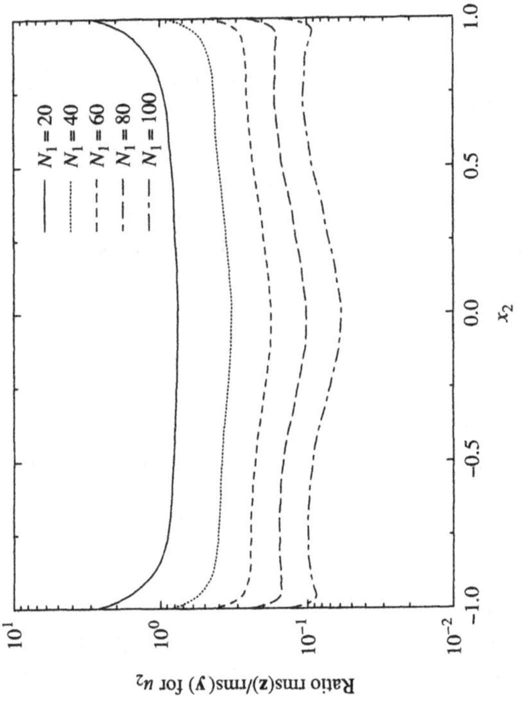

Ratio rms(z)/rms(y) for u_2
x_2
$N_1 = 20$
$N_1 = 40$
$N_1 = 60$
$N_1 = 80$
$N_1 = 100$

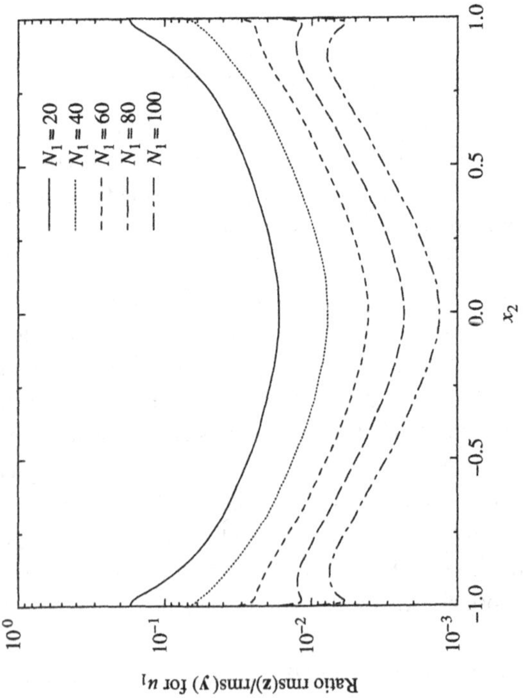

Ratio rms(z)/rms(y) for u_1
x_2
$N_1 = 20$
$N_1 = 40$
$N_1 = 60$
$N_1 = 80$
$N_1 = 100$

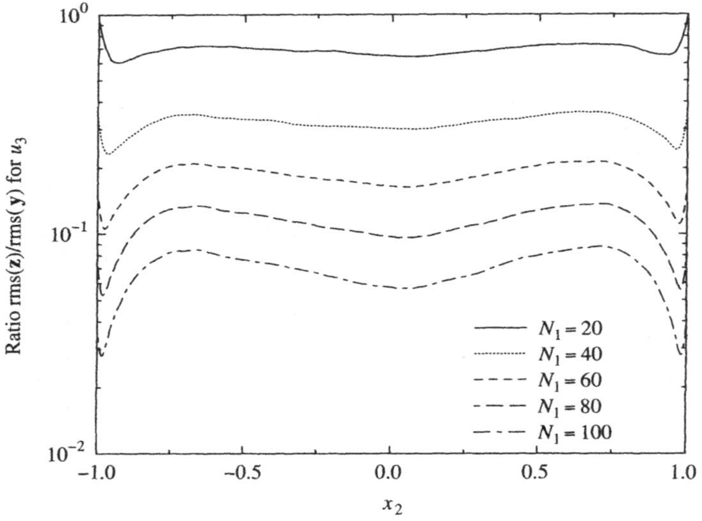

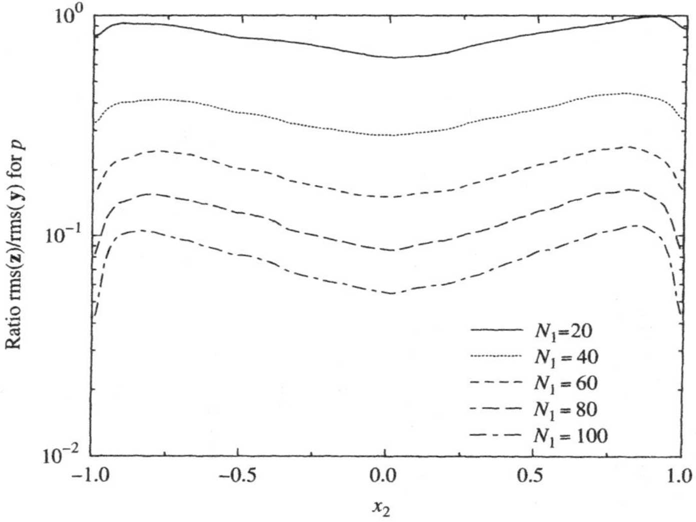

Figure 8.5. Ratio $\langle |\mathbf{z}_{i,N_1}^N|^2\rangle^{1/2}/\langle|\mathbf{y}_{i,N_1}|^2\rangle^{1/2}$ as a function of x_2, for each component u_i of the velocity, $i = 1, 2, 3$, and for the pressure.

Now we consider the nonlinear interaction terms appearing in the large-scale equations (6.51). We set

$$\begin{aligned} H_{\omega_2}\left(\mathbf{y}_{N_1},\mathbf{y}_{N_1}\right) &= \frac{\partial}{\partial x_3}H_1\left(\mathbf{y}_{N_1},\mathbf{y}_{N_1}\right) - \frac{\partial}{\partial x_1}H_3\left(\mathbf{y}_{N_1},\mathbf{y}_{N_1}\right), \\ H_{\mathrm{int},\omega_2}\left(\mathbf{y}_{N_1},\mathbf{z}^N_{N_1}\right) &= \frac{\partial}{\partial x_3}H_{\mathrm{int},1}\left(\mathbf{y}_{N_1},\mathbf{z}^N_{N_1}\right) - \frac{\partial}{\partial x_1}H_{\mathrm{int},3}\left(\mathbf{y}_{N_1},\mathbf{z}^N_{N_1}\right), \end{aligned} \tag{8.33}$$

the nonlinear terms of the equation of ω_2 (3.22), and

$$\begin{aligned} H_{\Delta u_2}\left(\mathbf{y}_{N_1},\mathbf{y}_{N_1}\right) &= \Delta H_2\left(\mathbf{y}_{N_1},\mathbf{y}_{N_1}\right) - \frac{\partial}{\partial x_2}\left(\boldsymbol{\nabla}\cdot\mathbf{H}\right)\left(\mathbf{y}_{N_1},\mathbf{y}_{N_1}\right), \\ H_{\mathrm{int},\Delta u_2}\left(\mathbf{y}_{N_1},\mathbf{z}^N_{N_1}\right) &= \Delta H_{\mathrm{int},2}\left(\mathbf{y}_{N_1},\mathbf{z}^N_{N_1}\right) - \frac{\partial}{\partial x_2}\left(\boldsymbol{\nabla}\cdot\mathbf{H}_{\mathrm{int}}\right)\left(\mathbf{y}_{N_1},\mathbf{z}^N_{N_1}\right), \end{aligned} \tag{8.34}$$

the nonlinear terms of the equation of Δu_2 (3.23). In Figure 8.6, we have represented, as functions of x_2, the ratios $\langle|H_{\mathrm{int},\Delta u_2}(\mathbf{y}_{N_1},\mathbf{z}^N_{N_1})|^2\rangle^{1/2}/\langle|H_{\Delta u_2}(\mathbf{y}_{N_1},\mathbf{y}_{N_1})|^2\rangle^{1/2}$ and $\langle|H_{\mathrm{int},\omega_2}(\mathbf{y}_{N_1},\mathbf{z}^N_{N_1})|^2\rangle^{1/2}/\langle|H_{\omega_2}(\mathbf{y}_{N_1},\mathbf{y}_{N_1})|^2\rangle^{1/2}$. We can see that these ratios increase when $|x_2|$ varies from 0 to 1 and they decrease when N_1 is increased. Furthermore, they are less than one for every cutoff level chosen, even near the walls. So, as for the homogeneous case, the nonlinear interaction terms represent a small correction to the large-scale evolution of ω_2 and Δu_2.

The previous quantities being global in x_1 and x_3, a more detailed analysis can be achieved by considering the one-dimensional energy spectra for each velocity component. We define the one-dimensional energy spectra in the streamwise direction as

$$E_{u_l u_l}\left(k_{x_1},x_2\right) = \sum_{k_3=1-N/2}^{N/2-1} |\tilde{u}_l(k_1,k_3,x_2)|^2, \tag{8.35}$$

with $l=1,2,3$ and $k_{x_1}=2\pi k_1/L_1$, $k_1=1,\ldots,N/2-1$. In the same manner, the one-dimensional energy spectra in the spanwise direction is defined by

$$E_{u_l u_l}\left(k_{x_3},x_2\right) = \sum_{k_1=1-N/2}^{N/2-1} |\tilde{u}_l(k_1,k_3,x_2)|^2, \tag{8.36}$$

with $l=1,2,3$ and $k_{x_3}=2\pi k_3/L_3$, $k_3=1,\ldots,N/2-1$. In Figure 8.7 (Figure 8.8), we have represented the energy spectra in the streamwise direction x_1 (spanwise direction x_3) for two values of x_2 (one value near the upper wall $x_2=0.97$ and one value near the centerline of the channel, $x_2=0.17$). We can see that the spectra in the streamwise direction decrease when the wavenumbers

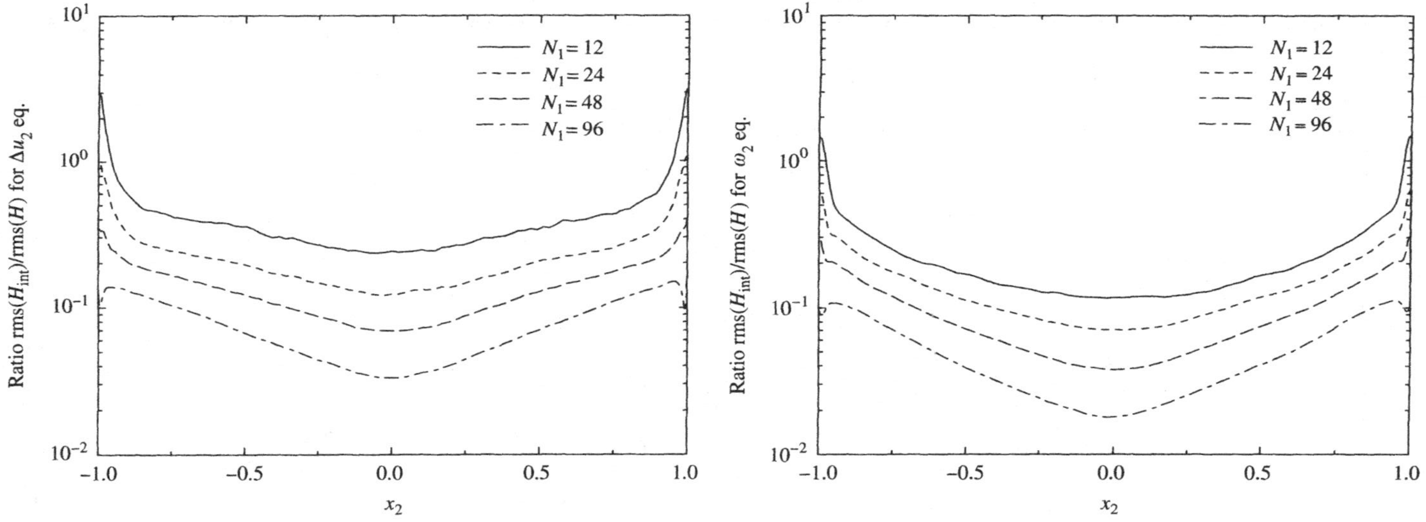

Figure 8.6. Ratios $\langle | H_{\mathrm{int},\Delta u_2}(\mathbf{y}_{N_1}, \mathbf{z}_{N_1}^N) |^2 \rangle^{1/2} / \langle | H_{\Delta u_2}(\mathbf{y}_{N_1}, \mathbf{y}_{N_1}) |^2 \rangle^{1/2}$ (left) and $\langle | H_{\mathrm{int},\omega_2}(\mathbf{y}_{N_1}, \mathbf{z}_{N_1}^N) |^2 \rangle^{1/2} / \langle | H_{\omega_2}(\mathbf{y}_{N_1}, \mathbf{y}_{N_1}) |^2 \rangle^{1/2}$ (right) as functions of x_2.

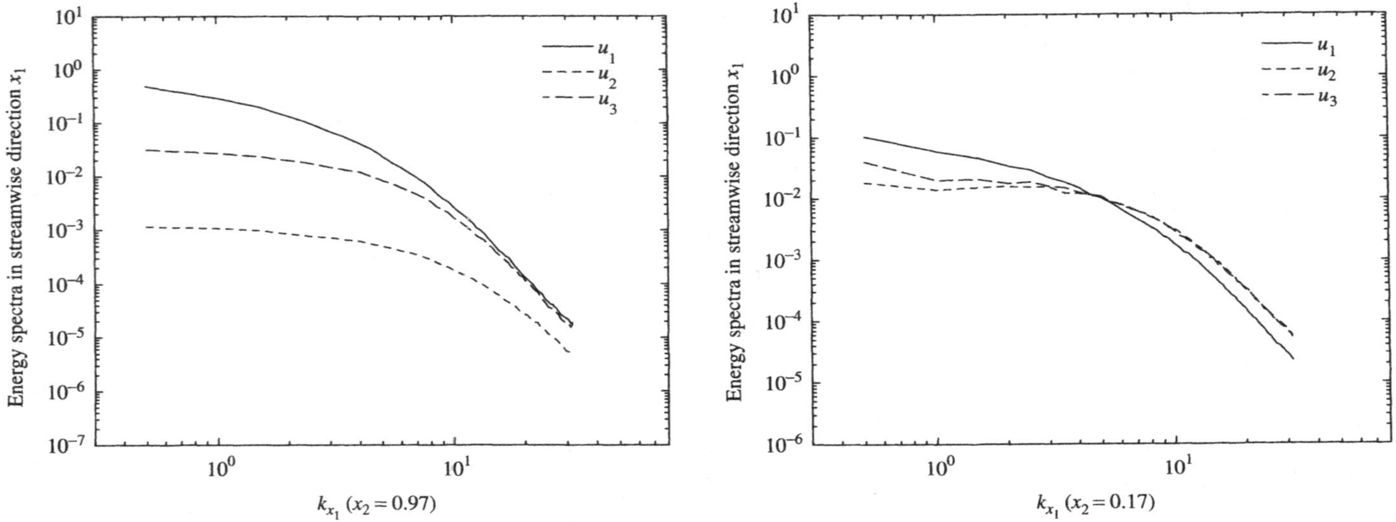

Figure 8.7. One-dimensional energy spectrum functions in the streamwise direction x_1 for each component of the velocity $u_i, i = 1, 2, 3$, and for two values of $x_2 : x_2 = 0.97$ (near the upper wall) and $x_2 = 0.17$ (near the centerline of the channel).

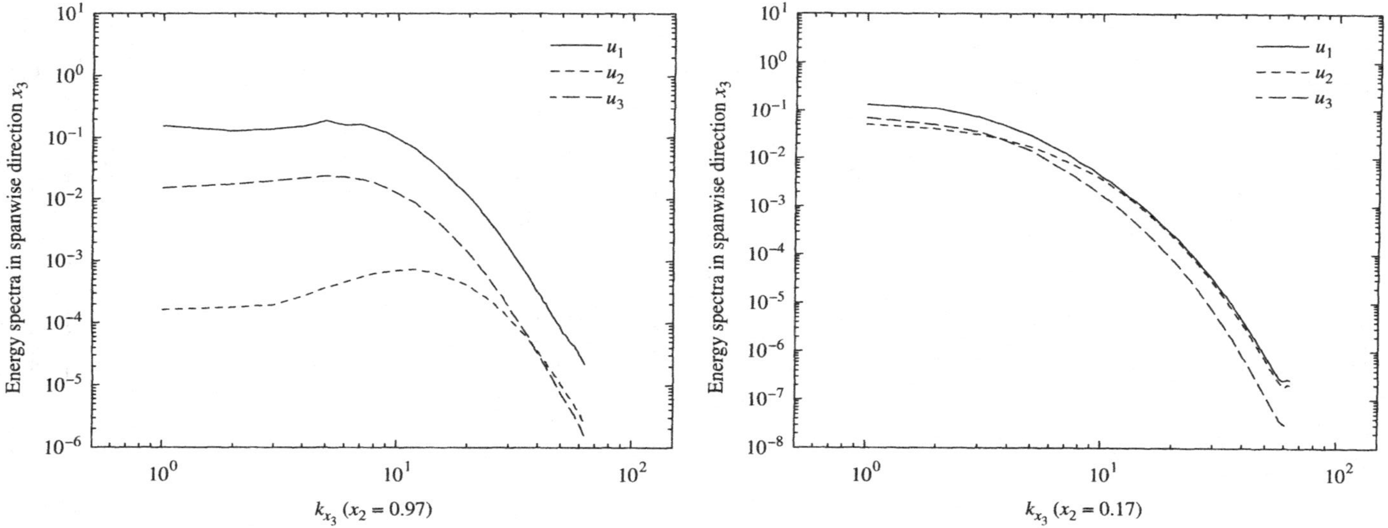

Figure 8.8. One-dimensional energy spectrum functions in the spanwise direction x_3 for each component of the velocity u_i, $i=1, 2, 3$, and for two values of $x_2 : x_2=0.97$ (near the upper wall) and $x_2=0.17$ (near the centerline of the channel).

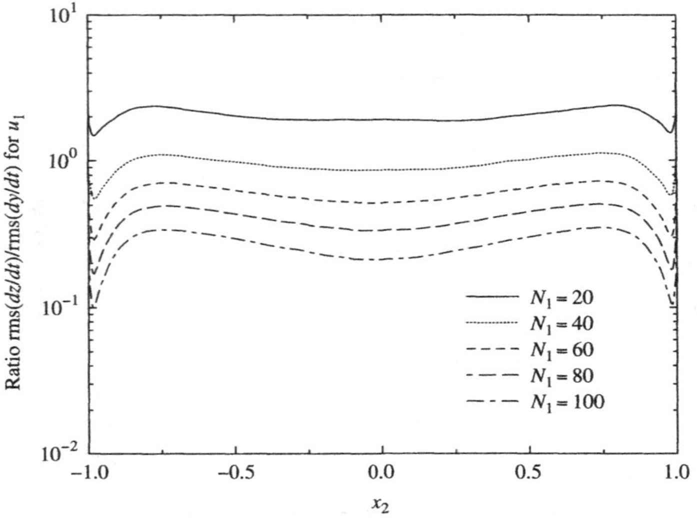

10^1
10^0
10^{-1}
10^{-2}
Ratio rms(dz/dt)/rms(dy/dt) for u_1
−1.0
−0.5
0.0
0.5
1.0
x_2
$N_1 = 20$
$N_1 = 40$
$N_1 = 60$
$N_1 = 80$
$N_1 = 100$

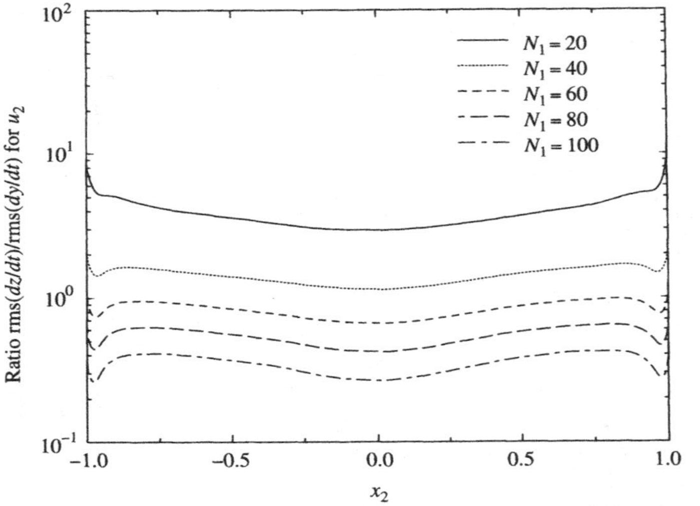

10^2
10^1
10^0
10^{-1}
Ratio rms(dz/dt)/rms(dy/dt) for u_2
−1.0
−0.5
0.0
0.5
1.0
x_2
$N_1 = 20$
$N_1 = 40$
$N_1 = 60$
$N_1 = 80$
$N_1 = 100$

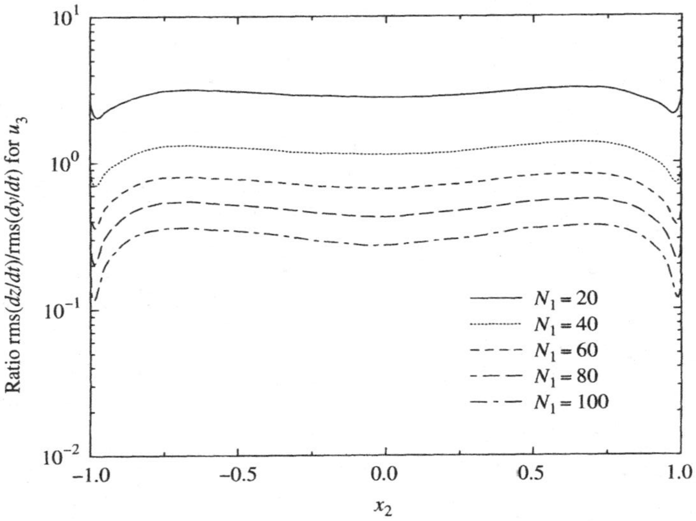

Figure 8.9. Ratio $\langle|\dot{z}^N_{i,N_1}|^2\rangle^{1/2}/\langle|\dot{y}_{i,N_1}|^2\rangle^{1/2}$ as a function of x_2, for each component u_i of the velocity, $i = 1, 2, 3$.

increase. For the spectra in the spanwise direction, near the centerline of the channel they decrease, but near the wall they increase (especially for u_2) until $k_{x_3} \in [8, 15]$, after which they decrease. This confirms the fact that the small scales, associated with high wavenumbers, are less dominating in the kinetic energy of the velocity field than the large scales associated with small wavenumbers.

Now we compare the time evolution of the small and large scales. In Figure 8.9, we have represented the ratio $\langle |\dot{z}^N_{i,N_1}|^2\rangle^{1/2} / \langle |\dot{y}_{i,N_1}|^2\rangle^{1/2}$ for the three components of the velocity field ($i = 1, 2, 3$). As previously, from Parseval's identity (8.5) and from the definitions of $\dot{\mathbf{y}}_{N_1}$ and $\dot{\mathbf{z}}^N_{N_1}$, we see that this ratio is a decreasing function of N_1. Moreover, this ratio becomes less than unity as soon as the cutoff level N_1 is greater than 60 for u_2 and 40 for u_1 and u_3.

In order to refine the previous observations on the time derivatives, we consider in Figure 8.10 (Figure 8.11) the one-dimensional energy spectra in the streamwise (spanwise) direction for each component of the time derivative of the velocity field $\dot{\mathbf{u}}_N$ and for three values of x_2. Firstly, we remark that the streamwise spectra increase with the wavenumber, in particular for u_2 and for x_2 far from the walls, until $k_{x_1} = 10$. Then they decrease. However, the spanwise spectra decrease when the wavenumbers are increased, especially near the centerline of the channel. So the behavior of the streamwise and of the spanwise spectra, for the time derivative, as functions of the wavenumbers and of x_2, are quite different. The spectral analysis in the direction x_1 shows that the Fourier coefficients associated with high wavenumbers k_{x_1} (small scales) have greater time derivatives, in particular near the middle of the channel, than the Fourier coefficients associated with high wave-numbers k_{x_3}.

Finally, as has been said before, the ratios, as functions of x_2, $\langle |z^N_{i,N_1}|^2\rangle^{1/2} / \langle |y_{i,N_1}|^2\rangle^{1/2}$ and $\langle |\dot{z}^N_{i,N_1}|^2\rangle^{1/2} / \langle |\dot{y}_{i,N_1}|^2\rangle^{1/2}$ become less than one for cutoff values greater than a minimum value. So, as for the homogeneous turbulence, it is necessary to control the error due to the quasistatic approximation on the small scales in order to estimate the lower values of N_1. An estimation of this error will be given in Chapter 9.

8.2. Multilevel Schemes

8.2.1. Periodic Flows

A Two-Level Scheme

We first consider the case of a two-level scheme defined by a fine level that corresponds to the fine grid (i.e. $k \leq k_N$) and a coarse level (i.e. $k \leq k_{N_1}$

($< k_N$)). The velocity field $\mathbf{u}_N$ is then decomposed as previously (see Section 8.1.1) into

$$\mathbf{u}_N = \mathbf{y}_{N_1} + \mathbf{z}_{N_1}^N. \tag{8.37}$$

The velocity components $(\mathbf{y}_{N_1}, \mathbf{z}_{N_1}^N)$ satisfy the following coupled system of ordinary differential equations (see (6.18), (6.20)):

$$\begin{aligned}\frac{\partial}{\partial t}\left(e^{\nu A t}\mathbf{y}_{N_1}(t)\right) &= e^{\nu A t}\left[P_{N_1}\mathbf{g}_N - P_{N_1}B\left(\mathbf{y}_{N_1}, \mathbf{y}_{N_1}\right)\right] \qquad (8.38)\\ &\quad - e^{\nu A t}P_{N_1}B_{\text{int}}\left(\mathbf{y}_{N_1},\ \mathbf{z}_{N_1}^N\right),\\ \frac{\partial}{\partial t}\left(e^{\nu A t}\mathbf{z}_{N1}^N(t)\right) &= -e^{\nu A t}Q_{N_1}^N B\left(\mathbf{y}_{N_1} + \mathbf{z}_{N_1}^N,\ \mathbf{y}_{N_1} + \mathbf{z}_{N_1}^N\right), \qquad (8.39)\end{aligned}$$

where A is the Stokes operator (sec Chapter 1). For the sake of simplicity, since in all the periodic numerical simulations presented in Chapter 10 the external force acts only on the first modes $k = 1$ (or 2) of the Fourier decomposition, we have assumed in (8.39) that $Q_{N_1}^N\mathbf{g}_N(t)$=0. Note that the case $Q_{N_1}^N\mathbf{g}_N(t) \neq 0$ can be considered as well (see for instance Debussche, Dubois, and Temam (1995)).

As was previously shown (see Section 8.1.1) the time variations over one time step of the small-scale components in the decomposition (8.37) are small compared to $|\mathbf{z}_{N_1}^N|_0$ and then much smaller when compared to $|\mathbf{y}_{N_1}|_0$. Hence, instead of computing $\mathbf{z}_{N_1}^N$ at each time iteration as in a full Galerkin method (see Section 3.1), we propose to locally neglect these time variations on time intervals of the form $(t_0, t_0 + \Delta t)$, where t_0 is any given time. We then approximate $\mathbf{z}_{N_1}^N(t_0 + \Delta t)$ by the value of these scales at time t_0, that is, we write

$$\mathbf{z}_{N_1}^N(t_0 + \Delta t) \simeq \mathbf{z}_{N_1}^N(t_0). \tag{8.40}$$

The approximation (8.40) corresponds to a quasistatic approximation on the time interval $(t_0, t_0 + \Delta t)$.

We now define a similar treatment for the nonlinear interaction term $P_{N_1}B_{\text{int}}(\mathbf{y}_{N_1}, \mathbf{z}_{N_1}^N)$ appearing in the right-hand side of (8.38). When the large-scale equation is integrated over the interval $(t_0, t_0 + \Delta t)$, a term coming from the right-hand side of (8.38) appears:

$$I(\Delta t, N_1) = \int_{t_0}^{t_0+\Delta t} e^{\nu A(s-t_0-\Delta t)}P_{N_1}B_{\text{int}}\left(\mathbf{y}_{N_1}(s),\ \mathbf{z}_{N_1}^N(s)\right)ds. \tag{8.41}$$

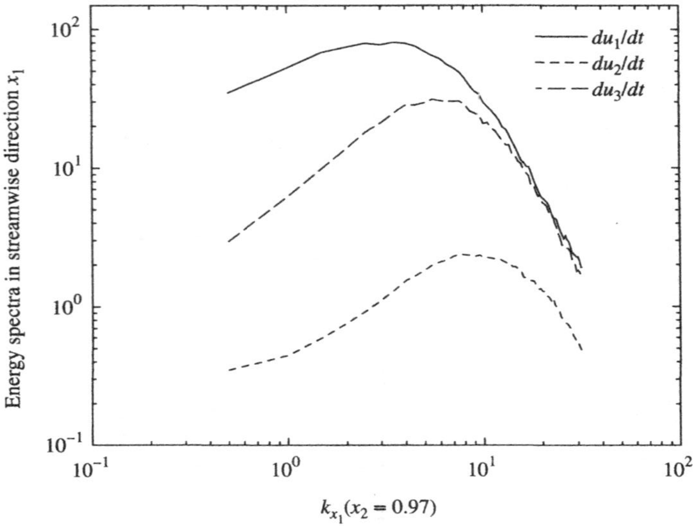

10^2
10^1
10^0
10^{-1}
10^{-1}
10^0
10^1
10^2
Energy spectra in streamwise direction x_1
$k_{x_1}(x_2 = 0.97)$
du_1/dt
du_2/dt
du_3/dt

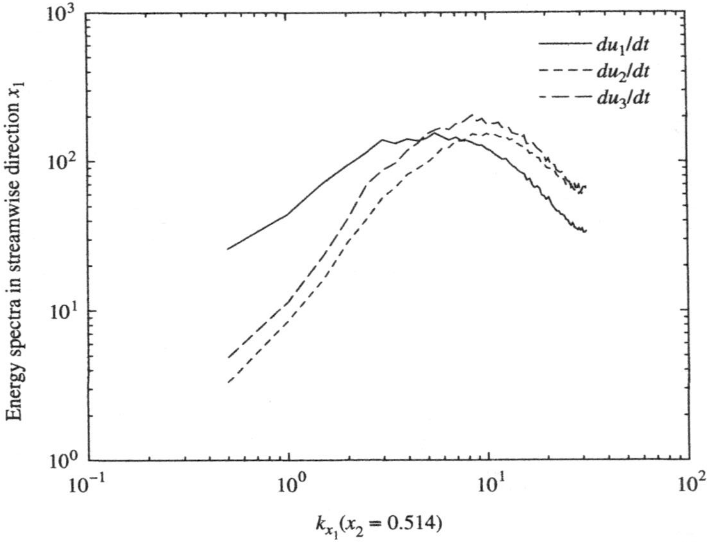

10^3
10^2
10^1
10^0
10^{-1}
10^0
10^1
10^2
Energy spectra in streamwise direction x_1
$k_{x_1}(x_2 = 0.514)$
du_1/dt
du_2/dt
du_3/dt

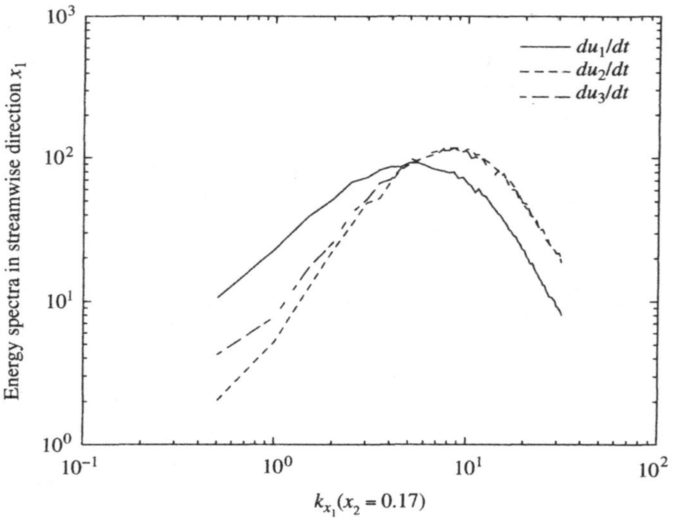

Figure 8.10. One-dimensional energy spectrum functions in the streamwise direction x_1 for each component of the time derivative velocity $\partial u_i/\partial t$, $i=1, 2, 3$, and for three values of x_2: $x_2=0.97$ (near the upper wall), $x_2=0.514$, and $x_2=0.17$ (near the centerline of the channel).

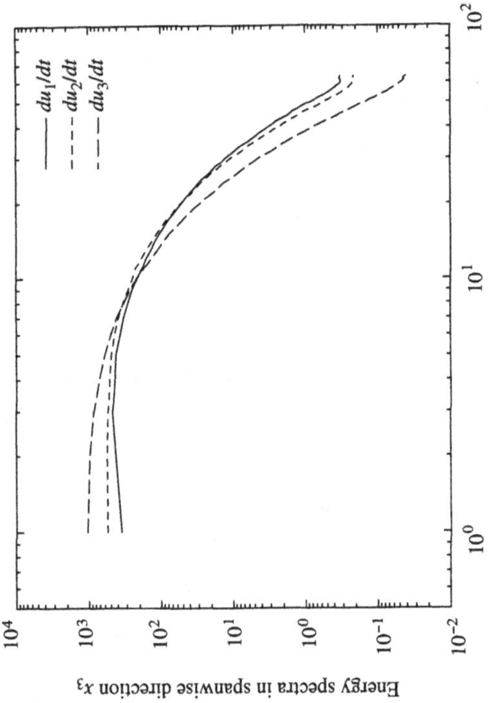
Energy spectra in spanwise direction x_3
$k_{x_3}(x_2 = 0.514)$
du_1/dt
du_2/dt
du_3/dt

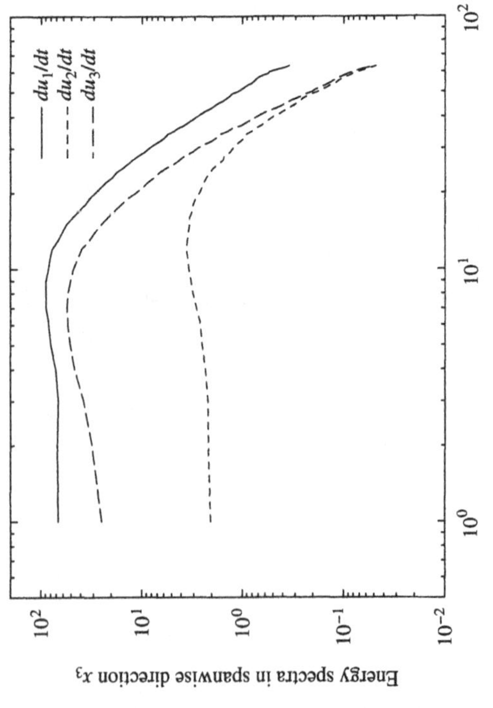
Energy spectra in spanwise direction x_3
$k_{x_3}(x_2 = 0.97)$
du_1/dt
du_2/dt
du_3/dt

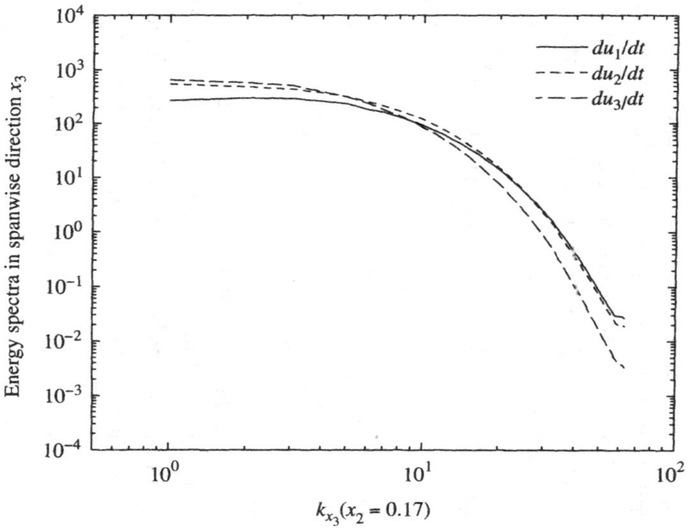

Figure 8.11. One-dimensional energy spectrum functions in the spanwise direction x_3 for each component of the time derivative velocity $\partial u_i/\partial t$, $i=1, 2, 3$ and for three values of x_2: $x_2 = 0.97$ (near the upper wall), $x_2 = 0.514$, and $x_2 = 0.17$ (near the centerline of the channel).

Recalling that the time variations of the interaction term are small (in $|\cdot|_0$) compared to $P_{N_1} B(\mathbf{y}_{N_1}, \mathbf{y}_{N_1})$, the above integral $I(\Delta t, N_1)$ can be rewritten as

$$I(\Delta t, N_1) = \int_{t_0}^{t_0+\Delta t} e^{\nu A(s-t_0-\Delta t)} P_{N_1} B_{\text{int}}\big(\mathbf{y}_{N_1}(t_0), \mathbf{z}_{N_1}^N(t_0)\big)\, ds \qquad (8.42)$$
$$+ \int_{t_0}^{t_0+\Delta t} e^{\nu A(s-t_0-\Delta t)} \big[P_{N_1} B_{\text{int}}\big(\mathbf{y}_{N_1}(s), \mathbf{z}_{N_1}^N(s)\big)$$
$$- P_{N_1} B_{\text{int}}\big(\mathbf{y}_{N_1}(t_0), \mathbf{z}_{N_1}^N(t_0)\big)\big]\, ds.$$

The second integral in the right-hand side of (8.42) depends on the time variations of the term $P_{N_1} B_{\text{int}}(\mathbf{y}_{N_1}, \mathbf{z}_{N_1}^N)$ on the interval $(t_0, t_0 + \Delta t)$ and is then small compared to the first term. The former term is neglected when a quasistatic approximation similar to (8.40) is performed on $P_{N_1} B_{\text{int}}(\mathbf{y}_{N_1}(t), \mathbf{z}_{N_1}^N(t))$, that is, when we assume that

$$P_{N_1} B_{\text{int}}\big(\mathbf{y}_{N_1}(s), \mathbf{z}_{N_1}^N(s)\big) \simeq P_{N_1} B_{\text{int}}\big(\mathbf{y}_{N_1}(t_0), \mathbf{z}_{N_1}^N(t_0)\big) \qquad (8.43)$$
$$\text{for all} \quad s \in (t_0, t_0 + \Delta t).$$

Hence, $I(\Delta t, N_1)$ is approximated by

$$I(\Delta t, N_1) \simeq e^{-\nu A\, \Delta t} P_{N_1} B_{\text{int}}\big(\mathbf{y}_{N_1}(t_0), \mathbf{z}_{N_1}^N(t_0)\big) \int_0^{\Delta t} e^{\nu A s}\, ds.$$

In order to implement (8.40) and (8.43) we have developed an adaptive scheme based on multigrid like V-cycles, that is, the number of unknowns which have to be integrated and hence the cutoff level vary in time. This scheme consists in computing a sequence of pairs $\{\mathbf{y}_{N_1}^m, \mathbf{z}_{N_1}^{N,m}\}$, $m \geq 0$, satisfying

$$\mathbf{y}_{N_1}^0 = P_{N_1} \mathbf{u}_N(t = 0), \qquad \mathbf{z}_{N_1}^{N,0} = Q_{N_1}^N \mathbf{u}_N(t = 0).$$

We write $t_m = m\, \Delta t$, and for the sake of simplicity we set $\mathbf{z}_1^m = \mathbf{z}_{N_1}^{N,m}$. We first assume that the whole time integration is divided into two-level V-cycles. The V-cycle j, $j \geq 0$, is constituted by two successive steps and corresponds to the time interval (t_{2j}, t_{2j+2}):

Step 1: $\mathbf{y}_{N1}^{2j+1}$ is computed by integrating (8.38) over the time interval (t_{2j}, t_{2j+1}), that is, we compute

$$\mathbf{y}_{N_1}^{2j+1} = e^{-\nu A\, \Delta t} \mathbf{y}_{N_1}^{2j} + \Delta t\, e^{-\nu A\, \Delta t} \Big[P_{N_1} \mathbf{g}_N^{2j} - P_{N_1} B\big(\mathbf{y}_{N_1}^{2j}, \mathbf{y}_{N_1}^{2j}\big)\Big] \qquad (8.44)$$
$$- \Delta t\, e^{-\nu A \Delta t} P_{N_1} B_{\text{int}}\big(\mathbf{y}_{N_1}^{2j}, \mathbf{z}_1^{2j}\big).$$

A quasistatic approximation on the interval (t_{2j}, t_{2j+1}) is used for the small scales, that is, we write

$$\mathbf{z}_1^{2j+1} = \mathbf{z}_1^{2j}. \tag{8.45}$$

Note that all integrals appearing in the right-hand side of (8.44) are approximated by an explicit scheme (see Section 3.1). For the sake of simplicity we have chosen an Euler scheme; (8.44) can be generalized as well to any other explicit scheme of higher order, such as the Runge–Kutta schemes. In practice, an explicit Runge–Kutta method of third order is used.

Step 2: $\{\mathbf{y}_{N_1}^{2j+2}, \mathbf{z}_1^{2j+2}\}$ are computed by integrating the full system (8.38), (8.39) on the time interval (t_{2j+1}, t_{2j+2}), leading to

$$\begin{aligned} \mathbf{y}_{N1}^{2j+2} &= e^{-\nu A\,\Delta t}\mathbf{y}_{N_1}^{2j+1} + \Delta t e^{-\nu A\,\Delta t} \\ &\quad\times \left[P_{N_1}\mathbf{g}_N^{2j+1} - P_{N_1}B\left(\mathbf{y}_{N_1}^{2j+1}, \mathbf{y}_{N_1}^{2j+1}\right)\right] \\ &\quad - \Delta t\, e^{-\nu A\,\Delta t} P_{N_1} B_{\text{int}}\left(\mathbf{y}_{N_1}^{2j}, \mathbf{z}_1^{2j}\right), \end{aligned} \tag{8.46}$$

$$\begin{aligned} \mathbf{z}_1^{2j+2} &= e^{-\nu A\,\Delta t}\mathbf{z}_1^{2j+1} - \Delta t\, e^{-\nu A\,\Delta t} \\ &\quad\times Q_{N_1}^N B\left(\mathbf{y}_{N_1}^{2j+1} + \mathbf{z}_1^{2j+1}, \mathbf{y}_{N_1}^{2j+1} + \mathbf{z}_1^{2j+1}\right) \end{aligned} \tag{8.47}$$

Note that according to (8.45), $\mathbf{z}_1^{2j+1}$ has to be replaced by $\mathbf{z}_1^{2j}$ in the right-hand side of (8.47).

During steps 1 and 2, the nonlinear interaction term in the large-scale equation is kept fixed, equal to its value at the time iteration t_{2j}. So $P_{N_1} B_{\text{int}}(\mathbf{y}_{N_1}, \mathbf{z}_{N_1}^N)$ has to be computed only once at the beginning of the V-cycle and has to be stored. The computation of the nonlinear interaction term at time t_{2j} is achieved as follows:

$$P_{N_1} B_{\text{int}}\left(\mathbf{y}_{N_1}^{2j}, \mathbf{z}_1^{2j}\right) = P_{N_1} B\left(\mathbf{y}_{N_1}^{2j} + \mathbf{z}_1^{2j}, \mathbf{y}_{N_1}^{2j} + \mathbf{z}_1^{2j}\right) - P_{N_1} B\left(\mathbf{y}_{N_1}^{2j}, \mathbf{y}_{N_1}^{2j}\right),$$

which requires $O(N^n \log_2 N + N_1^n \log_2 N_1)$ operations.

Hence, the small scales are kept constant during one time step, while the nonlinear interaction term is kept constant during the whole V-cycle, which corresponds to two time steps. In fact, we have noted that the error committed by using the quasistatic approximation on the interaction term is much smaller than the error corresponding to the small scales (see Section 9.3). As a consequence, the time variations of $P_{N_1} B_{\text{int}}(\mathbf{y}_{N_1}, \mathbf{z}_{N_1}^N)$ can be neglected on time intervals longer than $2\,\Delta t$. We have then defined cycles constituted by a succession of V-cycles. These cycles correspond to $2n_V$ iterations, where n_V is the number of V-cycles performed. Then, a cycle j $(j \geq 1)$ is constituted by the iterations $2(j-1)n_V+1$ to $2jn_V$, which correspond to the integration of the system (8.38), (8.39) on the time interval $(t_{2(j-1)n_V}, t_{2jn_V})$.

Each iteration $2(j-1)n_V+(2i+1)$, for $i \geq 0$, corresponds to the computation of the large-scale components $\mathbf{y}_{N_1}$ while the iterations $2(j-1)n_V+2i$, for $i \geq 1$, correspond to an evaluation of the large scales and an update of $\mathbf{z}_{N_1}^N$. During the whole cycle j, that is, during $2n_V$ iterations, the nonlinear interaction term is kept constant and equal to its value at the beginning of the cycle, that is,

$$P_{N_1} B_{\text{int}}\left(\mathbf{y}_{N_1}^{2(j-1)n_V+i}, \mathbf{z}_1^{2(j-1)n_V+i}\right) = P_{N_1} B_{\text{int}}\left(\mathbf{y}_{N_1}^{2(j-1)n_V}, \mathbf{z}_1^{2(j-1)n_V}\right), \quad \text{for } i=1,\ldots,2n_V. \tag{8.48}$$

Hence, when i is even, the number of operations needed to perform the iteration $2(j-1)n_V+i$ is $O(sN^n \log_2 N)$, while $O(sN_1^n \log_2 N_1)$ operations are needed to perform iteration $2(j-1)n_V+i$ when i is odd. As in Section 3.1, s denotes the number of sub-time-steps of the one-step explicit scheme used for the treatment of the nonlinearity. The total number of operations required to perform cycle j is $O((sn_V+1)N^n \log_2 N + sn_V N_1^n \log_2 N_1)$, while $O(2sn_V N^n \log_2 N)$ operations were needed for advancing a DNS on the same time interval. The reduction of the number of operations when a two-level DML is used can then be estimated by

$$\begin{aligned} \mathrm{R}(N_1) &= \frac{(sn_V-1)N^n \log_2 N - sn_V N_1^n \log_2 N_1}{2sn_V N^n \log_2 N} \\ &= 0.5\left(\frac{sn_V-1}{sn_V}\right) - 0.5\left(\frac{N_1}{N}\right)^n \frac{\log_2 N_1}{\log_2 N}. \end{aligned}$$

As an example, if we set $N_1 = N/2$, we obtain

$$\mathrm{R}(N/2) \simeq 0.5\left(\frac{sn_V-1}{sn_V} - \frac{1}{2^n}\right).$$

Hence, in the case $n=3$, $\mathrm{R}(N/2) \leq 0.4378$.

Generalization: A Multilevel Scheme with p Levels

According to (8.29), the variation over one time step of the small scales is a decreasing function of the cutoff level. Hence, if the variations of $\mathbf{z}_{N_1}^N$ can be neglected over two time steps, as is done in the preceding two-level scheme, then the variations of $\mathbf{z}_{N_2}^N$, for $N > N_2 > N_1$, can be neglected over a time interval larger than two time steps, and so on for other successive cutoff levels lying between N_1 and N. In order to take into account this behavior of the small scales, we have generalized the two-level scheme to a p-level scheme that we now describe.

We consider a sequence of p cutoff levels $\{N_i\}_{i=1,p}$ satisfying $N_{i+1} > N_i$, for $i = 1, \ldots, p-1$, and $N_p = N$. We decompose $\mathbf{u}_N$ into

$$\mathbf{u}_N = \mathbf{y}_{N_1} + \sum_{i=1}^{p-1} \mathbf{z}_{N_i}^{N_{i+1}}. \tag{8.49}$$

The velocity components $\{\mathbf{y}_{N_1}, \mathbf{z}_{N_i}^{N_{i+1}}\}_{i=1,p-1}$ satisfy the coupled system of ordinary differential equations

$$\frac{\partial}{\partial t}\left(e^{\nu A t}\mathbf{y}_{N_1}(t)\right) = e^{\nu A t}\left[P_{N_1}\mathbf{g}_N - P_{N_1}B\left(\mathbf{y}_{N_1}, \mathbf{y}_{N_1}\right)\right] \tag{8.50}$$

$$- e^{\nu A t} P_{N_1} B_{\text{int}}\left(\mathbf{y}_{N_1}, \sum_{i=1}^{p-1} \mathbf{z}_{N_i}^{N_{i+1}}\right),$$

$$\frac{\partial}{\partial t}\left(e^{\nu A t}\mathbf{z}_{N_i}^{N_{i+1}}(t)\right) \tag{8.51$_i$}$$

$$= -e^{\nu A t} Q_{N_i}^{N_{i+1}} B\left(\mathbf{y}_{N_1} + \sum_{i=1}^{p-1} \mathbf{z}_{N_i}^{N_{i+1}}, \mathbf{y}_{N_1} + \sum_{i=1}^{p-1} \mathbf{z}_{N_i}^{N_{i+1}}\right),$$

where $i = 1, \ldots, p-1$ and $Q_{N_i}^{N_{i+1}} = P_{N_{i+1}} - P_{N_i}$. By summing equations (8.50) and (8.51)$_i$ for $i=1, \ldots, k$ where $k \leq p-1$, and rewriting the nonlinear term, we deduce that $\mathbf{y}_{N_1} + \sum_{i=1}^{k} \mathbf{z}_{N_i}^{N_{i+1}}$ satisfies the following equation:

$$\frac{\partial}{\partial t}\left[e^{\nu A t}\left(\mathbf{y}_{N_1} + \sum_{i=1}^{k} \mathbf{z}_{N_i}^{N_{i+1}}\right)(t)\right] \tag{8.52$_k$}$$

$$= e^{\nu A t} P_{N_{k+1}}\mathbf{g}_N - e^{\nu A t} P_{N_{k+1}} B\left(\mathbf{y}_{N_1} + \sum_{i=1}^{k} \mathbf{z}_{N_i}^{N_{i+1}}, \mathbf{y}_{N_1} + \sum_{i=1}^{k} \mathbf{z}_{N_i}^{N_{i+1}}\right)$$

$$- e^{\nu A t} P_{N_{k+1}} B_{\text{int}}\left(\mathbf{y}_{N_1} + \sum_{i=1}^{k} \mathbf{z}_{N_i}^{N_{i+1}}, \sum_{i=k+1}^{p-1} \mathbf{z}_{N_i}^{N_{i+1}}\right).$$

The most natural way to implement the property mentioned above of the length of the quasistatic approximation is to generalize the two-level V-cycles previously introduced to p-level V-cycles. The p-level scheme consists in computing a sequence $\{\mathbf{y}_{N_1}^m, \mathbf{z}_{N_i}^{N_{i+1},m}\}_{i=1,p-1}$, $m \geq 0$, satisfying

$$\mathbf{y}_{N_1}^0 = P_{N_1}\mathbf{u}_N(t=0),$$

$$\mathbf{z}_{N_i}^{N_{i+1},0} = Q_{N_i}^{N_{i+1}}\mathbf{u}_N(t=0) \qquad \text{for} \quad i = 1, \ldots, p-1.$$

For the sake of simplicity, we set $\mathbf{z}_i^m = \mathbf{z}_{N_i}^{N_{i+1},m}$ and $q = p-1$. As in the two-level case, we first consider that a succession of p-level V-cycles are performed to advance the flow field computation. A p-level V-cycle j, $j \geq 0$, is constituted by $2q$ successive steps that we describe hereafter and corresponds to the time interval $(t_{2jq}, t_{(2j+2)q})$:

Step l, $1 \leq l \leq q-1$: The components $\{\mathbf{y}_{N_1}^{2jq+l},\ \mathbf{z}_i^{2jq+l}\}_{i=1,q-l}$ are computed by integrating $(8.52)_{q-l}$ over the time interval $(t_{2jq+l-1}, t_{2jq+l})$, that is, we compute

$$
\begin{aligned}
&\mathbf{y}_{N_1}^{2jq+l} + \sum_{i=1}^{q-l} \mathbf{z}_i^{2jq+l} \qquad (8.53)_{q-l}\\
&\quad = e^{-\nu A\,\Delta t}\left(\mathbf{y}_{N_1}^{2jq+l-1} + \sum_{i=1}^{q-l} \mathbf{z}_i^{2jq+l-1}\right)\\
&\qquad + \Delta t\, e^{-\nu A\,\Delta t}\, P_{N_{q-l+1}} \mathbf{g}_N^{2jq+l-1} - \Delta t\, e^{-\nu A\,\Delta t}\\
&\qquad \times P_{N_{q-l+1}} B\left(\mathbf{y}_{N_1}^{2jq+l-1} + \sum_{i=1}^{q-l} \mathbf{z}_i^{2jq+l-1},\ \mathbf{y}_{N_1}^{2jq+l-1} + \sum_{i=1}^{q-l} \mathbf{z}_i^{2jq+l-1}\right)\\
&\qquad - \Delta t\, e^{-\nu A\,\Delta t}\, P_{N_{q-l+1}} B_{\text{int}}\left(\mathbf{y}_{N_1}^{2jq} + \sum_{i=1}^{q-l} \mathbf{z}_i^{2jq},\ \sum_{i=q-l+1}^{q} \mathbf{z}_i^{2jq}\right).
\end{aligned}
$$

A quasistatic approximation on the time interval $(t_{2jq+l-1}, t_{2jq+l})$ is applied to the small scales:

$$
\mathbf{z}_i^{2jq+l} = \mathbf{z}_i^{2jq+l-1} \qquad \text{for} \quad i = q-l+1, \ldots, q. \tag{8.54}
$$

Step q: The large-scale components $\mathbf{y}_{N_1}^{(2j+1)q}$ are integrated over the time interval $(t_{(2j+1)q-1}, t_{(2j+1)q})$, that is, we compute

$$
\begin{aligned}
\mathbf{y}_{N_1}^{(2j+1)q} &= e^{-\nu A\,\Delta t}\, \mathbf{y}_{N_1}^{(2j+1)q-1} \qquad (8.55)\\
&\quad + \Delta t\, e^{-\nu A\,\Delta t}\left[P_{N_1} \mathbf{g}_N^{(2j+1)q-1} - P_{N_1} B\left(\mathbf{y}_{N_1}^{(2j+1)q-1}, \mathbf{y}_{N_1}^{(2j+1)q-1}\right)\right]\\
&\quad - \Delta t\, e^{-\nu A\,\Delta t}\, P_{N_1} B_{\text{int}}\left(\mathbf{y}_{N_1}^{2jq}, \sum_{i=1}^{q} \mathbf{z}_i^{2jq}\right).
\end{aligned}
$$

A quasistatic approximation on the time interval $(t_{(2j+1)q-1}, t_{(2j+1)q})$ is applied to the small scales, that is,

$$
\mathbf{z}_i^{(2j+1)q} = \mathbf{z}_i^{(2j+1)q-1} \qquad \text{for} \quad i = 1, \ldots, q. \tag{8.56}
$$

Step l, $q+1 \leq l \leq 2q-1$: The components $\{\mathbf{y}_{N_1}^{2jq+l}, \mathbf{z}_i^{2jq+l}\}_{i=1,l-q}$ are computed by integrating $(8.52)_{l-q}$ over the time interval $(t_{2jq+l-1}, t_{2jq+l})$, that is, they satisfy $(8.53)_{l-q}$, and a quasistatic approximation is applied to the small scales:

$$\mathbf{z}_i^{2jq+l} = \mathbf{z}_i^{2jq+l-1}, \qquad \text{for} \quad i = l-q+1, \ldots, q. \tag{8.57}$$

Note that with such treatment of the small scales, we have in the right-hand side of $(8.53)_{l-q}$

$$\mathbf{z}_{l-q}^{2jq+l-1} = \mathbf{z}_{l-q}^{(2j+2)q-l}.$$

Step $2q$: This last step consists in computing all the velocity components $\{\mathbf{y}_{N_1}^{(2j+2)q}, \mathbf{z}_i^{(2j+2)q}\}_{i=1,q}$ by integrating $(8.52)_q$ on the time interval $(t_{2(j+1)q-1}, t_{2(j+1)q})$, that is,

$$\begin{aligned}
&\mathbf{y}_{N_1}^{(2j+2)q} + \sum_{i=1}^{q} \mathbf{z}_i^{(2j+2)q} \\
&= e^{-\nu A\,\Delta t}\left(\mathbf{y}_{N_1}^{(2j+2)q-1} + \sum_{i=1}^{q} \mathbf{z}_i^{(2j+2)q-1}\right) \\
&\quad + \Delta t\, e^{-\nu A\,\Delta t} P_N \mathbf{g}_N^{(2j+2)q-1} - \Delta t\, e^{-\nu A\,\Delta t} \\
&\quad \times P_N B\left(\mathbf{y}_{N_1}^{(2j+2)q-1} + \sum_{i=1}^{q} \mathbf{z}_i^{(2j+2)q-1}, \mathbf{y}_{N_1}^{(2j+2)q-1} + \sum_{i=1}^{q} \mathbf{z}_i^{(2j+2)q-1}\right).
\end{aligned} \tag{8.58}$$

The treatments (8.54) and (8.56) of the small scales imply that

$$\mathbf{z}_i^{(2j+1)q} = \mathbf{z}_i^{(2j+1)q-i} \qquad \text{for} \quad i=1, \ldots, q.$$

Moreover, the treatment (8.57) implies that

$$\mathbf{z}_i^{(2j+1)q+i-1} = \mathbf{z}_i^{(2j+1)q} \qquad \text{for} \quad i=1, \ldots, q.$$

We finally obtain

$$\mathbf{z}_i^{(2j+1)q+i-1} = \mathbf{z}_i^{(2j+1)q-i} \qquad \text{for} \quad i=1, \ldots, q,$$

so that the scales $\mathbf{z}_{N_i}^{N_{i+1}}$ are kept constant on a time interval of length $(2i-1)\,\Delta t$ for $i = 1, \ldots, p-1$. Hence, with such a scheme, the length of the time interval of the quasistatic approximation of the small scales increases when the cutoff level increases.

One can argue that this numerical treatment is in contradiction with the Kolmogorov theory (see Section 2.2.4), which assumes that the characteristic time scale $\tau(k)$ is a decreasing function of k. However, we can expect that $(2p-3)\,\Delta t$, which is the largest time length of the quasistatic approximation, remains smaller than $\tau_d \simeq \mathrm{Re}_L^{1/4}\,\Delta t$ (see (8.32)).

As a direct consequence of the above scheme, the nonlinear interaction terms in $(8.53)_l$, for $l=-1,0,\ldots,q-1$, that is, $P_{N_l} B_{\mathrm{int}}(\mathbf{y}_{N_1}+\sum_{i=1}^{l-1}\mathbf{z}_{N_i}^{N_{i+1}}, \sum_{i=l}^{q}\mathbf{z}_{N_i}^{N_{i+1}})$, are kept constant and equal to their values at time t_{2jq} during the whole V-cycle:

$$
\begin{aligned}
P_{N_l} B_{\mathrm{int}} &\left(\mathbf{y}_{N_1}^{2jq+k} + \sum_{i=1}^{l-1}\mathbf{z}_i^{2jq+k}, \sum_{i=l}^{q}\mathbf{z}_i^{2jq+k}\right) \\
&= P_{N_l} B_{\mathrm{int}} \left(\mathbf{y}_{N_1}^{2jq} + \sum_{i=1}^{l-1}\mathbf{z}_i^{2jq}, \sum_{i=l}^{q}\mathbf{z}_i^{2jq}\right),
\end{aligned}
$$

for $k=1,\ldots,2q$. Note that the term $\sum_{i=1}^{l-1}\mathbf{z}_i$ vanishes for $l=1$. Hence, the nonlinear terms

$$
\left\{P_{N_l} B_{\mathrm{int}} \left(\mathbf{y}_{N_1}^{2jq} + \sum_{i=1}^{l-1}\mathbf{z}_i^{2jq}, \sum_{i=l}^{q}\mathbf{z}_i^{2jq}\right)\right\}_{l=1,q}
$$

have to be computed and stored at the beginning of the V-cycle. Their computation is done as follows:

$$
\begin{aligned}
P_{N_l} B_{\mathrm{int}} &\left(\mathbf{y}_{N_1}^{2jq} + \sum_{i=1}^{l-1}\mathbf{z}_i^{2jq}, \sum_{i=l}^{q}\mathbf{z}_i^{2jq}\right) \\
&= P_{N_l} B \left(\mathbf{y}_{N_1}^{2jq} + \sum_{i=1}^{q}\mathbf{z}_i^{2jq}, \mathbf{y}_{N_1}^{2jq} + \sum_{i=1}^{q}\mathbf{z}_i^{2jq}\right) \\
&\quad - P_{N_l} B \left(\mathbf{y}_{N_1}^{2jq} + \sum_{i=1}^{l-1}\mathbf{z}_i^{2jq}, \mathbf{y}_{N_1}^{2jq} + \sum_{i=1}^{l-1}\mathbf{z}_i^{2jq}\right),
\end{aligned}
$$

which represents an overhead of $O(N^n \log_2 N + \sum_{i=1}^{q} N_i^n \log_2 N_i)$ operations.

As in the two-level case, we define cycles constituted by several p-level V-cycles. Then, a cycle j ($j \geq 1$) corresponds to the time interval $(t_{2q(j-1)n_V}, t_{2qjn_V})$, where n_V is the number of V-cycles performed. During these $2qn_V$ iterations, the nonlinear interaction terms on the levels $\{N_i\}_{i=1,q}$ are kept constant and equal to their value at $t_{2q(j-1)n_V}$. The total number of operations needed to complete these $2qn_V$ iterations is of the order of

$$(sn_V+1)N^n\log_2 N + (sn_V+1)N_1^n\log_2 N_1 + (2sn_V+1)\sum_{i=2}^{q-1} N_i^n\log_2 N_i + 2sn_V N_q^n\log_2 N_q,$$

compared to $O(2qsn_V N^n\log_2 N)$ operations with a Galerkin method.

As an example, if we set $p=5$ and $N_i=(3+i)N/8$, $i=1,\dots,5$, the total number of operations of the corresponding p-level V-cycle is of the order of

$$(sn_V+1)N^n\log_2 N + (sn_V+1)\left(\frac{N}{2}\right)^n\log_2\left(\frac{N}{2}\right) + 2sn_V\left(\frac{7}{8}\right)^n N^n\log_2\left(\frac{7N}{8}\right) + (2sn_V+1) \times\left[\left(\frac{3}{4}\right)^n N^n\log_2\left(\frac{3N}{4}\right) + \left(\frac{5}{8}\right)^n N^n\log_2\left(\frac{5N}{8}\right)\right],$$

which is asymptotically of the order of

$$\left((sn_V+1)(1+2^{-n}) + 2sn_V\left(\tfrac{7}{8}\right)^n + (2sn_V+1)\left[\left(\tfrac{5}{8}\right)^n+\left(\tfrac{3}{4}\right)^n\right]\right)N^n\log_2 N.$$

In the three-dimensional case ($n=3$), the number of operations for this p-level cycle is then of the order of $(3.797sn_V+1.79)N^n\log_2 N$ instead of $8sn_V N^n\log_2 N$ for a Galerkin method. A reduction factor of the order of 0.5 is then obtained with the p-level method.

To derive the multilevel scheme described above, we have assumed that the small scales as well as the nonlinear interaction terms, for any level N_i above a given one N_1, can be treated via a quasistatic approximation. Such numerical treatment induces perturbations in the large-scale equation, which have to be controlled. A control strategy must then be developed in order to determine the proper cutoff levels[3] $\{N_i\}_{i=1,p}$ and the number of V-cycles n_V which can be performed. This implementation issue will be addressed in Chapter 9.

Remark 8.1. At each time iteration of a p-level V-cycle, the equation $(8.53)_{|q-l|}$, for $l=1,\dots,2q$, is solved. Hence, by projecting $(8.53)_{|q-l|}$ with

[3] For the sake of simplicity, we have assumed that $N_p=N$ in the description of the multilevel scheme; however, $N_p<N$ can be considered as well. In such a case, a specific treatment has to be defined for the scales $\mathbf{z}_{N_p}^N$. In the implementation of the multilevel schemes, N_p is always less or equal to N (see Chapter 9).

respect to P_{N_1}, we obtain the equation satisfied by the large scales at each time iteration

$$\begin{aligned}
\mathbf{y}_{N_1}^{2jq+l} &= e^{-\nu A\,\Delta t}\,\mathbf{y}_{N_1}^{2jq+l-1} \\
&\quad + \Delta t\, e^{-\nu A\,\Delta t}\Bigg[P_{N_1}\mathbf{g}_N^{2jq+l-1} \\
&\quad - P_{N_1}B\left(\mathbf{y}_{N_1}^{2jq+l-1}+\sum_{i=1}^{|q-l|}\mathbf{z}_i^{2jq+l-1},\ \mathbf{y}_{N_1}^{2jq+l-1}+\sum_{i=1}^{|q-l|}\mathbf{z}_i^{2jq+l-1}\right)\Bigg] \\
&\quad - \Delta t\, e^{-\nu A\,\Delta t}P_{N_1}B_{\text{int}}\left(\mathbf{y}_{N_1}^{2jq}+\sum_{i=1}^{|q-l|}\mathbf{z}_i^{2jq},\ \sum_{i=|q-l|+1}^{q}\mathbf{z}_i^{2jq}\right),
\end{aligned}$$

where $l = 1, \ldots, 2q$. Note that the nonlinear term can be rewritten as

$$\begin{aligned}
&P_{N_1}B\left(\mathbf{y}_{N_1}^{2jq+l-1}+\sum_{i=1}^{|q-l|}\mathbf{z}_i^{2jq+l-1},\ \mathbf{y}_{N_1}^{2jq+l-1}+\sum_{i=1}^{|q-l|}\mathbf{z}_i^{2jq+l-1}\right) \\
&= P_{N_1}B\left(\mathbf{y}_{N_1}^{2jq+l-1},\ \mathbf{y}_{N_1}^{2jq+l-1}\right) + P_{N_1}B_{\text{int}}\left(\mathbf{y}_{N_1}^{2jq+l-1},\ \sum_{i=1}^{|q-l|}\mathbf{z}_i^{2jq+l-1}\right).
\end{aligned}$$

Hence, we note that all interactions between the large scales $\mathbf{y}_{N_1}$ and the smaller scales $\{\mathbf{z}_{N_i}^{N_{i+1}}\}_{i=1,q}$ are not kept constant during the whole time interval $(t_{2jq}, t_{2(j+1)q})$ corresponding to the jth V-cycle. Indeed, the above relation shows that only some part of this interaction, varying with the cutoff level, is kept constant at each time iteration. This implies a transition between a quasistatic and a fully dynamic treatment of the nonlinear interaction terms. □

8.2.2. Wall-Bounded Flows

In this section, we describe a multilevel scheme with p levels in the periodic directions of the channel flow problem.

As has been said in Section 8.1.2, the contribution of $\mathbf{z}_{N_1}^N$ to the velocity field $\mathbf{u}_N$ is reduced when the cutoff level N_1 is increased. So the quantity $\mathbf{z}_{N_1}^N$ is computed with less accuracy when N_1 increases. Moreover, the ratio $\langle|\dot{z}_{i,N_1}^N|^2\rangle^{1/2}/\langle|\dot{y}_{i,N_1}|^2\rangle^{1/2}$ is a decreasing function of the cutoff level, $i=1,\ldots 3$.

In order to take into account this behavior of the small scales, as for the homogeneous case, we have developed a p-level scheme.

We consider a sequence of p cutoff levels $\{N_i\}_{i=1,p}$ satisfying $N_{i+1} > N_i$, for $i=1, \ldots, p-1$, and $N_p=N$. We decompose $\mathbf{u}_N$ into

$$\mathbf{u}_N = \mathbf{y}_{N_1} + \sum_{i=1}^{p-1} \mathbf{z}_{N_i}^{N_{i+1}}. \tag{8.59}$$

This decomposition of $\mathbf{u}_N$ induces a similar decomposition on Δu_2 and ω_2:

$$\Delta u_{2,N} = \Delta y_{2,N_1} + \sum_{i=1}^{p-1} \Delta z_{2,N_i}^{N_{i+1}} \tag{8.60}$$

and

$$\omega_{2,N} = y_{\omega_{2,N_1}} + \sum_{i=1}^{p-1} z_{\omega_{2,N_i}}^{N_{i+1}}. \tag{8.61}$$

The components $\{\Delta y_{2,N_1}, \Delta z_{2,N_i}^{N_{i+1}}\}$ satisfy the coupled systems of ordinary differential equations (see (6.51) and (6.53))

$$\begin{aligned} &\frac{\partial \Delta y_{2,N_1}}{\partial t} - \nu\, \Delta^2 y_{2,N_1} + \Delta P_{N_1} H_2(\mathbf{y}_{N_1}, \mathbf{y}_{N_1}) \\ &\quad - \frac{\partial}{\partial x_2} P_{N_1}(\nabla \cdot \mathbf{H})(\mathbf{y}_{N_1}, \mathbf{y}_{N_1}) \\ &= -\Delta P_{N_1} H_{\text{int},2}\left(\mathbf{y}_{N_1}, \sum_{i=1}^{p-1} \mathbf{z}_{N_i}^{N_{i+1}}\right) + \frac{\partial}{\partial x_2} P_{N_1}(\nabla \cdot \mathbf{H}_{\text{int}})\left(\mathbf{y}_{N_1}, \sum_{i=1}^{p-1} \mathbf{z}_{N_i}^{N_{i+1}}\right), \end{aligned} \tag{8.62}$$

$$\begin{aligned} &\frac{\partial \Delta z_{2,N_i}^{N_{i+1}}}{\partial t} - \nu \Delta^2 z_{2,N_i}^{N_{i+1}} \\ &\quad + \Delta Q_{N_i}^{N_{i+1}} H_2\left(\mathbf{y}_{N_1} + \sum_{j=1}^{p-1} \mathbf{z}_{N_j}^{N_{j+1}}, \mathbf{y}_{N_1} + \sum_{j=1}^{p-1} \mathbf{z}_{N_j}^{N_{j+1}}\right) \\ &\quad - \frac{\partial}{\partial x_2} Q_{N_i}^{N_{i+1}} (\nabla \cdot \mathbf{H})\left(\mathbf{y}_{N_1} + \sum_{j=1}^{p-1} \mathbf{z}_{N_j}^{N_{j+1}}, \mathbf{y}_{N_1} + \sum_{j=1}^{p-1} \mathbf{z}_{N_j}^{N_{j+1}}\right) = 0, \end{aligned} \tag{8.63$_i$}$$

and the components $\{y_{\omega_{2,N_1}}, z_{\omega_{2,N_i}}^{N_{i+1}}\}$ are solutions of the following equations (see

(6.51) and (6.53)):

$$\frac{\partial y_{\omega 2,N_1}}{\partial t} - \nu \Delta y_{\omega 2,N_1} + \frac{\partial}{\partial x_3} P_{N_1} H_1(\mathbf{y}_{N_1}, \mathbf{y}_{N_1}) - \frac{\partial}{\partial x_1} P_{N_1} H_3(\mathbf{y}_{N_1}, \mathbf{y}_{N_1}) \tag{8.64}$$

$$= -\frac{\partial}{\partial x_3} P_{N_1} H_{\text{int},1}\left(\mathbf{y}_{N_1}, \sum_{i=1}^{p-1} \mathbf{z}_{N_i}^{N_{i+1}}\right) + \frac{\partial}{\partial x_1} P_{N_1} H_{\text{int},3}\left(\mathbf{y}_{N_1}, \sum_{i=1}^{p-1} \mathbf{z}_{N_i}^{N_{i+1}}\right),$$

$$\frac{\partial z_{\omega 2,N_i}^{N_{i+1}}}{\partial t} - \nu \Delta z_{\omega 2,N_i}^{N_{i+1}} + \frac{\partial}{\partial x_3} Q_{N_i}^{N_{i+1}} H_1\left(\mathbf{y}_{N_1} + \sum_{j=1}^{p-1} \mathbf{z}_{N_j}^{N_{j+1}}, \mathbf{y}_{N_1} + \sum_{j=1}^{p-1} \mathbf{z}_{N_j}^{N_{j+1}}\right) \tag{8.65$_i$}$$

$$- \frac{\partial}{\partial x_1} Q_{N_i}^{N_{i+1}} H_3\left(\mathbf{y}_{N_1} + \sum_{j=1}^{p-1} \mathbf{z}_{N_j}^{N_{j+1}}, \mathbf{y}_{N_1} + \sum_{j=1}^{p-1} \mathbf{z}_{N_j}^{N_{j+1}}\right) = 0,$$

where $i = 1, \ldots, p-1$ and $Q_{N_i}^{N_{i+1}} = P_{N_{i+1}} - P_{N_i}$. By summing equations (8.62) and $(8.63)_i$ for $i = 1, \ldots, k$, where $k \leq p-1$, we deduce that $y_{2,N_1} + \sum_{i=1}^{k} z_{2,N_i}^{N_{i+1}}$ satisfies the following equation:

$$\frac{\partial}{\partial t} \Delta \left(y_{2,N_1} + \sum_{i=1}^{k} z_{2,N_i}^{N_{i+1}}\right) - \nu \Delta^2 \left(y_{2,N_1} + \sum_{i=1}^{k} z_{2,N_i}^{N_{i+1}}\right) \tag{8.66$_k$}$$

$$+ \Delta P_{N_{k+1}} H_2\left(\mathbf{y}_{N_1} + \sum_{i=1}^{k} \mathbf{z}_{N_i}^{N_{i+1}}, \mathbf{y}_{N_1} + \sum_{i=1}^{k} \mathbf{z}_{N_i}^{N_{i+1}}\right)$$

$$- \frac{\partial}{\partial x_2} P_{N_{k+1}} (\nabla \cdot \mathbf{H}) \left(\mathbf{y}_{N_1} + \sum_{i=1}^{k} \mathbf{z}_{N_i}^{N_{i+1}}, \mathbf{y}_{N_1} + \sum_{i=1}^{k} \mathbf{z}_{N_i}^{N_{i+1}}\right)$$

$$= -\Delta P_{N_{k+1}} H_{\text{int},2}\left(\mathbf{y}_{N_1} + \sum_{i=1}^{k} \mathbf{z}_{N_i}^{N_{i+1}}, \sum_{i=k+1}^{p-1} \mathbf{z}_{N_i}^{N_{i+1}}\right)$$

$$+ \frac{\partial}{\partial x_2} P_{N_{k+1}} (\nabla \cdot \mathbf{H}_{\text{int}}) \left(\mathbf{y}_{N_1} + \sum_{i=1}^{k} \mathbf{z}_{N_i}^{N_{i+1}}, \sum_{i=k+1}^{p-1} \mathbf{z}_{N_i}^{N_{i+1}}\right).$$

In the same manner, by summing equations (8.64) and $(8.67)_i$ for $i = 1, \ldots, k$, where $k \leq p-1$, we deduce that $y_{\omega 2,N_1} + \sum_{i=1}^{k} z_{\omega 2,N_i}^{N_{i+1}}$ satisfies the following equation:

$$\begin{aligned}
&\frac{\partial}{\partial t}\left(y_{\omega_2,N_1}+\sum_{i=1}^{k} z_{\omega_2,N_i}^{N_{i+1}}\right)-\nu\Delta\left(y_{\omega_2,N_1}+\sum_{i=1}^{k} z_{\omega_2,N_i}^{N_{i+1}}\right) \\
&\quad+\frac{\partial}{\partial x_3}P_{N_{k+1}}H_1\left(\mathbf{y}_{N_1}+\sum_{i=1}^{k}\mathbf{z}_{N_i}^{N_{i+1}},\ \mathbf{y}_{N_1}+\sum_{i=1}^{k}\mathbf{z}_{N_i}^{N_{i+1}}\right) \\
&\quad-\frac{\partial}{\partial x_1}P_{N_{k+1}}H_3\left(\mathbf{y}_{N_1}+\sum_{i=1}^{k}\mathbf{z}_{N_i}^{N_{i+1}},\ \mathbf{y}_{N_1}+\sum_{i=1}^{k}\mathbf{z}_{N_i}^{N_{i+1}}\right) \\
&=-\frac{\partial}{\partial x_3}P_{N_{k+1}}H_{\text{int},1}\left(\mathbf{y}_{N_1}+\sum_{i=1}^{k}\mathbf{z}_{N_i}^{N_{i+1}},\ \sum_{i=k+1}^{p-1}\mathbf{z}_{N_i}^{N_{i+1}}\right) \\
&\quad+\frac{\partial}{\partial x_1}P_{N_{k+1}}H_{\text{int},3}\left(\mathbf{y}_{N_1}+\sum_{i=1}^{k}\mathbf{z}_{N_i}^{N_{i+1}},\ \sum_{i=k+1}^{p-1}\mathbf{z}_{N_i}^{N_{i+1}}\right).
\end{aligned} \tag{8.67$_k$}$$

We set $\mathbf{z}_i^m = \mathbf{z}_{N_i}^{N_{i+1},m}$, $z_{\omega_2,i}^m = z_{\omega_2,N_i}^{N_{i+1},m}$, and $q = p-1$. As for the homogeneous case (see Section 8.2.1), a succession of p-level V-cycles are performed to advance the flow field computation. A p-level V-cycle consists in computing a sequence $\{\mathbf{y}_{N_1}^m, \mathbf{z}_{N_i}^{N_{i+1},m}\}_{i=1,p-1}$, $m \geq 0$, using $(8.66)_i$, $(8.67)_i$, and the incompressibility constraint. A p-level V-cycle j, $j \geq 0$, is constituted by $2q$ successive steps that we describe hereafter, and it corresponds to the time interval $(t_{2jq}, t_{(2j+2)q})$:

Step l, $1 \leq l \leq q-1$: The components $\{y_{2,N_1}^{2jq+l}, z_{2,i}^{2jq+l}\}$ and $\{y_{\omega_2,N_1}^{2jq+l}, z_{\omega_2,i}^{2jq+l}\}$, for $i = 1, \dots, q-l$, are respectively computed by integrating $(8.66)_{q-l}$ and $(8.67)_{q-l}$ over the time interval $(t_{2jq+l-1}, t_{2jq+l})$. Knowing $\{y_{2,N_1}^{2jq+l}, z_{2,i}^{2jq+l}\}$ and $\{y_{\omega_2,N_1}^{2jq+l}, z_{\omega_2,i}^{2jq+l}\}$ for $i = 1, \dots, q-l$, we deduce $\{y_{1,N_1}^{2jq+l}, z_{1,i}^{2jq+l}\}$ and $\{y_{3,N_1}^{2jq+l}, z_{3,i}^{2jq+l}\}$ for $i = 1, \dots, q-l$, using the incompressibility constraint. A quasistatic approximation on the time interval $(t_{2jq+l-1}, t_{2jq+l})$ is applied to the small scales:

$$\mathbf{z}_i^{2jq+l} = \mathbf{z}_i^{2jq+l-1}, \qquad \text{for} \quad i = q-l+1, \dots, q \tag{8.68}$$

and

$$z_{\omega_2,i}^{2jq+l} = z_{\omega_2,i}^{2jq+l-1}, \qquad \text{for} \quad i = q-l+1, \dots, q. \tag{8.69}$$

Step q: The large-scale components $y_{2,N_1}^{(2j+1)q}$ $(y_{\omega_2,N_1}^{(2j+1)q})$ are computed, integrating (8.62), (8.64) over the time interval $(t_{(2j+1)q-1}, t_{(2j+1)q})$. Knowing

$y_{2,N_1}^{(2j+1)q}$ and $y_{\omega_2,N_1}^{(2j+1)q}$, we deduce $y_{1,N_1}^{(2j+1)q}$ and $y_{3,N_1}^{(2j+1)q}$ using the incompressibility constraint. A quasistatic approximation on the time interval $(t_{(2j+1)q-1}, t_{(2j+1)q})$ is applied to the small scales:

$$\mathbf{z}_i^{(2j+1)q} = \mathbf{z}_i^{(2j+1)q-1}, \qquad \text{for} \quad i = 1, \ldots, q \tag{8.70}$$

and

$$z_{\omega_2,i}^{(2j+1)q} = z_{\omega_2,i}^{(2j+1)q-1} \qquad \text{for} \quad i = 1, \ldots, q. \tag{8.71}$$

Step l, $q+1 \leq l \leq 2q-1$: The components $\{y_{2,N_1}^{2jq+l}, z_{2,i}^{2jq+l}\}$ and $\{y_{\omega_2,N_1}^{2jq+l}, z_{\omega_2,i}^{2jq+l}\}$, for $i = 1, \ldots, l-q$, are respectively computed by integrating $(8.66)_{l-q}$ and $(8.67)_{l-q}$ over the time interval $(t_{2jq+l-1}, t_{2jq+l})$. Knowing $\{y_{2,N_1}^{2jq+l}, z_{2,i}^{2jq+l}\}$ and $\{y_{\omega_2,N_1}^{2jq+l}, z_{\omega_2,i}^{2jq+l}\}$ for $i = 1, \ldots, l-q$, we deduce $\{y_{1,N_1}^{2jq+l}, z_{1,i}^{2jq+l}\}$ and $\{y_{3,N_1}^{2jq+l}, z_{3,i}^{2jq+l}\}$ for $i = 1, \ldots, l-q$ using the incompressibility constraint. A quasistatic approximation on the time interval $(t_{2jq+l-1}, t_{2jq+l})$ is applied to the small scales:

$$\mathbf{z}_i^{2jq+l} = \mathbf{z}_i^{2jq+l-1} \qquad \text{for} \quad i = l-q+1, \ldots, q \tag{8.72}$$

and

$$z_{\omega_2,i}^{2jq+l} = z_{\omega_2,i}^{2jq+l-1} \qquad \text{for} \quad i = l-q+1, \ldots, q. \tag{8.73}$$

Step $2q$: This last step consists in computing all velocity components $\{y_{2,N_1}^{(2j+2)q}, z_{2,i}^{(2j+2)q}\}$ and $\{y_{\omega_2,N_1}^{(2j+2)q}, z_{\omega_2,i}^{(2j+2)q}\}$ for $i = 1, \ldots, q$ by integrating respectively $(8.66)_q$ and $(8.67)_q$ on the time interval $(t_{2(j+1)q-1}, t_{2(j+1)q})$. Knowing $\{y_{2,N_1}^{(2j+2)q}, z_{2,i}^{(2j+2)q}\}$ and $\{y_{\omega_2,N_1}^{(2j+2)q}, z_{\omega_2,i}^{(2j+2)q}\}$ for $i = 1, \ldots, q$, we deduce $\{y_{1,N_1}^{(2j+2)q}, z_{1,i}^{(2j+2)q}\}$ and $\{y_{3,N_1}^{(2j+2)q}, z_{3,i}^{(2j+2)q}\}$ for $i = 1, \ldots, q$, using the incompressibility constraint.

Note that the treatments (8.68), (8.69) and (8.70), (8.71) of the small scales imply that

$$\mathbf{z}_i^{(2j+1)q} = \mathbf{z}_i^{(2j+1)q-i} \qquad \text{for} \quad i = 1, \ldots, q$$

and

$$z_{\omega_2,i}^{(2j+1)q} = z_{\omega_2,i}^{(2j+1)q-i} \qquad \text{for} \quad i = 1, \ldots, q.$$

Moreover, the treatments (8.72), (8.73) imply that

$$\mathbf{z}_i^{(2j+1)q+i-1} = \mathbf{z}_i^{(2j+1)q} \qquad \text{for} \quad i = 1, \ldots, q$$

and

$$z_{\omega_2,i}^{(2j+1)q+i-1} = z_{\omega_2,i}^{(2j+1)q} \quad \text{for} \quad i = 1, \ldots, q.$$

We finally obtain

$$\mathbf{z}_i^{(2j+1)q+i-1} = \mathbf{z}_i^{(2j+1)q-i} \quad \text{for} \quad i = 1, \ldots, q$$

and

$$z_{\omega_2,i}^{(2j+1)q+i-1} = z_{\omega_2,i}^{(2j+1)q-i} \quad \text{for} \quad i = 1, \ldots, q,$$

so that the scales $\mathbf{z}_{N_i}^{N_{i+1}}$ are kept constant on a time interval of length $(2i-1)\,\Delta t$, for $i = 1, \ldots, p-1$. Hence, with such a scheme, the length of the time interval of the quasistatic approximation of the small scales increases when the cutoff level increases.

As a direct consequence of the above scheme, the nonlinear interaction terms in $(8.66)_{l=1,\ldots,q}$ and $(8.67)_{l=1,\ldots,q}$ are kept constant and equal to their values at time t_{2jq} during the whole V-cycle:

$$\begin{aligned} &P_{N_l} H_{\text{int},\Delta u_2}\left(\mathbf{y}_{N_1}^{2jq+k} + \sum_{i=1}^{l-1} \mathbf{z}_i^{2jq+k}, \sum_{i=l}^{q} \mathbf{z}_i^{2jq+k}\right) \\ &\quad = P_{N_l} H_{\text{int},\Delta u_2}\left(\mathbf{y}_{N_1}^{2jq} + \sum_{i=1}^{l-1} \mathbf{z}_i^{2jq}, \sum_{i=l}^{q} \mathbf{z}_i^{2jq}\right) \end{aligned}$$

for $k = 1, \ldots, 2q$,

$$\begin{aligned} &P_{N_l} H_{\text{int},\omega_2}\left(\mathbf{y}_{N_1}^{2jq+k} + \sum_{i=1}^{l-1} \mathbf{z}_i^{2jq+k}, \sum_{i=l}^{q} \mathbf{z}_i^{2jq+k}\right) \\ &\quad = P_{N_l} H_{\text{int},\omega_2}\left(\mathbf{y}_{N_1}^{2jq} + \sum_{i=1}^{l-1} \mathbf{z}_i^{2jq}, \sum_{i=l}^{q} \mathbf{z}_i^{2jq}\right) \end{aligned}$$

for $k = 1, \ldots, 2q$, using the notation introduced in (8.33) and (8.34). Hence, the nonlinear interaction terms have to be computed and stored at the beginning of the V-cycle. Their computation is done as follows:

$$\begin{aligned} &P_{N_l} H_{\text{int},\Delta u_2}\left(\mathbf{y}_{N_1}^{2jq} + \sum_{i=1}^{l-1} \mathbf{z}_i^{2jq}, \sum_{i=l}^{q} \mathbf{z}_i^{2jq}\right) \\ &\quad = P_{N_l} H_{\Delta u_2}\left(\mathbf{y}_{N_1}^{2jq} + \sum_{i=1}^{q} \mathbf{z}_i^{2jq}, \mathbf{y}_{N_1}^{2jq} + \sum_{i=1}^{q} \mathbf{z}_i^{2jq}\right) \\ &\qquad - P_{N_l} H_{\Delta u_2}\left(\mathbf{y}_{N_1}^{2jq} + \sum_{i=1}^{l-1} \mathbf{z}_i^{2jq}, \mathbf{y}_{N_1}^{2jq} + \sum_{i=1}^{l-1} \mathbf{z}_i^{2jq}\right), \end{aligned}$$

$$P_{N_l} H_{\text{int},\omega_2}\left(\mathbf{y}_{N_1}^{2jq} + \sum_{i=1}^{l-1} \mathbf{z}_i^{2jq}, \sum_{i=l}^{q} \mathbf{z}_i^{2jq}\right)$$

$$= P_{N_l} H_{\omega_2}\left(\mathbf{y}_{N_1}^{2jq} + \sum_{i=1}^{q} \mathbf{z}_i^{2jq}, \mathbf{y}_{N_1}^{2jq} + \sum_{i=1}^{q} \mathbf{z}_i^{2jq}\right)$$

$$- P_{N_l} H_{\omega_2}\left(\mathbf{y}_{N_1}^{2jq} + \sum_{i=1}^{l-1} \mathbf{z}_i^{2jq}, \mathbf{y}_{N_1}^{2jq} + \sum_{i=1}^{l-1} \mathbf{z}_i^{2jq}\right).$$

As in the homogeneous case, we define cycles constituted by several p-level V-cycles. Then, a cycle $j\,(j \geq 1)$ corresponds to the time interval $(t_{2q(j-1)n_V}, t_{2qjn_V})$, where n_V is the number of V-cycles performed. During these $2qn_V$ iterations, the nonlinear interaction terms on the levels $\{N_i\}_{i=1,q}$ are kept constant and equal to their value at $t_{2q(j-1)n_V}$.

Orientation

To derive the multilevel scheme described above, we have assumed that the small scales as well as the nonlinear interaction terms, for any level N_i above a given one N_1, can be treated via a quasistatic approximation. Such numerical treatment induces perturbations in the large-scale equation, which have to be controlled. A control strategy must then be developed in order to define the proper cutoff levels $\{N_i\}_{i=1,p}$. This implementation issue will be detailed in Chapter 9.

9

Computational Implementation of the Dynamic Multilevel Methods

In Sections 9.1 to 9.4, the implementation of the DML methods applied to the homogeneous problem is described in details. In Sections 9.1 to 9.3 we describe and motivate different parts of the algorithm, and in Section 9.4 we present the synthesis of the DML algorithm, which we actually implement in Chapter 10. In Section 9.5, the nonhomogeneous case is considered.

9.1. General Description of the DML Methods

The dynamic multilevel schemes which have been implemented and applied to the simulation of homogeneous turbulence are based on the p-level schemes described in Section 8.2.1. As in the previous chapters, we denote by N the total number of modes retained in each direction and we assume that the corresponding wavenumber k_N is of the order of the Kolmogorov wavenumber k_d. First of all, we choose a sequence of m cutoff levels $\{N_i\}_{i=1,m}$ satisfying

$$4 \leq N_1 < N_2 < \cdots < N_i < N_{i+1} < \cdots < N_{m-1} \leq N_m = N,$$

and such that $\Delta_i = N_{i+1} - N_i$ remains approximately constant for $i = 1, \ldots, m-1$. This sequence will provide possible choices of p different values for the cutoff levels.

Since nonlinear terms depending on the coarse levels N_i, $i = 1, \ldots, m$, are computed via a pseudospectral method (see Section 3.1), n-dimensional FFTs at N_i^n modes are used in the algorithm. Restrictions on the values of the cutoff levels are imposed via the FFT algorithm. Indeed, the levels N_i have to be of the form $2^p 3^q 5^r 7^s$, $p, q, r, s \in \mathbb{N}$, with $p \geq 1$.

At the beginning of each cycle j, two distinct levels $N_{i_1(j)}$, $N_{i_2(j)}$, with $1 \leq i_1(j) < i_2(j) \leq m$, and the parameter $n_V(j)$, fixing the number of

V-cycles performed, are computed according to some criteria that we will describe hereafter. We set $q(j) = i_2(j) - i_1(j)$ and introduce

$$n_j = \sum_{i=1}^{j} 2q(i)n_V(i) \qquad \text{for } j \geq 1, \tag{9.1}$$

with $n_0 = 0$; hence, $t_{n_0} = 0$ corresponds to the initial condition and $t_{n_j} = n_j\,\Delta t$. Each cycle j corresponds to the time interval $(t_{n_{j-1}}, t_{n_j})$ and then to $2q(j)n_V(j)$ time iterations. The levels $N_{i_1(j)}$, $N_{i_2(j)}$ define a subsequence of $q(j)+1$ values lying in $\{N_i\}_{i=1,m}$, where $q(j) = i_2(j) - i_1(j)$, which satisfy

$$N_{i_1(j)} < N_{i_1(j)+1} < \cdots < N_{i_1(j)+q(j)-1} < N_{i_2(j)} \leq N.$$

We denote by $\mathbf{u}_N^{n_j}$ the approximation (3.11) of the velocity field (3.4) at time t_{n_j}, that is, $\mathbf{u}_N(t_{n_j})$, computed by the DML method and satisfying

$$\mathbf{u}_N^{n_0} = \mathbf{u}_N(t = 0).$$

Based on the sequence of levels defined above, at time t_{n_j} we decompose $\mathbf{u}_N^{n_j}$ into

$$\mathbf{u}_N^{n_j} = \mathbf{y}_{N_{i_1}}^{n_j} + \sum_{i=i_1}^{i_2-1} \mathbf{z}_i^{n_j} + \mathbf{z}_{N_{i_2}}^{N,n_j}, \tag{9.2}$$

where we denote $\mathbf{z}_i^{n_j} = \mathbf{z}_i(t_{n_j}) = \mathbf{z}_{N_i}^{N_{i+1}}(t_{n_j})$.

We consider and describe hereafter the cycle $j+1$ leading to the computation of $\mathbf{u}_N^{n_{j+1}}$. In order to simplify the notation we omit in the following the dependence of i_1, i_2, q, and n_V on $j+1$. As in Section 8.2.1, a p-level ($p = q+1$) scheme on the time interval $(t_{n_j}, t_{n_{j+1}})$ can be implemented. Let $l \in [1, 2qn_V]$; then $n_j + l$ denotes the current iteration. In order to simplify the notation, we introduce

$$i(l) = i_1 + |k - q| \qquad \text{with } k = l - 2q\left[\frac{l-1}{2q}\right], \tag{9.3}$$

where $[\cdot]$ denotes the integer part. From (9.3), $k \in [1, 2q]$, so that $i(l) \in [i_1, i_2]$. Then, $N_{i(l)}$ denotes the current cutoff level, and the time iteration $n_j + l$ consists in computing the scales $\mathbf{y}_{N_{i_1}}^{n_j+l} + \sum_{i=i_1}^{i(l)-1} \mathbf{z}_i^{n_j+l}$ by solving

$$\mathbf{y}_{N_{i_1}}^{n_j+l} + \sum_{i=i_1}^{i(l)-1} \mathbf{z}_i^{n_j+l} \tag{9.4$_l$}$$

$$= e^{-\nu A\,\Delta t}\left(\mathbf{y}_{N_{i_1}}^{n_j+l-1} + \sum_{i=i_1}^{i(l)-1} \mathbf{z}_i^{n_j+l-1}\right) + \Delta t e^{-\nu A\,\Delta t}\, P_{N_{i(l)}} \mathbf{g}_N^{n_j+l-1}$$

$$- \Delta t\, e^{-\nu A \Delta t} P_{N_{i(l)}} B \left(\mathbf{y}_{N_{i_1}}^{n_j+l-1} + \sum_{i=i_1}^{i(l)-1} \mathbf{z}_i^{n_j+l-1}, \mathbf{y}_{N_{i_1}}^{n_j+l-1} + \sum_{i=i_1}^{i(l)-1} \mathbf{z}_i^{n_j+l-1} \right)$$

$$- \Delta t\, e^{-\nu A \Delta t} P_{N_{i(l)}} B_{\text{int}} \left(\mathbf{y}_{N_{i_1}}^{n_j} + \sum_{i=i_1}^{i(l)-1} \mathbf{z}_i^{n_j}, \sum_{i=i(l)}^{i_2-1} \mathbf{z}_i^{n_j} + \mathbf{z}_{N_{i_2}}^{N,n_j} \right).$$

Note that when $l = (2r+1)q$, with $r = 0, \ldots, n_V - 1$, then $i(l) = i_1$, so that the intermediate scales $\sum_{i=i_1}^{i(l)-1} \mathbf{z}_i^{n_j+s}$, $s = 0$ and $l-1$, vanish in $(9.4)_l$. The cutoff level corresponding to time iteration $n_j + (2r+1)q$ is then N_{i_1}, and only the large scales $\mathbf{y}_{N_{i_1}}$ are updated.

A quasistatic approximation on the time interval (t_{n_j+l-1}, t_{n_j+l}) is applied to the small-scale components $\mathbf{z}_i(t)$ for $i = i(l), \ldots, i_2 - 1$ and $\mathbf{z}_{N_{i_2}}^N$:

$$\mathbf{z}_i^{n_j+l} = \mathbf{z}_i^{n_j+l-1} \qquad \text{for } i = i(l), \ldots, i_2 - 1,$$

$$\mathbf{z}_{N_{i_2}}^{N,n_j+l} = \mathbf{z}_{N_{i_2}}^{N,n_j+l-1} \qquad \text{for } l \neq 2qr,\ r = 1, \ldots, n_V.$$

Note that the above treatment implies that

$$\mathbf{z}_i^{n_j+(2r-1)q+i-i_1} = \mathbf{z}_i^{n_j+(2r-1)q+i_1-i-1} \qquad \text{for} \qquad (9.5)$$

$$i = i_1, \ldots, i_2 - 1 \quad \text{and} \quad r = 1, \ldots, n_V,$$

so that the intermediate scales $\{\mathbf{z}_i(t)\}_{i=i_1, i_2-1}$ are kept constant on time intervals of length $[2(i - i_1) + 1]\,\Delta t$, for $i = 1, \ldots, q$.

At the final time t_{n_j+2qr} of each V-cycle ($r = 1, \ldots, n_V$), the small scales $\mathbf{z}_{N_{i_2}}^N$ are updated either by integrating all the scales $\mathbf{y}_{N_{i_1}} + \sum_{i=i_1}^{i_2-1} \mathbf{z}_i + \mathbf{z}_{N_{i_2}}^N$ on the time interval $(t_{n_j+2qr-1}, t_{n_j+2qr})$ or by projecting the velocity field onto an approximate inertial manifold (see Chapter 5); namely we write

$$\mathbf{z}_{N_{i_2}}^{N,n_j+2qr} = \Phi \left(\mathbf{y}_{N_{i_1}}^{n_j+2qr}, \mathbf{z}_{i_1}^{n_j+2qr}, \ldots, \mathbf{z}_{i_2-1}^{n_j+2qr}, \mathbf{z}_{N_{i_2}}^{N,n_j+2q(r-1)} \right),$$

where Φ is a suitable functional. In practice, $\mathbf{z}_{N_{i_2}}^{N,n_j+2qr}$ is computed as follows:

$$\mathbf{z}_{N_{i_2}}^{N,n_j+2qr} = e^{-2q\nu A \Delta t} \mathbf{z}_{N_{i_2}}^{N,n_j+2q(r-1)} \qquad (9.6)$$

$$- q\,\Delta t \Bigg[Q_{N_{i_2}}^N B \left(\mathbf{y}_{N_{i_1}}^{n_j+2qr} + \sum_{i=i_1}^{i_2-1} \mathbf{z}_i^{n_j+2qr}, \mathbf{y}_{N_{i_1}}^{n_j+2qr} + \sum_{i=i_1}^{i_2-1} \mathbf{z}_i^{n_j+2qr} \right)$$

$$+ e^{-2q\nu A\,\Delta t} Q_{N_{i_2}}^{N} B\left(\mathbf{y}_{N_{i_1}}^{n_j+2q(r-1)} + \sum_{i=i_1}^{i_2-1} \mathbf{z}_i^{n_j+2q(r-1)},\right.$$

$$\left.\left. \mathbf{y}_{N_{i_1}}^{n_j+2q(r-1)} + \sum_{i=i_1}^{i_2-1} \mathbf{z}_i^{n_j+2q(r-1)}\right)\right].$$

Note that in (9.6), the nonlinear term is treated via a second-order semi-implicit scheme while the nonlinear interaction term is neglected. The computation of $\mathbf{z}_{N_{i_2}}^{N,n_j+2qr}$ then requires the evaluation of a nonlinear term on the fine grid. Moreover, we note that as the small scales $\mathbf{z}_{N_{i_2}}^{N,n_j+2q(r-1)}$ are not used during the V-cycle $(t_{n_j+2q(r-1)}, t_{n_j+2qr})$, we can compute a part of the right-hand side of (9.6) at $t_{n_j+2q(r-1)}$, so that

$$\begin{aligned}
&\mathbf{z}_{N_{i_2}}^{N,n_j+2qr}\\
&\quad = e^{-2q\nu A\,\Delta t}\,\mathbf{z}_{N_{i_2}}^{N,n_j+2q(r-1)} - q\,\Delta t\, e^{-2q\nu A\,\Delta t}\\
&\qquad \times Q_{N_{i_2}}^{N} B\left(\mathbf{y}_{N_{i_1}}^{n_j+2q(r-1)} + \sum_{i=i_1}^{i_2-1} \mathbf{z}_i^{n_j+2q(r-1)}, \mathbf{y}_{N_{i_1}}^{n_j+2q(r-1)} + \sum_{i=i_1}^{i_2-1} \mathbf{z}_i^{n_j+2q(r-1)}\right).
\end{aligned}$$

We then avoid any extra storage.

The nonlinear interaction terms appearing in $(9.4)_{l=1,2qn_V}$ are kept constant during the whole cycle $j+1$; they are computed at time t_{n_j} as follows:

$$\begin{aligned}
&P_{N_{i(l)}} B_{\text{int}}\left(\mathbf{y}_{N_{i_1}}^{n_j} + \sum_{i=i_1}^{i(l)-1} \mathbf{z}_i^{n_j}, \sum_{i=i(l)}^{i_2-1} \mathbf{z}_i^{n_j} + \mathbf{z}_{N_{i_2}}^{N,n_j}\right)\\
&\quad = P_{N_{i(l)}} B\left(\mathbf{u}_N^{n_j}, \mathbf{u}_N^{n_j}\right) - P_{N_{i(l)}} B\left(\mathbf{y}_{N_{i_1}}^{n_j} + \sum_{i=i_1}^{i(l)-1} \mathbf{z}_i^{n_j}, \mathbf{y}_{N_{i_1}}^{n_j} + \sum_{i=i_1}^{i(l)-1} \mathbf{z}_i^{n_j}\right)
\end{aligned}$$

for $i(l) = i_1+1, \ldots, i_2-1$. The nonlinear interaction term corresponding to the level N_{i_1} is computed similarly:

$$P_{N_{i_1}} B_{\text{int}}\left(\mathbf{y}_{N_{i_1}}^{n_j}, \sum_{i=i_1}^{i_2-1} \mathbf{z}_i^{n_j} + \mathbf{z}_{N_{i_2}}^{N,n_j}\right) = P_{N_{i_1}} B\left(\mathbf{u}_N^{n_j}, \mathbf{u}_N^{n_j}\right) - P_{N_{i_1}} B\left(\mathbf{y}_{N_{i_1}}^{n_j}, \mathbf{y}_{N_{i_1}}^{n_j}\right).$$

As in Section 8.2.1, we can estimate the total number of operations required by the cycle $j+1$, namely

$$\begin{aligned}
&(sn_V+1)N^n \log_2 N + (sn_V+1)N_{i_1}^n \log_2 N_{i_1} \qquad (9.7)\\
&\quad + (2sn_V+1)\sum_{i=i_1+1}^{i_2-1} N_i^n \log_2 N_i + 2sn_V N_{i_2}^n \log_2 N_{i_2}.
\end{aligned}$$

Here we have assumed that the smallest scales $\mathbf{z}^N_{N_{i_2}}$ are updated via an integration of the full system on the fine grid on the time interval $(t_{n_j+2qr-1}, t_{n_j+2qr})$, $r = 1, \ldots, n_V$, corresponding to the end of each V-cycle. When an approximate inertial manifold is used, the factor $sn_V N^n \log_2 N$ has to be replaced by $n_V N^n \log_2 N$ in (9.7). We note that the number of operations is largely dependent on the values of i_1 and i_2, and an important saving in CPU time can be obtained when N_{i_2} is small compared to N.

As was previously said, the parameters defining the multilevel scheme (i_1, i_2, and n_V) depend on j and are computed at each time t_{n_j}. The multilevel scheme is based on the fact that the time variations of the small scales and of the nonlinear interaction terms can be neglected during each cycle. However, these time variations can change from one cycle to another, so that the cutoff levels have to be adjusted. On two-dimensional simulations (see Debussche, Dubois, and Temam (1995) for instance), we have noted very strong variations on the time derivative of the small scales and of the nonlinear interaction terms.

9.2. Survey of Related DML Methods

We now describe several methods that have been previously defined and implemented for computing the parameters $i_1(j)$, $i_2(j)$, and $n_V(j)$. We recall from the previous section that the approximated field at time t_{n_j}, $j \geq 0$, is known and satisfies the decomposition (9.2). Let us consider a cutoff level N_i, $1 \leq i \leq m$. We introduce the following decomposition:

$$\mathbf{u}_N^{n_j} = \mathbf{y}_{N_i}^{n_j} + \mathbf{z}_{N_i}^{N,n_j}, \tag{9.8}$$

where $\mathbf{y}_{N_i}^{n_j} = P_{N_i}\mathbf{u}_N^{n_j}$ and $\mathbf{z}_{N_i}^{N,n_j} = Q_{N_i}^N \mathbf{u}_N^{n_j}$.

In an earlier version of the DML method, the lowest coarse level $N_{i_1(j)}$ was defined by comparing the kinetic energy of the small scales $|\mathbf{y}_{N_i}^{n_j}|_0$ with that of the large scales $|\mathbf{z}_{N_i}^{N,n_j}|_0$. Similarly, the highest coarse level $N_{i_2(j)}$ was defined by comparing the time derivative of the small scales $|\dot{\mathbf{z}}_{N_i}^{N,n_j}|_0$ with $|\dot{\mathbf{y}}_{N_i}^{n_j}|_0$.

In a later version, the definition of the level N_{i_2} was modified and was based on the comparison of $|\mathbf{z}_{N_{i_2}}^{N,n_j}|_0$ with the accuracy of the numerical scheme (here, of the third order in time).

In Jauberteau, Rosier, and Temam (1989, 1990), Jauberteau (1990), and Dubois, Jauberteau, and Temam (1993), the number of V-cycles performed, $n_V(j)$, was fixed at once and remained constant during the whole computation. In Dubois and Temam (1993), $n_V(j)$ was updated during the computation and estimated using the characteristic relaxation time for the viscous term on the finest coarse level $N_{i_2(j)}$, that is, $n_V(j) = [2q(j)\,\Delta t\,(\nu k^2_{N_{i_2}})]^{-1}$, where $k_{N_{i_2(j)}}$ is the wavenumber associated to $N_{i_2(j)}$.

Several two-dimensional simulations have been performed with the previous algorithms. Firstly, exact (analytical) solutions were simulated in order to measure the accuracy and the validity of the methods. Then, computations of physically more realistic flows (homogeneous turbulent flows) have been conducted. For more details about these simulations, the reader is referred to the references cited above.

In Debussche, Dubois, and Temam (1995), the level $N_{i_1(j)}$ was determined by comparing $\|\mathbf{z}_{N_i}^{N,n_j}\|$ with $\|\mathbf{y}_{N_i}^{n_j}\|$, that is, by comparing the enstrophy contained in the small scales with the enstrophy contained in the large scales. The level $N_{i_2(j)} > N_{i_1(j)}$ was determined by comparing the kinetic energy $|\mathbf{z}_{N_i}^{N,n_j}|_0$ contained in the small scales with that contained in the large ones, $|\mathbf{y}_{N_i}^{n_j}|_0$.

The test used to define the cutoff level $N_{i_1(j)}$ allows us to control the size of the nonlinear interaction term $P_{N_i} B_{\text{int}}(\mathbf{y}_{N_i}^{n_j}, \mathbf{z}_{N_i}^{N,n_j})$ compared with the nonlinear term $P_{N_i} B(\mathbf{y}_{N_i}^{n_j}, \mathbf{y}_{N_i}^{n_j})$. Indeed, in the two-dimensional case, we have (see Dubois (1993))

$$\frac{\left|P_{N_i} B_{\text{int}}\left(\mathbf{y}_{N_i}, \mathbf{z}_{N_i}^{N}\right)\right|_0}{|P_{N_i} B(\mathbf{y}_{N_i}, \mathbf{y}_{N_i})|_0} \simeq \frac{\|\mathbf{z}_{N_i}^{N}\|}{\|\mathbf{y}_{N_i}\|}$$

for any cutoff level $N_i < N$.

In order to validate the choice of the levels $N_{i_1(j)}$ and $N_{i_2(j)}$ obtained from the previous estimates and to evaluate $n_V(j)$, Debussche, Dubois, and Temam (1995) derived several estimates on the time variations of the small scales and of the nonlinear interaction terms by using the time analyticity of the solution of the two-dimensional Navier–Stokes equations, which was established in Foias, Manley, and Temam (1988). These estimates allowed them to control the perturbation errors introduced in the scheme by the quasistatic approximation. This algorithm has been successfully used for the simulation of two-dimensional homogeneous turbulent flows on long intervals of time (see Dubois (1993) and Debussche, Dubois, and Temam (1995)). Improvements, due to the dynamic control of the perturbations, were shown by comparing the current version of the DML method with the earlier versions cited above.

In Dubois and Jauberteau (1998), a new algorithm similar to, but improving on, the previous one has been proposed. Some simplifications have been made in that the definitions of the levels $N_{i_1(j)}$ and $N_{i_2(j)}$ are now based only on the control of the errors due to the quasistatic treatment. This is the algorithm that was applied to the simulations of three-dimensional homogeneous turbulent flows, described in Chapter 10. The following sections (Sections 9.3 and 9.4) are devoted to a detailed description of this algorithm.

9.3. Dynamic Control of the Numerical Parameters

In order to determine the parameters i_1, i_2, and n_V of the multilevel scheme described in Section 9.1, we first derive estimates of the perturbations that we are injecting in the system by using a quasistatic approximation for the small scales on several time steps and for the nonlinear interaction terms during a cycle j. Using these estimates, we then propose a dynamic procedure for the evaluation of i_1, i_2, and n_V based on the control of the perturbations.

Let us consider a time t_{n_j+l} with $1 \leq l \leq 2qn_V$, so that $t_{n_j+l} \in (t_{n_j}, t_{n_{j+1}})$. According to Section 9.1, the scales $\mathbf{y}_{N_{i_1}} + \sum_{i=i_1}^{i(l)-1} \mathbf{z}_i$, where $i(l)$ is defined by (9.3), are computed at t_{n_j+l}. These scales satisfy the following evolution equation:

$$\frac{d}{dt}\left[e^{\nu At}\left(\mathbf{y}_{N_{i_1}} + \sum_{i=i_1}^{i(l)-1} \mathbf{z}_i\right)\right] \tag{9.9$_l$}$$

$$= e^{\nu At} P_{N_{i(l)}} \mathbf{g}_N - e^{\nu At} P_{N_{i(l)}} B\left(\mathbf{y}_{N_{i_1}} + \sum_{i=i_1}^{i(l)-1} \mathbf{z}_i, \mathbf{y}_{N_{i_1}} + \sum_{i=i_1}^{i(l)-1} \mathbf{z}_i\right)$$

$$- e^{\nu At} P_{N_{i(l)}} B_{\text{int}}\left(\mathbf{y}_{N_{i_1}} + \sum_{i=i_1}^{i(l)-1} \mathbf{z}_i, \sum_{i=i(l)}^{i_2-1} \mathbf{z}_i + \mathbf{z}_{N_{i_2}}^N\right),$$

where A is the Stokes operator (see Chapter 1). In order to derive (9.4)$_l$, the equation (9.9)$_l$ is integrated over the time interval (t_{n_j+l-1}, t_{n_j+l}), leading to

$$\mathbf{y}_{N_{i_1}}(t_{n_j+l}) + \sum_{i=i_1}^{i(l)-1} \mathbf{z}_i(t_{n_j+l}) \tag{9.10$_l$}$$

$$= e^{-\nu A\Delta t}\left(\mathbf{y}_{N_{i_1}}(t_{n_j+l-1}) + \sum_{i=i_1}^{i(l)-1} \mathbf{z}_i(t_{n_j+l-1})\right) + \int_{t_{n_j+l-1}}^{t_{n_j+l}} e^{\nu A(s-t_{n_j+l})}$$

$$\times \left[P_{N_{i(l)}} \mathbf{g}_N(s) - P_{N_{i(l)}} B\left(\mathbf{y}_{N_{i_1}}(s) + \sum_{i=i_1}^{i(l)-1} \mathbf{z}_i(s), \mathbf{y}_{N_{i_1}}(s)\right.\right.$$

$$\left.\left. + \sum_{i=i_1}^{i(l)-1} \mathbf{z}_i(s)\right)\right] ds - \int_{t_{n_j+l-1}}^{t_{n_j+l}} e^{\nu A(s-t_{n_j+l})}$$

$$\times P_{N_{i(l)}} B_{\text{int}}\left(\mathbf{y}_{N_{i_1}}(s) + \sum_{i=i_1}^{i(l)-1} \mathbf{z}_i(s), \sum_{i=i(l)}^{i_2-1} \mathbf{z}_i(s) + \mathbf{z}_{N_{i_2}}^N(s)\right) ds.$$

The last integral in the right-hand side of $(9.10)_l$ is approximated by

$$-e^{-\nu A\Delta t}$$

$$\times P_{N_{i(l)}} B_{\text{int}}\left(\mathbf{y}_{N_{i_1}}(t_{n_j}) + \sum_{i=i_1}^{i(l)-1} \mathbf{z}_i(t_{n_j}), \sum_{i=i(l)}^{i_2-1} \mathbf{z}_i(t_{n_j}) + \mathbf{z}^N_{N_{i_2}}(t_{n_j})\right) \int_0^{\Delta t} e^{\nu A s}\, ds.$$

Hence, the error induced in the system at time t_{n_j+l} (consistency error) can be estimated by

$$\varepsilon^{\mathbf{u}}_{\text{int},i(l)}(\mathbf{x}) = \int_{t_{n_j+l-1}}^{t_{n_j+l}} e^{\nu A(s-t_{n_j+l})} \tag{9.11}$$

$$\times \left[P_{N_{i(l)}} B_{\text{int}}\left(\mathbf{y}_{N_{i_1}}(s) + \sum_{i=i_1}^{i(l)-1} \mathbf{z}_i(s), \sum_{i=i(l)}^{i_2-1} \mathbf{z}_i(s) + \mathbf{z}^N_{N_{i_2}}(s)\right)\right.$$

$$\left. - P_{N_{i(l)}} B_{\text{int}}\left(\mathbf{y}_{N_{i_1}}(t_{n_j}) + \sum_{i=i_1}^{i(l)-1} \mathbf{z}_i(t_{n_j}), \sum_{i=i(l)}^{i_2-1} \mathbf{z}_i(t_{n_j}) + \mathbf{z}^N_{N_{i_2}}(t_{n_j})\right)\right] ds.$$

This error can be bounded as follows:

$$\left|\varepsilon^{\mathbf{u}}_{\text{int},i(l)}(\mathbf{x})\right|_0 \le l\, \Delta t^2 \tag{9.12}$$

$$\times \operatorname*{Max}_{s\in[t_{n_j}, t_{n_j+l}]} \left| P_{N_{i(l)}} \dot{B}_{\text{int}}\left(\mathbf{y}_{N_{i_1}}(s) + \sum_{i=i_1}^{i(l)-1} \mathbf{z}_i(s), \sum_{i=i(l)}^{i_2-1} \mathbf{z}_i(s) + \mathbf{z}^N_{N_{i_2}}(s)\right)\right|_0$$

for $l = 1, \ldots, 2qn_V$. From the definition (9.3) of $i(l)$ we can easily show that

$$\begin{aligned} i_{k+2qr} &= i_{(2r+2)q-k} = i(k) \qquad \text{for } k = 1, \ldots, q-1, \\ i_{(2r+1)q} &= i_1, \\ i_{(2r+2)q} &= i_2 \end{aligned}$$

for $r = 0, \ldots, n_V - 1$. Bounds on $|\varepsilon^{\mathbf{u}}_{\text{int},i(l)}(\mathbf{x})|_0$ for $l = 1, \ldots, 2qn_V$ can then be easily derived, namely

$$\left|\varepsilon^{\mathbf{u}}_{\text{int},i(l)}(\mathbf{x})\right|_0 \le [2qn_V + i(l) - i_2]$$

$$\times \Delta t^2 \operatorname*{Max}_{s\in[t_{n_j}, t_{n_{j+1}}]} \left| P_{N_{i(l)}} \dot{B}_{\text{int}}\left(\mathbf{y}_{N_{i_1}}(s) + \sum_{i=i_1}^{i(l)-1} \mathbf{z}_i(s), \sum_{i=i(l)}^{i_2-1} \mathbf{z}_i(s) + \mathbf{z}^N_{N_{i_2}}(s)\right)\right|_0$$

for $l = 2qr + k$ or $l = 2(r+1)q - k$, with $k = 1, \dots, q-1$,

$$\left|\varepsilon^{\mathbf{u}}_{\text{int},i((2r+1)q)}(\mathbf{x})\right|_0$$
$$\leq (2n_V - 1)q\ \Delta t^2 \operatorname*{Max}_{s\in[t_{n_j}, t_{n_{j+1}}]} \left| P_{N_{i_1}} \dot{B}_{\text{int}}\left(\mathbf{y}_{N_{i_1}}(s), \sum_{i=i_1}^{i_2-1} \mathbf{z}_i(s) + \mathbf{z}^N_{N_{i_2}}(s)\right)\right|_0,$$

and

$$\left|\varepsilon^{\mathbf{u}}_{\text{int},i(2(r+1)q)}(\mathbf{x})\right|_0$$
$$\leq 2qn_V\ \Delta t^2 \operatorname*{Max}_{s\in[t_{n_j}, t_{n_{j+1}}]} \left| P_{N_{i_2}} \dot{B}_{\text{int}}\left(\mathbf{y}_{N_{i_1}}(s) + \sum_{i=i_1}^{i_2-1} \mathbf{z}_i(s),\ \mathbf{z}^N_{N_{i_2}}(s)\right)\right|_0$$

for $r = 0, \dots, n_V - 1$. Then we can derive the following uniform bound:

$$\left|\varepsilon^{\mathbf{u}}_{\text{int},i(l)}(\mathbf{x})\right|_0 \leq \varepsilon_1(i) \qquad \text{for } l = 1, \dots, 2qn_V, \tag{9.13}$$

where

$$\varepsilon_1(i) = 2qn_V\ \Delta t^2 \tag{9.14}$$
$$\times \operatorname*{Max}_{s\in[t_{n_j}, t_{n_{j+1}}]} \left| P_{N_i} \dot{B}_{\text{int}}\left(\mathbf{y}_{N_{i_1}}(s) + \sum_{k=i_1}^{i-1} \mathbf{z}_k(s), \sum_{k=i}^{i_2-1} \mathbf{z}_k(s) + \mathbf{z}^N_{N_{i_2}}(s)\right)\right|_0.$$

As we can see in Figure 9.1, $\varepsilon_1(i)$ decreases when the cutoff level N_i is increased.

We now aim to derive an estimate of the error due to the quasistatic approximation of the intermediate scales $\{\mathbf{z}_i\}_{i=i_1,i_2-1}$. By recalling (9.6), we note that the scales $\mathbf{z}_i$ are kept constant on time intervals

$$I_{i,r} = (t_{n_j+(2r-1)q+i_1-i-1}, t_{n_j+(2r-1)q+i-i_1})$$

for $r = 1, \dots, n_V$. The quasistatic approximation on such intervals consists in neglecting

$$\varepsilon^{\mathbf{u}}_{i,r}(\mathbf{x}) = \int_{I_{i,r}} \dot{\mathbf{z}}_i(s)\, ds,$$

which can be bounded as follows:

$$\left|\varepsilon^{\mathbf{u}}_{i,r}(\mathbf{x})\right|_0 \leq [2(i - i_1) + 1]\ \Delta t \operatorname*{Max}_{s\in I_{i,r}} |\dot{\mathbf{z}}_i(s)|_0 \tag{9.15}$$

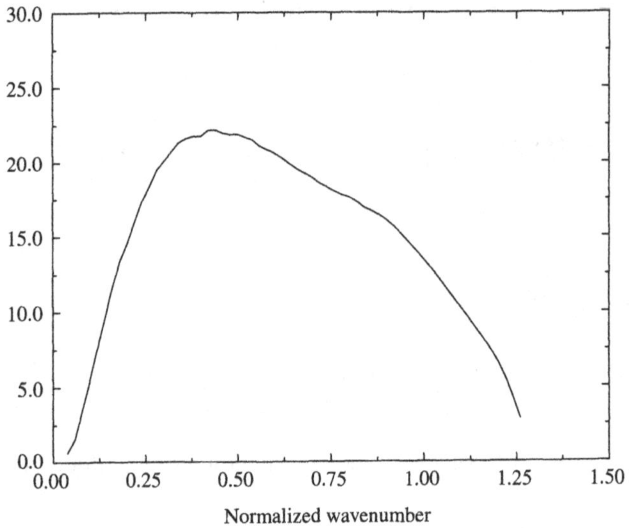

Figure 9.1. $|P_{N_1}\dot{B}_{\text{int}}(\mathbf{y}_{N_1}, \mathbf{z}_{N_1}^N)|_0$ as a function of the normalized wavenumber ηk_{N_1}.

for $i = i_1, \dots, i_2 - 1$ and $r = 1, \dots, n_V$. A bound independent of r can be derived, namely

$$\left|\varepsilon_{i,r}^{\mathbf{u}}(\mathbf{x})\right|_0 \le \varepsilon_2(i) \tag{9.16}$$

$$= [2(i - i_1) + 1]\, \Delta t \max_{s\in[t_{n_j}, t_{n_{j+1}}]} \left| \sum_{k=i}^{i_2-1} \dot{\mathbf{z}}_k(s) + \dot{\mathbf{z}}_{N_{i_2}}^N(s) \right|_0$$

for $i = i_1, \dots, i_2 - 1$. We can note that $\varepsilon_2(i)$ is decreased when the cutoff level N_i is increased.

We still have to derive an estimate of the error induced by the treatment of the small scales $\mathbf{z}_{N_{i_2}}^N$ during the whole cycle. Let us first recall the evolution equation satisfied by these scales:

$$\frac{d}{dt}\left(e^{\nu At}\, \mathbf{z}_{N_{i_2}}^N\right) = -e^{\nu At}\, Q_{N_{i_2}}^N B\left(\mathbf{y}_{N_{i_1}} + \sum_{i=i_1}^{i_2-1} \mathbf{z}_i, \mathbf{y}_{N_{i_1}} + \sum_{i=i_1}^{i_2-1} \mathbf{z}_i\right) \tag{9.17}$$

$$- e^{\nu At}\, Q_{N_{i_2}}^N B_{\text{int}}\left(\mathbf{y}_{N_{i_1}} + \sum_{i=i_1}^{i_2-1} \mathbf{z}_i, \mathbf{z}_{N_{i_2}}^N\right).$$

In order to derive (9.6), the equation (9.17) is integrated on the time interval

$(t_{n_j+2q(r-1)}, t_{n_j+2qr})$, leading to

$$\mathbf{z}_{N_{i_2}}^{N,n_j+2qr} \tag{9.18}$$
$$= e^{-2q\nu A\,\Delta t}\,\mathbf{z}_{N_{i_2}}^{N,n_j+2q(r-1)} - \int_{t_{n_j+2q(r-1)}}^{t_{n_j+2qr}} e^{\nu A(s-t_{n_j+2qr})}$$
$$\times\, Q_{N_{i_2}}^N B\left(\mathbf{y}_{N_{i_1}}(s) + \sum_{i=i_1}^{i_2-1}\mathbf{z}_i(s), \mathbf{y}_{N_{i_1}}(s) + \sum_{i=i_1}^{i_2-1}\mathbf{z}_i(s)\right) ds$$
$$- \int_{t_{n_j+2q(r-1)}}^{t_{n_j+2qr}} e^{\nu A(s-t_{n_j+2qr})} Q_{N_{i_2}}^N B_{\text{int}}\left(\mathbf{y}_{N_{i_1}}(s) + \sum_{i=i_1}^{i_2-1}\mathbf{z}_i(s), \mathbf{z}_{N_{i_2}}^N(s)\right) ds.$$

The first integral in (9.18) is approximated by a second-order semi-implicit scheme (the Crank–Nicholson scheme), while the latter term is neglected. Hence, the major error (consistency error) made by using (9.6) can be estimated by

$$\varepsilon_{i_2(l),r}^{\mathbf{u}}(\mathbf{x}) = \int_{t_{n_j+2q(r-1)}}^{t_{n_j+2qr}} e^{\nu A(s-t_{n_j+2qr})} Q_{N_{i_2}}^N B_{\text{int}}\left(\mathbf{y}_{N_{i_1}}(s) + \sum_{i=i_1}^{i_2-1}\mathbf{z}_i(s), \mathbf{z}_{N_{i_2}}^N(s)\right) ds,$$

which can be bounded as follows:

$$\left|\varepsilon_{i_2(l),r}^{\mathbf{u}}(\mathbf{x})\right|_0 \le 2q\,\Delta t$$
$$\times \operatorname*{Max}_{s\in[t_{n_j+2q(r-1)}, t_{n_j+2qr}]} \left| Q_{N_{i_2}}^N B_{\text{int}}\left(\mathbf{y}_{N_{i_1}}(s) + \sum_{i=i_1}^{i_2-1}\mathbf{z}_i(s), \mathbf{z}_{N_{i_2}}^N(s)\right)\right|_0,$$

for $r = 1, \dots, n_V$. An uniform bound can be derived (see Figure 8.2):

$$\left|\varepsilon_{i_2(l),r}^{\mathbf{u}}(\mathbf{x})\right|_0 \le \varepsilon_3 = 2q\,\Delta t \operatorname*{Max}_{s\in[t_{n_j}, t_{n_{j+1}}]} \left|\dot{\mathbf{z}}_{N_{i_2}}^N(s)\right|_0. \tag{9.19}$$

Note that ε_3 is an estimate of the same type as $\varepsilon_2(i)$ given by (9.16).

We can note that, for a similar error on a cutoff level N_i, the ratio between the allowable number of iterations of the quasistatic approximation of $P_{N_i} B_{\text{int}}(\mathbf{y}_{N_i}, \mathbf{z}_{N_i}^N)$ (see (9.14)) and the allowable number of iterations of the quasistatic approximation of $\mathbf{z}_{N_i}^N$ (see (9.16)) can be estimated by

$$\frac{k_N U_\infty(t_j)\left|\dot{\mathbf{z}}_{N_i}^N(t_j)\right|_0}{\left|P_{N_i}\dot{B}_{\text{int}}\left(\mathbf{y}_{N_i}(t_j), \mathbf{z}_{N_i}^N(t_j)\right)\right|_0},$$

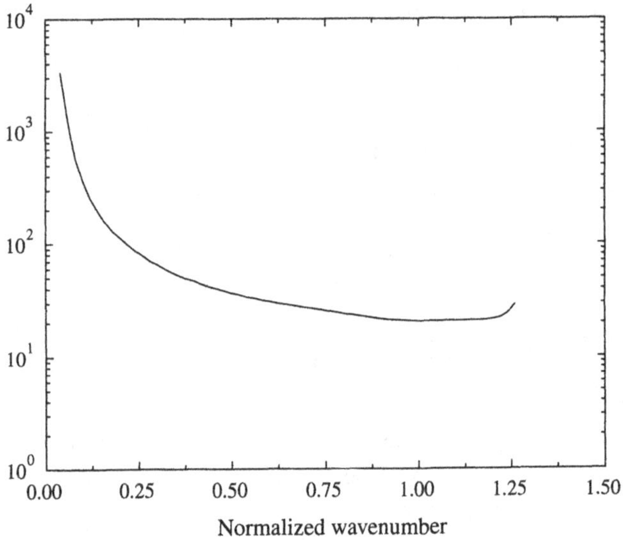

Figure 9.2. Ratio $(k_N U_\infty)|\dot{\mathbf{z}}^N_{N_1}|_0/|P_{N_1}\dot{B}_{\text{int}}(\mathbf{y}_{N_1},\mathbf{z}^N_{N_1})|_0$ as a function of the normalized wavenumber ηk_{N_1}.

where we have used the CFL condition (4.5). This ratio is much larger than unity (see Figure 9.2). Hence it appears that the nonlinear interaction term can be frozen over a much longer time interval than the corresponding small scales.

In summary, the error $\varepsilon_1(i)$ induced in the computation of the velocity field by the quasistatic approximation of the nonlinear interaction terms on the level N_i during a cycle j is of order Δt^2 (see (9.14)). The error $\varepsilon_2(i)$ (respectively, ε_3) induced in the computation of the velocity field, by the quasistatic approximation of the small scales on the level N_i (respectively, N_{i_2}), $N_{i_1} \leq N_i < N_{i_2}$, is of order Δt (see (9.16) and (9.19)). In the following section, we use these error estimates to compute the parameters i_1, i_2, and n_V of the multilevel scheme (see Section 9.1), in order to control each error ε_1, ε_2, and ε_3 (and so the total error) injected in the computation of the velocity field.

9.4. Synthesis of the Algorithm

In this section, we summarize the DML method in the way it has been implemented in order to obtain the numerical results presented in Chapter 10. As has been previously said, the DML algorithm consists in a succession of cycles based on multilevel V-cycles (see Figure 9.3). A cycle j, $j \geq 1$, corresponds to a time interval $(t_{n_{j-1}}, t_{n_j})$, where n_j (see (9.1)) is the number of time

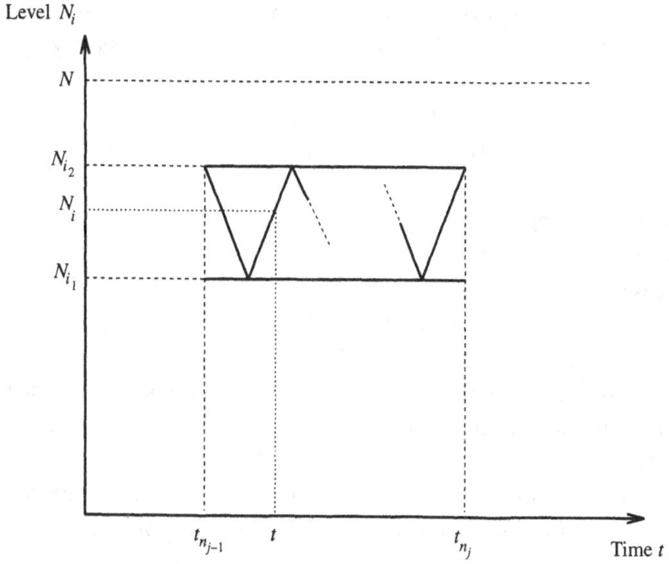

Figure 9.3. The jth cycle of the multilevel strategy; $t_{n_j} = t_{n_{j-1}} + 2q(j)n_V(j)$.

iterations for the cycle and $t_{n_j} = n_j\,\Delta t$. At time $t_{n_{j-1}}$ the following numerical approximation of $\mathbf{u}_N(t_{n_{j-1}})$ has been obtained:

$$\mathbf{u}_N^{n_{j-1}} = \mathbf{y}_{N_{i_1(j-1)}}^{n_{j-1}} + \sum_{i=i_1(j-1)}^{i_2(j-1)-1} \mathbf{z}_i^{n_{j-1}} + \mathbf{z}_{N_{i_2(j-1)}}^{N,n_{j-1}}. \tag{9.20}$$

Let ξ_r be a given nondimensional parameter corresponding to a relative error on the total kinetic energy. The DML method on the time interval $(t_{n_{j-1}}, t_{n_j})$ can then be summarized as follows:

1. *Computation of the absolute error* $\xi(j) = \xi_r |\mathbf{u}_N^{n_{j-1}}|_0$.
2. *Computation of* $i_1(j)$ by imposing that the error $\varepsilon_2(i_1(j))$, estimated in (9.16), is less than or equal to $\xi(j) : i_1(j)$ is the lowest value $i \in [1, m]$ satisfying

$$\frac{\xi(j)}{\left|\mathbf{z}_{N_i}^{N,n_{j-1}}\right|_0} \geq K_1\,\Delta t, \tag{9.21}$$

where Δt is the time step and K_1 a given (tolerance) parameter; in practice K_1 is of the order of 1 or 2.

3. *Computation of* $i_2(j)$ by imposing that the error $\varepsilon_2(i_2(j))$ is less than or equal to $\xi(j) : i_2(j)$ is the lowest value $i \in [i_1(j)+2, m]$ satisfying

$$\frac{\xi(j)}{\left|\dot{\mathbf{z}}_{N_i}^{N,n_{j-1}}\right|_0} \geq 2(i-i_1)\,\Delta t. \tag{9.22}$$

   Note that we impose $i_2(j) \geq i_1(j)+2$, so that $p(j)$ is at least equal to 3. If this inequality on $i_2(j)$ cannot be verified, the cutoff level $N_{i_1}(j)$ is decreased.
4. *The value of* $i_1(j)$ *is adjusted so that* $i_1(j) = \max(i_1(j), i_2(j) - \Delta_{\text{level}} + 1)$, where Δ_{level} is a given parameter. We impose that $q(j) = i_2(j) - i_1(j)$ is bounded by Δ_{level}. Since $|\dot{\mathbf{z}}_{N_i}^N|_0$ decreases when the cutoff level N_i increases, (9.21) is still satisfied if i_1 is increased with the above condition. In practice, the parameter Δ_{level} is of the order of 5.
5. *Computation of* $n_V(j)$ by imposing that the error $\varepsilon_1(i_1(j))$, computed in (9.14), is less than or equal to $\xi(j)$: the number of V-cycles is estimated by

$$n_V(j) \simeq \frac{\xi(j)}{2qK_2\,\Delta t^2\left|P_{N_{i_1}}\dot{B}_{\text{int}}\left(\mathbf{y}_{N_{i_1}}^{n_{j-1}}, \mathbf{z}_{N_{i_1}}^{N,n_{j-1}}\right)\right|_0}, \tag{9.23}$$

   where K_2 is a given (tolerance) parameter. In practice the nonlinear interaction term $|P_{N_{i_1}}\dot{B}_{\text{int}}(\mathbf{y}_{N_{i_1}}^{n_{j-1}}, \mathbf{z}_{N_{i_1}}^{N,n_{j-1}})|_0$ is not computed directly, since it is expensive in CPU time ($O(N^n \log_2 N)$ operations). We have noted in the numerical results (see Figure 9.4) that

$$\begin{aligned}&\left|P_{N_{i_1}}\dot{B}_{\text{int}}\left(\mathbf{y}_{N_{i_1}}^{n_{j-1}}, \mathbf{z}_{N_{i_1}}^{N,n_{j-1}}\right)\right|_0 \\ &\quad\simeq \frac{\left|\dot{\mathbf{z}}_{N_{i_1}}^{N,n_{j-1}}\right|_0\left\|\mathbf{z}_{N_{i_1}}^{N,n_{j-1}}\right\|}{4\left|\mathbf{z}_{N_{i_1}}^{N,n_{j-1}}\right|_0\left\|\mathbf{y}_{N_{i_1}}^{n_{j-1}}\right\|}\left|P_{N_{i_1}}B\left(\mathbf{y}_{N_{i_1}}^{n_{j-1}}, \mathbf{y}_{N_{i_1}}^{n_{j-1}}\right)\right|_0.\end{aligned} \tag{9.24}$$

   In the test (9.23), the above estimate for $|P_{N_{i_1}}\dot{B}_{\text{int}}(\mathbf{y}_{N_{i_1}}^{n_{j-1}})\mathbf{z}_{N_{i_1}}^{N,n_{j-1}}|_0$ is then used. The evaluation of (9.24) requires in this case only $O(N_{i_1}^n \log_2 N_{i_1})$ operations, in order to compute $P_{N_{i_1}}B(\mathbf{y}_{N_{i_1}}^{n_{j-1}}, \mathbf{y}_{N_{i_1}}^{n_{j-1}})$. Since $\varepsilon_1(i)$ is decreased when the cutoff level N_i is increased (see Section 9.3), the control of this error is done on the lowest coarse level N_{i_1}.
6. *Computation of the scales* $\mathbf{y}_{N_{i_1(j)}}^{n_{j-1}+l}$, $\{\mathbf{z}_i^{n_{j-1}+l}\}_{i=i_1(j),i_2(j)-1}$, $\mathbf{z}_{N_{i_2(j)}}^{N,n_{j-1}+l}$ for $l = 1, \ldots, 2q(j)n_V(j)$: these scales are computed according to the scheme described in Section 9.1. We recall that the nonlinear terms are treated with

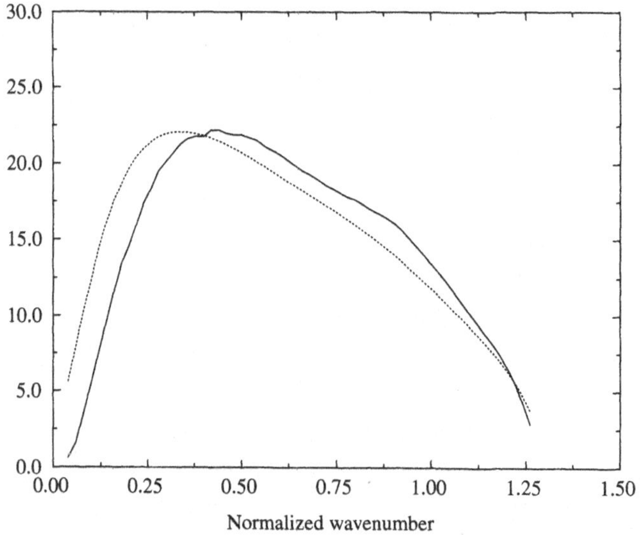

Figure 9.4. $|P_{N_i}\dot{B}_{\mathrm{int}}(\mathbf{y}_{N_i}, \mathbf{z}_{N_i}^N)|_0$ (solid line) and $|P_{N_i}B(\mathbf{y}_{N_i}, \mathbf{y}_{N_i})|_0|\dot{\mathbf{z}}_{N_i}^N|_0\|\mathbf{z}_{N_i}^N\|/(4|\mathbf{z}_{N_i}^N|_0 \|\mathbf{y}_{N_i}\|)$ (dotted line) as functions of the normalized wavenumber ηk_{N_1}.

a third-order explicit Runge–Kutta scheme. The quasistatic approximation is performed on the nonlinear interaction terms at each sub-time-step of the Runge–Kutta calculation.

7. *At time* $t_{n_j} = t_{n_{j-1}} + 2q(j)n_V(j)$ the approximate velocity field $\mathbf{u}_N^{n_j}$ is obtained according to (9.20) where $j-1$ is replaced by j. We then update j in $j+1$ and return to step 1.

This algorithm is repeated until the final time T of the integration is reached.

Remark 9.1. The memory size required by the DML codes is larger than for the DNS code. Indeed, the DML method previously described needs two additional arrays of complex numbers in order to store the spectral coefficients of the nonlinear interaction terms at the beginning of each cycle j, $j \geq 0$. This storage has to be done for the different cutoff levels $N_i \in [N_{i_1}, N_{i_2}]$. According to step 4 described above, the maximum number of levels is bounded by $\Delta_{\mathrm{level}} + 1$. This induces an overhead of the memory, which can be estimated as $\sum_{l=1}^{\Delta_{\mathrm{level}}}(N_{m-l}+1)^2(N_{m-l}+2)$. For comparison, we have listed in Table 9.1 the memory size of a three-dimensional DNS code (see Table 3.1) and estimates of the overhead DML/DNS. Note that this overhead is independent of the resolution; it depends of the sequence of coarse levels retained. □

Table 9.1. *Estimation of the memory size of a three-dimensional DML code in Mega-Octets (real quantities are assumed to be stored in double precision). The overhead DML/DNS is estimated as the quotient (memory size*$_{\text{DML}}$ *− memory size*$_{\text{DNS}}$*)/(memory size*$_{\text{DNS}}$*)*

Resolution	64^3	128^3	256^3
Memory size (DNS) (in Mega-Octets)	21	168	1,341
Overhead (DML/DNS, %)	18	21	32

9.5. The Channel Flow Problem

In this section, we firstly present a dynamic multilevel method applied in the two periodic directions x_1, x_3 and error estimates (Section 9.5.1). Then we consider a dynamic multilevel method in the direction x_2 normal to the walls (Section 9.5.2).

9.5.1. The DML Method Applied in the Homogeneous Directions

By comparison with the homogeneous case (see Section 9.4), we propose the following dynamic multilevel algorithm. This algorithm consists in a succession of cycles based on multilevel V-cycles. A cycle j, $j \geq 1$, corresponds to a time interval $(t_{n_{j-1}}, t_{n_j})$. At time $t_{n_{j-1}}$ the following numerical approximation of $\mathbf{u}_N(t_{n_{j-1}})$ has been obtained:

$$\mathbf{u}_N^{n_{j-1}} = \mathbf{y}_{N_{i_1(j-1)}}^{n_{j-1}} + \sum_{i=i_1(j-1)}^{i_2(j-1)-1} \mathbf{z}_i^{n_{j-1}} + \mathbf{z}_{N_{i_2(j-1)}}^{N,n_{j-1}}. \tag{9.25}$$

We recall (see Section 9.1) that $\mathbf{z}_i^{n_{j-1}} = \mathbf{z}_i(t_{n_{j-1}}) = \mathbf{z}_{N_i}^{N_{i+1}}(t_{n_{j-1}})$ and $z_{2,i}^{n_{j-1}}$ is the second component of the vector $\mathbf{z}_i^{n_{j-1}}$. We have (see (6.49) and (6.50)) that (9.25) induces a similar decomposition on Δu_2 and ω_2:

$$\Delta u_{2,N} = \Delta y_{2N_{i_1(j-1)}}^{n_{j-1}} + \sum_{i=i_1(j-1)}^{i_2(j-1)-1} \Delta z_{2,i}^{n_{j-1}} + \Delta z_{2,N_{i_2(j-1)}}^{N,n_{j-1}}, \tag{9.26}$$

$$\omega_{2,N} = y_{\omega 2,N_{i_1(j-1)}}^{n_{j-1}} + \sum_{i=i_1(j-1)}^{i_2(j-1)-1} z_{\omega 2,i}^{n_{j-1}} + z_{\omega 2,N_{i_2(j-1)}}^{N,n_{j-1}}, \tag{9.27}$$

with $z_{\omega_2,i} = z_{\omega_2,N_i}^{N_{i+1}}$.

Let ξ_r be a given nondimensional parameter corresponding to a relative error on the total kinetic energy. The DML method on the time interval $(t_{n_{j-1}}, t_{n_j})$ can then be summarized as follows:

1. *Computation of the absolute error* $\xi(j) = \xi_r |\mathbf{u}_N^{n_{j-1}}|_0$.
2. *Computation of* $i_1(j)$ associated with the lowest coarse level $N_{i_1(j)}$.
3. *Computation of* $i_2(j)$ associated with the highest coarse level $N_{i_2(j)}$.
4. *The value of* $i_1(j)$ *is adjusted so that* $i_1(j) = \max(i_1(j), i_2(j) - \Delta_{\text{level}} + 1)$, where Δ_{level} is a given parameter. So we impose that $q(j) = i_2(j) - i_1(j)$ is bounded by Δ_{level}.
5. *Computation of* $n_V(j)$ (the number of V-cycles in the cycle j).
6. *Computation of the scale components* $\{y_{2,N_{i_1}}^{n_{j-1}+l}, z_{2,i}^{n_{j-1}+l}\}_{i=i_1,i(l)-1}$ *and* $\{y_{\omega_2,N_{i_1}}^{n_{j-1}+l}, z_{\omega_2,i}^{n_{j-1}+l}\}_{i=i_1,i(l)-1}$ by integrating the equations

$$\begin{aligned}
&\frac{\partial}{\partial t}\Delta\left(y_{2,N_{i_1}} + \sum_{i=i_1}^{i(l)-1} z_{2,i}\right) - \nu\Delta^2\left(y_{2,N_{i_1}} + \sum_{i=i_1}^{i(l)-1} z_{2,i}\right) \\
&\quad + \Delta P_{N_{i(l)}} H_2\left(\mathbf{y}_{N_{i_1}} + \sum_{i=i_1}^{i(l)-1} \mathbf{z}_i, \mathbf{y}_{N_{i_1}} + \sum_{i=i_1}^{i(l)-1} \mathbf{z}_i\right) \\
&\quad - \frac{\partial}{\partial x_2} P_{N_{i(l)}} (\boldsymbol{\nabla}\cdot\mathbf{H})\left(\mathbf{y}_{N_{i_1}} + \sum_{i=i_1}^{i(l)-1} \mathbf{z}_i, \mathbf{y}_{N_{i_1}} + \sum_{i=i_1}^{i(l)-1} \mathbf{z}_i\right) \\
&= -\Delta P_{N_{i(l)}} H_{\text{int},2}\left(\mathbf{y}_{N_{i_1}} + \sum_{i=i_1}^{i(l)-1} \mathbf{z}_i, \sum_{i=i(l)}^{i_2-1} \mathbf{z}_i + \mathbf{z}_{N_{i_2}}^N\right) \\
&\quad + \frac{\partial}{\partial x_2} P_{N_{i(l)}} (\boldsymbol{\nabla}\cdot\mathbf{H}_{\text{int}})\left(\mathbf{y}_{N_{i_1}} + \sum_{i=i_1}^{i(l)-1} \mathbf{z}_i, \sum_{i=i(l)}^{i_2-1} \mathbf{z}_i + \mathbf{z}_{N_{i_2}}^N\right)
\end{aligned} \tag{9.28$_l$}$$

and

$$\begin{aligned}
&\frac{\partial}{\partial t}\left(y_{\omega_2,N_{i_1}} + \sum_{i=i_1}^{i(l)-1} z_{\omega_2,i}\right) - \nu\Delta\left(y_{\omega_2,N_{i_1}} + \sum_{i=i_1}^{i(l)-1} z_{\omega_2,i}\right) \\
&\quad + \frac{\partial}{\partial x_3} P_{N_{i(l)}} H_1\left(\mathbf{y}_{N_{i_1}} + \sum_{i=i_1}^{i(l)-1} \mathbf{z}_i, \mathbf{y}_{N_{i_1}} + \sum_{i=i_1}^{i(l)-1} \mathbf{z}_i\right) \\
&\quad - \frac{\partial}{\partial x_1} P_{N_{i(l)}} H_3\left(\mathbf{y}_{N_{i_1}} + \sum_{i=i_1}^{i(l)-1} \mathbf{z}_i, \mathbf{y}_{N_{i_1}} + \sum_{i=i_1}^{i(l)-1} \mathbf{z}_i\right)
\end{aligned} \tag{9.29$_l$}$$

$$= -\frac{\partial}{\partial x_3} P_{N_{i(l)}} H_{\text{int},1}\left(\mathbf{y}_{N_{i_1}} + \sum_{i=i_1}^{i(l)-1} \mathbf{z}_i, \sum_{i=i(l)}^{i_2-1} \mathbf{z}_i + \mathbf{z}_{N_{i_2}}^N\right)$$

$$+ \frac{\partial}{\partial x_1} P_{N_{i(l)}} H_{\text{int},3}\left(\mathbf{y}_{N_{i_1}} + \sum_{i=i_1}^{i(l)-1} \mathbf{z}_i, \sum_{i=i(l)}^{i_2-1} \mathbf{z}_i + \mathbf{z}_{N_{i_2}}^N\right)$$

over the time interval $(t_{n_{j-1}+(l-1)}, t_{n_{j-1}+l})$, where $n_{j-1}+l$ denotes the current iteration and $N_{i(l)}$ the current cutoff level, $l \in [1, 2qn_V]$ (see (9.3)). We recall that the nonlinear terms are treated with a third-order explicit Runge–Kutta scheme. The quasistatic approximation is performed on the nonlinear interaction terms at each sub-time step of Runge–Kutta. Knowing $\{y_{2,N_{i_1}}^{n_{j-1}+l}, z_{2,i}^{n_{j-1}+l}\}$ and $\{y_{\omega_2,N_{i_1}}^{n_{j-1}+l}, z_{\omega_2,i}^{n_{j-1}+l}\}$ for $i = i_1, \ldots, i(l)-1$, we deduce $\{y_{1,N_{i_1}}^{n_{j-1}+l}, z_{1,i}^{n_{j-1}+l}\}$ and $\{y_{3,N_{i_1}}^{n_{j-1}+l}, z_{3,i}^{n_{j-1}+l}\}$ for $i = i_1, \ldots, i(l)-1$, by using the incompressibility constraint, except for the mode $\mathbf{0}$, which does not vanish in the channel case. To compute $\tilde{u}_j^{n_{j-1}+l}(\mathbf{0})$, $j = 1, 3$, we use the following equations (see (3.31)):

$$\frac{\partial \tilde{u}_j}{\partial t}(\mathbf{0}) - \nu \frac{\partial^2 \tilde{u}_j}{\partial x_2^2}(\mathbf{0}) + \hat{H}_{j,\mathbf{0}}\left(\mathbf{y}_{N_{i_1}} + \sum_{i=i_1}^{i(l)-1} \mathbf{z}_i, \mathbf{y}_{N_{i_1}} + \sum_{i=i_1}^{i(l)-1} \mathbf{z}_i\right) \tag{9.30$_l$}$$

$$+ \hat{H}_{\text{int},j,\mathbf{0}}\left(\mathbf{y}_{N_{i_1}} + \sum_{i=i_1}^{i(l)-1} \mathbf{z}_i, \sum_{i=i(l)}^{i_2-1} \mathbf{z}_i + \mathbf{z}_{N_{i_2}}^N\right) + K_p \delta_{1j} = 0$$

$$\text{for } j = 1, 3.$$

A quasistatic approximation on the time interval $(t_{n_{j-1}+(l-1)}, t_{n_{j-1}+l})$ is applied to the small scales:

$$\mathbf{z}_i^{n_{j-1}+l} = \mathbf{z}_i^{n_{j-1}+(l-1)} \qquad \text{for } i = i(l), \ldots, m-1, \tag{9.31}$$

and

$$z_{\omega_2,i}^{n_{j-1}+l} = z_{\omega_2,i}^{n_{j-1}+(l-1)} \qquad \text{for } i = i(l), \ldots, m-1. \tag{9.32}$$

Note that when $l = (2r+1)q$, with $r = 0, \ldots, n_V - 1$, then $i(l) = i_1$, so that the intermediate scales $\sum_{i=i_1}^{i(l)-1} z_{2,i}$ $(\sum_{i=i_1}^{i(l)-1} z_{\omega_2,i})$ vanish in $(9.28)_l$ $((9.29)_l)$. The cutoff level corresponding to the time iteration $n_{j-1} + (2r+1)q$ is N_{i_1}, and only the large scales $\mathbf{y}_{N_{i_1}}$ and $y_{\omega_2,N_{i_1}}$ are updated.

7. *At time* $t_{n_j} = t_{n_{j-1}} + 2q(j)n_V(j)$ the approximate velocity field $\mathbf{u}_N^{n_j}$ is obtained according to (9.25) where $j-1$ is replaced by j. We then update j in $j+1$ and return to step 1.

This algorithm is repeated until the final time T of the integration is reached.

We will now derive error estimates. Then these estimates will be used to derive appropriate tests for computing the parameters $i_1(j)$, $i_2(j)$ and $n_V(j)$ of the algorithm.

Let $\varepsilon^{\omega_2}_{\mathrm{int},i(l)}$ be the error injected, at time $t_{n_{j-1}+l}$, in the computation of ω_2 (consistency error) namely the error due to the quasistatic approximation on the nonlinear interaction term in Equation $(9.29)_l$. We have

$$\begin{aligned}
&\hat{\varepsilon}^{\omega_2}_{\mathrm{int},i(l)}(\mathbf{k}, x_2)\\
&= \int_{t_{n_{j-1}+(l-1)}}^{t_{n_{j-1}+l}} \Bigg[\left(\frac{\partial}{\partial x_3}\hat{H}_{\mathrm{int},1,\mathbf{k}}\left(\mathbf{y}_{N_{i(l)}}(t), \mathbf{z}^N_{N_{i(l)}}(t)\right)\right.\\
&\qquad \left. - \frac{\partial}{\partial x_1}\hat{H}_{\mathrm{int},3,\mathbf{k}}\left(\mathbf{y}_{N_{i(l)}}(t), \mathbf{z}^N_{N_{i(l)}}(t)\right)\right)\\
&\qquad - \left(\frac{\partial}{\partial x_3}\hat{H}_{\mathrm{int},1,\mathbf{k}}\left(\mathbf{y}_{N_{i(l)}}\left(t_{n_{j-1}}\right), \mathbf{z}^N_{N_{i(l)}}\left(t_{n_{j-1}}\right)\right)\right.\\
&\qquad \left.\left. - \frac{\partial}{\partial x_1}\hat{H}_{\mathrm{int},3,\mathbf{k}}\left(\mathbf{y}_{N_{i(l)}}\left(t_{n_{j-1}}\right), \mathbf{z}^N_{N_{i(l)}}\left(t_{n_{j-1}}\right)\right)\right)\right] dt\\
&= \int_{t_{n_{j-1}+(l-1)}}^{t_{n_{j-1}+l}} i\underline{\mathbf{k}}^{\perp}\cdot\left[\hat{\mathbf{H}}_{\mathrm{int},\mathbf{k}}\left(\mathbf{y}_{N_{i(l)}}(t), \mathbf{z}^N_{N_{i(l)}}(t)\right)\right.\\
&\qquad \left. - \hat{\mathbf{H}}_{\mathrm{int},\mathbf{k}}\left(\mathbf{y}_{N_{i(l)}}\left(t_{n_{j-1}}\right), \mathbf{z}^N_{N_{i(l)}}\left(t_{n_{j-1}}\right)\right)\right] dt
\end{aligned}$$

for $\mathbf{k} \in S_{N_{i(l)}}\backslash\{\mathbf{0}\}$, where $\underline{\mathbf{k}}^{\perp} = (2\pi k_3/L_3, 0, -2\pi k_1/L_1)$ (see Section 3.2.2). We deduce the following upper bound, using the Cauchy–Schwarz inequality:

$$\begin{aligned}
\left|\hat{\varepsilon}^{\omega_2}_{\mathrm{int},i(l)}(\mathbf{k}, x_2)\right| &\le \int_{t_{n_{j-1}+(l-1)}}^{t_{n_{j-1}+l}} |i\underline{\mathbf{k}}^{\perp}|\\
&\times \left|\hat{\mathbf{H}}_{\mathrm{int},\mathbf{k}}\left(\mathbf{y}_{N_{i(l)}}(t), \mathbf{z}^N_{N_{i(l)}}(t)\right) - \hat{\mathbf{H}}_{\mathrm{int},\mathbf{k}}\left(\mathbf{y}_{N_{i(l)}}\left(t_{n_{j-1}}\right), \mathbf{z}^N_{N_{i(l)}}\left(t_{n_{j-1}}\right)\right)\right| dt,
\end{aligned}$$

which can be bounded as follows:

$$\begin{aligned}
&\left|\hat{\varepsilon}^{\omega_2}_{\mathrm{int},i(l)}(\mathbf{k}, x_2)\right|\\
&\le |i\underline{\mathbf{k}}^{\perp}| \int_{t_{n_{j-1}+(l-1)}}^{t_{n_{j-1}+l}} \operatorname*{Max}_{t\in[t_{n_{j-1}}, t_{n_j}]} \left|\dot{\hat{\mathbf{H}}}_{\mathrm{int},\mathbf{k}}\left(\mathbf{y}_{N_{i(l)}}(t), \mathbf{z}^N_{N_{i(l)}}(t)\right)\right| (t - t_{n_j-1})\, dt\\
&\le 2qn_V\Delta t^2\, |i\underline{\mathbf{k}}^{\perp}| \cdot \operatorname*{Max}_{t\in[t_{n_{j-1}}, t_{n_j}]} \left|\dot{\hat{\mathbf{H}}}_{\mathrm{int},\mathbf{k}}\left(\mathbf{y}_{N_{i(l)}}(t), \mathbf{z}^N_{N_{i(l)}}(t)\right)\right|.
\end{aligned}$$

So we obtain

$$\left|\hat{\varepsilon}^{\omega_2}_{\mathrm{int},i(l)}(\mathbf{k}, x_2)\right| \le 2qn_V\Delta t^2\, |\underline{\mathbf{k}}^{\perp}| \operatorname*{Max}_{t\in[t_{n_{j-1}}, t_{n_j}]} \left|\dot{\hat{\mathbf{H}}}_{\mathrm{int},\mathbf{k}}\left(\mathbf{y}_{N_{i(l)}}(t), \mathbf{z}^N_{N_{i(l)}}(t)\right)\right|.$$

Finally, we have the following majoration of the error injected in equation $(9.29)_l$ associated with ω_2, by the quasistatic approximation of the nonlinear interaction terms:

$$\hat{\varepsilon}^{\omega_2}_{\text{int},i(l)}(\mathbf{k}) = \left|\hat{\varepsilon}^{\omega_2}_{\text{int},i(l)}(\mathbf{k}, x_2)\right|_{L^2(-1,+1)} \tag{9.33}$$
$$\leq 4\pi q n_V \Delta t^2 \; |\mathbf{k}| \operatorname*{Max}_{t\in[t_{n_{j-1}},t_{n_j}]} \left|\hat{\mathbf{H}}_{\text{int},\mathbf{k}}\left(\mathbf{y}_{N_{i(l)}}(t), \mathbf{z}^N_{N_{i(l)}}(t)\right)\right|_{L^2(-1,+1)}$$

for $\mathbf{k} \in S_{N_{i(l)}} \backslash \{\mathbf{0}\}$.

Now we consider the error introduced in the computation of Δu_2 by the quasistatic approximation of the nonlinear interaction terms in equation $(9.28)_l$ (consistency error). We decompose this error into $\eta^{i(l)}_{\text{int},1}$ (the error introduced in Δu_2 due to the quasistatic approximation of the first nonlinear interaction term in $(9.28)_l$) and $\eta^{i(l)}_{\text{int},2}$ (the error introduced in Δu_2 due to the quasistatic approximation of the second nonlinear interaction term in $(9.28)_l$). Estimates of the errors $\eta^{i(l)}_{\text{int},j}$, $j = 1, 2$, are as follows:

$$\hat{\eta}^{i(l)}_{\text{int},1}(\mathbf{k}, x_2) = \int_{t_{n_{j-1}+(l-1)}}^{t_{n_{j-1}+l}} \left[D^2_{\mathbf{k}} \hat{H}_{\text{int},2,\mathbf{k}}\left(\mathbf{y}_{N_{i(l)}}(t), \mathbf{z}^N_{N_{i(l)}}(t)\right)\right.$$
$$\left. - D^2_{\mathbf{k}} \hat{H}_{\text{int},2,\mathbf{k}}\left(\mathbf{y}_{N_{i(l)}}(t_{n_{j-1}}), \mathbf{z}^N_{N_{i(l)}}(t_{n_{j-1}})\right)\right] dt,$$

and

$$\hat{\eta}^{i(l)}_{\text{int},2}(\mathbf{k}, x_2) = \int_{t_{n_{j-1}+(l-1)}}^{t_{n_{j-1}+l}} \left[\frac{\partial}{\partial x_2} D_{\mathbf{k}} \cdot \hat{\mathbf{H}}_{\text{int},\mathbf{k}}\left(\mathbf{y}_{N_{i(l)}}(t), \mathbf{z}^N_{N_{i(l)}}(t)\right)\right.$$
$$\left. - \frac{\partial}{\partial x_2} D_{\mathbf{k}} \cdot \hat{\mathbf{H}}_{\text{int},\mathbf{k}}\left(\mathbf{y}_{N_{i(l)}}(t_{n_{j-1}}), \mathbf{z}^N_{N_{i(l)}}(t_{n_{j-1}})\right)\right] dt,$$

for $\mathbf{k} \in S_{N_{i(l)}} \backslash \{\mathbf{0}\}$. We have used the notation introduced in (3.31). We deduce the following upper bound:

$$\hat{\eta}^{i(l)}_{\text{int},1}(\mathbf{k}) = \left|\hat{\eta}^{i(l)}_{\text{int},1}(\mathbf{k}, x_2)\right|_{L^2(-1,+1)} \tag{9.34}$$
$$\leq 2 q n_V \Delta t^2 \operatorname*{Max}_{t\in[t_{n_{j-1}},t_{n_j}]} \left|D^2_{\mathbf{k}} \hat{H}_{\text{int},2,\mathbf{k}}\left(\mathbf{y}_{N_{i(l)}}(t), \mathbf{z}^N_{N_{i(l)}}(t)\right)\right|_{L^2(-1,+1)},$$

and

$$\hat{\eta}^{i(l)}_{\text{int},2}(\mathbf{k}) = \left|\hat{\eta}^{i(l)}_{\text{int},2}(\mathbf{k}, x_2)\right|_{L^2(-1,+1)} \tag{9.35}$$
$$\leq 2 q n_V \Delta t^2 \operatorname*{Max}_{t\in[t_{n_{j-1}},t_{n_j}]} \left|\frac{\partial}{\partial x_2} D_{\mathbf{k}} \cdot \hat{\mathbf{H}}_{\text{int},\mathbf{k}}\left(\mathbf{y}_{N_{i(l)}}(t), \mathbf{z}^N_{N_{i(l)}}(t)\right)\right|_{L^2(-1,+1)}.$$

Now we consider the error $\varepsilon^{j,u_2}_{\text{int},i(l)}$ introduced, in the computation of u_2, by the error $\eta^{i(l)}_{\text{int},j}$ on Δu_2. We have the following relation:

$$\eta^{i(l)}_{\text{int},j} = \Delta \varepsilon^{j,u_2}_{\text{int},i(l)}, \tag{9.36}$$

with $j = 1, 2$. Using (9.36) we obtain the following relation

$$-\,|\underline{\mathbf{k}}|^2\, \hat{\varepsilon}^{j,u_2}_{\text{int},i(l)}(\mathbf{k}, x_2) + \frac{\partial^2}{\partial x_2^2} \hat{\varepsilon}^{j,u_2}_{\text{int},i(l)}(\mathbf{k}, x_2) = \hat{\eta}^{i(l)}_{\text{int},j}(\mathbf{k}, x_2). \tag{9.37}$$

Taking the scalar product of (9.37) with $\hat{\varepsilon}^{j,u_2}_{\text{int},i(l)}(\mathbf{k}, x_2)$ in $L^2(-1, +1)$, we deduce that

$$\begin{aligned} &-\,|\underline{\mathbf{k}}|^2 \left|\hat{\varepsilon}^{j,u_2}_{\text{int},i(l)}(\mathbf{k}, x_2)\right|^2_{L^2(-1,+1)} - \left|\frac{\partial}{\partial x_2} \hat{\varepsilon}^{j,u_2}_{\text{int},i(l)}(\mathbf{k}, x_2)\right|^2_{L^2(-1,+1)} \\ &= \left(\hat{\eta}^{i(l)}_{\text{int},j}(\mathbf{k}, x_2), \hat{\varepsilon}^{j,u_2}_{\text{int},i(l)}(\mathbf{k}, x_2)\right)_{L^2(-1,+1)}. \end{aligned} \tag{9.38}$$

Then, applying the Cauchy–Schwarz inequality in (9.38), we find

$$\begin{aligned} &|\underline{\mathbf{k}}|^2 \left|\hat{\varepsilon}^{j,u_2}_{\text{int},i(l)}(\mathbf{k}, x_2)\right|^2_{L^2(-1,+1)} + \left|\frac{\partial}{\partial x_2} \hat{\varepsilon}^{j,u_2}_{\text{int},i(l)}(\mathbf{k}, x_2)\right|^2_{L^2(-1,+1)} \\ &\leq \left|\hat{\eta}^{i(l)}_{\text{int},j}(\mathbf{k}, x_2)\right|_{L^2(-1,+1)} \left|\hat{\varepsilon}^{j,u_2}_{\text{int},i(l)}(\mathbf{k}, x_2)\right|_{L^2(-1,+1)}. \end{aligned} \tag{9.39}$$

Finally, using (9.34) and (9.35), we obtain

$$\begin{aligned} \hat{\varepsilon}^{1,u_2}_{\text{int},i(l)}(\mathbf{k}) &= \left|\hat{\varepsilon}^{1,u_2}_{\text{int},i(l)}(\mathbf{k}, x_2)\right|_{L^2(-1,+1)} \leq \frac{1}{|\underline{\mathbf{k}}|^2} \hat{\eta}^{i(l)}_{\text{int},1}(\mathbf{k}) \leq 2qn_V \Delta t^2 \\ &\times \frac{1}{|\underline{\mathbf{k}}|^2} \operatorname*{Max}_{t\in[t_{n_{j-1}}, t_{n_j}]} \left|D^2_{\mathbf{k}} \hat{H}_{\text{int},2,\mathbf{k}}(\mathbf{y}_{N_{i(l)}}(t), \mathbf{z}^N_{N_{i(l)}}(t))\right|_{L^2(-1,+1)} \end{aligned} \tag{9.40}$$

and

$$\begin{aligned} \hat{\varepsilon}^{2,u_2}_{\text{int},i(l)}(\mathbf{k}) &= \left|\hat{\varepsilon}^{2,u_2}_{\text{int},i(l)}(\mathbf{k}, x_2)\right|_{L^2(-1,+1)} \leq \frac{1}{|\underline{\mathbf{k}}|^2} \hat{\eta}^{i(l)}_{\text{int},2}(\mathbf{k}) \leq 2qn_V \Delta t^2 \\ &\times \frac{1}{|\underline{\mathbf{k}}|^2} \operatorname*{Max}_{t\in[t_{n_{j-1}} t_{n_j}]} \left|\frac{\partial}{\partial x_2} D_{\mathbf{k}} \cdot \hat{\mathbf{H}}_{\text{int},\mathbf{k}} \left(\mathbf{y}_{N_{i(l)}}(t), \mathbf{z}^N_{N_{i(l)}}(t)\right)\right|_{L^2(-1,+1)}, \end{aligned} \tag{9.41}$$

for $\mathbf{k} \in S_{N_{i(l)}} \backslash \{\mathbf{0}\}$. From the previous estimates (9.40) and (9.41), we obtain the following majoration of the error $\varepsilon^{u_2}_{\text{int},i(l)}$ injected, at time $t_{n_{j-1}+l}$, in the computation of u_2 by the quasistatic approximation of the nonlinear interaction terms in equation $(9.28)_l$:

$$\hat{\varepsilon}^{u_2}_{\text{int},i(l)}(\mathbf{k}) = \left|\hat{\varepsilon}^{u_2}_{\text{int},i(l)}(\mathbf{k}, x_2)\right|_{L^2(-1,+1)} \leq \hat{\varepsilon}^{1,u_2}_{\text{int},i(l)}(\mathbf{k}) + \hat{\varepsilon}^{2,u_2}_{\text{int},i(l)}(\mathbf{k}) \tag{9.42}$$

$$\leq 2qn_V\,\Delta t^2 \frac{1}{|\underline{\mathbf{k}}|^2}\left[\operatorname*{Max}_{t\in[t_{n_{j-1}},t_{n_j}]}\left|D^2_{\mathbf{k}}\hat{\dot{H}}_{\text{int},2,\mathbf{k}}\left(\mathbf{y}_{N_{i(l)}}(t), \mathbf{z}^N_{N_{i(l)}}(t)\right)\right|_{L^2(-1,+1)}\right.$$

$$\left. + \operatorname*{Max}_{t\in[t_{n_{j-1}},t_{n_j}]}\left|\frac{\partial}{\partial x_2}D_{\mathbf{k}}\cdot\hat{\dot{\mathbf{H}}}_{\text{int},\mathbf{k}}\left(\mathbf{y}_{N_{i(l)}}(t), \mathbf{z}^N_{N_{i(l)}}(t)\right)\right|_{L^2(-1,+1)}\right]$$

for $\mathbf{k} \in S_{N_{i(l)}}\setminus\{\mathbf{0}\}$.

For the mode $\mathbf{k} = \mathbf{0}$, we impose for the DML methodology, as for DNS, that $\hat{u}_2(\mathbf{k} = \mathbf{0}, x_2, t) = U_2(x_2, t) = 0$ (see (3.25)). So no error is induced on the computation of the averaged value in the periodic directions x_1 and x_3. It follows that $\hat{\varepsilon}^{u_2}_{\text{int},i(l)}(\mathbf{k} = \mathbf{0}, x_2) = 0$.

From the Parseval identity (8.5) and from (9.42), we deduce that

$$\left|\varepsilon^{u_2}_{\text{int},i(l)}\right|_{L^2(\Omega)} = \sqrt{L_1L_3}\times\sqrt{\sum_{\mathbf{k}\in S_{N_{i(l)}}}\left|\hat{\varepsilon}^{u_2}_{\text{int},i(l)}(\mathbf{k})\right|^2} \leq 2qn_V\,\Delta t^2\sqrt{L_1L_3} \tag{9.43}$$

$$\times\left[\operatorname*{Max}_{t\in[t_{n_{j-1}},t_{n_j}]}\left|P_{N_{i(l)}}\,\Delta\dot{H}_{\text{int},2}\left(\mathbf{y}_{N_{i(l)}}(t), \mathbf{z}^N_{N_{i(l)}}(t)\right)\right|_{L^2(\Omega)}\right.$$

$$\left. + \operatorname*{Max}_{t\in[t_{n_{j-1}},t_{n_j}]}\left|\frac{\partial}{\partial x_2}P_{N_{i(l)}}(\nabla\cdot\dot{\mathbf{H}}_{\text{int}})\left(\mathbf{y}_{N_{i(l)}}(t), \mathbf{z}^N_{N_{i(l)}}(t)\right)\right|_{L^2(\Omega)}\right],$$

where Ω is the domain defined in Section 3.2, with $h = 1$.

Let us now consider the error $\varepsilon^f_{\text{int},i(l)}$ induced in the computation of $f = -\partial u_2/\partial x_2$ by the error $\varepsilon^{u_2}_{\text{int},i(l)}$ on u_2. We have

$$\hat{\varepsilon}^f_{\text{int},i(l)}(\mathbf{k}, x_2) = -\frac{\partial}{\partial x_2}\hat{\varepsilon}^{u_2}_{\text{int},i(l)}(\mathbf{k}, x_2).$$

We derive that

$$\hat{\varepsilon}^f_{\text{int},i(l)}(\mathbf{k}) = \left|\hat{\varepsilon}^f_{\text{int},i(l)}(\mathbf{k}, x_2)\right|_{L^2(-1,+1)} = \left|\frac{\partial}{\partial x_2}\hat{\varepsilon}^{u_2}_{\text{int},i(l)}(\mathbf{k}, x_2)\right|_{L^2(-1,+1)}$$

$$\leq \left|\frac{\partial}{\partial x_2}\hat{\varepsilon}^{1,u_2}_{\text{int},i(l)}(\mathbf{k}, x_2)\right|_{L^2(-1,+1)} + \left|\frac{\partial}{\partial x_2}\hat{\varepsilon}^{2,u_2}_{\text{int},i(l)}(\mathbf{k}, x_2)\right|_{L^2(-1,+1)}.$$

From (9.39), (9.40), (9.41) and from (9.34), (9.35), we deduce that

$$\left|\frac{\partial}{\partial x_2}\hat{\varepsilon}^{1,u_2}_{\text{int},i(l)}(\mathbf{k},x_2)\right|_{L^2(-1,+1)} \leq \frac{1}{|\mathbf{k}|}\hat{\eta}^{i(l)}_{\text{int},1}(\mathbf{k})$$
$$\leq 2qn_V\,\Delta t^2\frac{1}{|\mathbf{k}|}\underset{t\in[t_{n_{j-1}},t_{n_j}]}{\text{Max}}\left|D^2_{\mathbf{k}}\hat{H}_{\text{int},2,\mathbf{k}}\big(\mathbf{y}_{N_{i(l)}}(t),\mathbf{z}^N_{N_{i(l)}}(t)\big)\right|_{L^2(-1,+1)},$$

and

$$\left|\frac{\partial}{\partial x_2}\hat{\varepsilon}^{2,u_2}_{\text{int},i(l)}(\mathbf{k},x_2)\right|_{L^2(-1,+1)} \leq \frac{1}{|\mathbf{k}|}\hat{\eta}^{i(l)}_{\text{int},2}(\mathbf{k})$$
$$\leq 2qn_V\,\Delta t^2\frac{1}{|\mathbf{k}|}\underset{t\in[t_{n_{j-1}},t_{n_j}]}{\text{Max}}\left|\frac{\partial}{\partial x_2}D_{\mathbf{k}}\cdot\hat{\mathbf{H}}_{\text{int},\mathbf{k}}\big(\mathbf{y}_{N_{i(l)}}(t),\mathbf{z}^N_{N_{i(l)}}(t)\big)\right|_{L^2(-1,+1)}.$$

Finally we obtain the following majoration of the error $\hat{\varepsilon}^{f}_{\text{int},i(l)}(\mathbf{k})$ injected in the computation of f, at time $t_{n_{j-1}+l}$, by the quasistatic approximation of the nonlinear interaction terms of equation $(9.28)_l$:

$$\begin{aligned}\hat{\varepsilon}^{f}_{\text{int},i(l)}(\mathbf{k}) \leq{}& 2qn_V\,\Delta t^2 \\ &\times\frac{1}{|\mathbf{k}|}\left[\underset{t\in[t_{n_{j-1}},t_{n_j}]}{\text{Max}}\left|D^2_{\mathbf{k}}\hat{H}_{\text{int},2,\mathbf{k}}\left(\mathbf{y}_{N_{i(l)}}(t),\mathbf{z}^N_{N_{i(l)}}(t)\right)\right|_{L^2(-1,+1)}\right. \\ &+\left.\underset{t\in[t_{n_{j-1}},t_{n_j}]}{\text{Max}}\left|\frac{\partial}{\partial x_2}D_{\mathbf{k}}\cdot\hat{\mathbf{H}}_{\text{int},\mathbf{k}}\left(\mathbf{y}_{N_{i(l)}}(t),\mathbf{z}^N_{N_{i(l)}}(t)\right)\right|_{L^2(-1,+1)}\right],\end{aligned} \tag{9.44}$$

for $\mathbf{k}\in S_{N_{i(l)}}\backslash\{\mathbf{0}\}$.

Now, we can establish estimates on the error $\varepsilon^{u_i}_{\text{int},i(l)}$ in the computation of u_i, $i = 1,3$, resulting from the quasistatic approximation of the nonlinear interaction terms in equation $(9.28)_l$ (Δu_2) and in equation $(9.29)_l$ (ω_2). We have the following relations:

$$\begin{aligned}&\mathbf{i}\frac{2\pi}{L_1}k_1\hat{\varepsilon}^{u_1}_{\text{int},i(l)}(\mathbf{k},x_2)+\mathbf{i}\frac{2\pi}{L_3}k_3\hat{\varepsilon}^{u_3}_{\text{int},i(l)}(\mathbf{k},x_2)=\hat{\varepsilon}^{f}_{\text{int},i(l)}(\mathbf{k},x_2),\\ &\mathbf{i}\frac{2\pi}{L_3}k_3\hat{\varepsilon}^{u_1}_{\text{int},i(l)}(\mathbf{k},x_2)-\mathbf{i}\frac{2\pi}{L_1}k_1\hat{\varepsilon}^{u_3}_{\text{int},i(l)}(\mathbf{k},x_2)=\hat{\varepsilon}^{\omega_2}_{\text{int},i(l)}(\mathbf{k},x_2).\end{aligned} \tag{9.45}$$

From (9.45), we can deduce that

$$\hat{\varepsilon}^{u_1}_{\text{int},i(l)}(\mathbf{k},x_2) = -\frac{\mathbf{i}}{|\mathbf{k}|^2}\left(\frac{2\pi}{L_1}k_1\hat{\varepsilon}^{f}_{\text{int},i(l)}(\mathbf{k},x_2)+\frac{2\pi}{L_3}k_3\hat{\varepsilon}^{\omega_2}_{\text{int},i(l)}(\mathbf{k},x_2)\right),$$
$$\hat{\varepsilon}^{u_3}_{\text{int},i(l)}(\mathbf{k},x_2) = -\frac{\mathbf{i}}{|\mathbf{k}|^2}\left(\frac{2\pi}{L_3}k_3\hat{\varepsilon}^{f}_{\text{int},i(l)}(\mathbf{k},x_2)-\frac{2\pi}{L_1}k_1\hat{\varepsilon}^{\omega_2}_{\text{int},i(l)}(\mathbf{k},x_2)\right).$$

Taking the norm in the space $L^2(-1,+1)$, we obtain the following estimates:

$$\begin{aligned}
\hat{\varepsilon}^{u_1}_{\text{int},i(l)}(\mathbf{k}) &= \left|\hat{\varepsilon}^{u_1}_{\text{int},i(l)}(\mathbf{k},x_2)\right|_{L^2(-1,+1)} \le \frac{1}{|\underline{\mathbf{k}}|}\left(\hat{\varepsilon}^{f}_{\text{int},i(l)}(\mathbf{k}) + \hat{\varepsilon}^{\omega_2}_{\text{int},i(l)}(\mathbf{k})\right), \\
\hat{\varepsilon}^{u_3}_{\text{int},i(l)}(\mathbf{k}) &= \left|\hat{\varepsilon}^{u_3}_{\text{int},i(l)}(\mathbf{k},x_2)\right|_{L^2(-1,+1)} \le \frac{1}{|\underline{\mathbf{k}}|}\left(\hat{\varepsilon}^{f}_{\text{int},i(l)}(\mathbf{k}) + \hat{\varepsilon}^{\omega_2}_{\text{int},i(l)}(\mathbf{k})\right).
\end{aligned} \tag{9.46}$$

Substituting (9.33) and (9.44) in (9.46), we finally obtain the following estimates for the error, at time $t_{n_{j-1}+l}$, induced in the computation of u_j, $j = 1, 3$, by the quasistatic approximation on the nonlinear intercation terms in $(9.28)_l$ and $(9.29)_l$:

$$\begin{aligned}
\hat{\varepsilon}^{u_j}_{\text{int},i(l)}(\mathbf{k}) &\le 2qn_V\,\Delta t^2 \\
&\times\Bigg[2\pi \operatorname*{Max}_{t\in[t_{n_{j-1}},t_{n_j}]}\left|\hat{\mathbf{H}}_{\text{int},\mathbf{k}}\left(\mathbf{y}_{N_{i(l)}}(t),\mathbf{z}^N_{N_{i(l)}}(t)\right)\right|_{L^2(-1,+1)^n} \\
&+\frac{1}{|\mathbf{k}|^2}\Bigg(\operatorname*{Max}_{t\in[t_{n_{j-1}},t_{n_j}]}\left|D^2_{\mathbf{k}}\hat{H}_{\text{int},2,\mathbf{k}}\left(\mathbf{y}_{N_{i(l)}}(t),\mathbf{z}^N_{N_{i(l)}}(t)\right)\right|_{L^2(-1,+1)} \\
&+\operatorname*{Max}_{t\in[t_{n_{j-1}},t_{n_j}]}\left|\frac{\partial}{\partial x_2}D_{\mathbf{k}}\cdot\hat{\mathbf{H}}_{\text{int},\mathbf{k}}\left(\mathbf{y}_{N_{i(l)}}(t),\mathbf{z}^N_{N_{i(l)}}(t)\right)\right|_{L^2(-1,+1)}\Bigg)\Bigg],
\end{aligned} \tag{9.47}$$

for $\mathbf{k}\in S_{N_{i(l)}}\backslash\{\mathbf{0}\}$ and $j = 1, 3$.

Concerning the error $\hat{\varepsilon}^{u_j}_{\text{int},i(l)}(\mathbf{k},x_2)$, for $\mathbf{k}=\mathbf{0}$ and $j = 1, 3$, since $\tilde{u}_j(\mathbf{0})$ is computed with $(9.30)_l$, we have the following estimates:

$$\begin{aligned}
\hat{\varepsilon}^{u_j}_{\text{int},i(l)}(\mathbf{k}=\mathbf{0},x_2) = \int_{t_{n_{j-1}+(l-1)}}^{t_{n_{j-1}+l}} &\left[\hat{H}_{\text{int},j,0}\left(\mathbf{y}_{N_{i(l)}}(t),\mathbf{z}^N_{N_{i(l)}}(t)\right)\right. \\
&\left.-\hat{H}_{\text{int},j,0}\left(\mathbf{y}_{N_{i(l)}}(t_{n_{j-1}}),\mathbf{z}^N_{N_{i(l)}}(t_{n_{j-1}})\right)\right]dt,
\end{aligned}$$

from which we deduce the following majoration:

$$\begin{aligned}
\hat{\varepsilon}^{u_j}_{\text{int},i(l)}(\mathbf{k}=\mathbf{0}) &= \left|\hat{\varepsilon}^{u_j}_{\text{int},i(l)}(\mathbf{k}=\mathbf{0},x_2)\right|_{L^2(-1,+1)} \\
&\le 2qn_V\,\Delta t^2 \operatorname*{Max}_{t\in[t_{n_{j-1}},t_{n_j}]}\left|\hat{H}_{\text{int},j,0}\left(\mathbf{y}_{N_{i(l)}}(t),\mathbf{z}^N_{N_{i(l)}}(t)\right)\right|_{L^2(-1,+1)}.
\end{aligned} \tag{9.48}$$

Using the Parseval identity (8.5) and (9.47) and (9.48), we obtain

$$\begin{aligned}
&\left|\varepsilon^{u_j}_{\text{int},i(l)}\right|_{L^2(\Omega)} \\
&= \sqrt{L_1L_3}\times\sqrt{\sum_{\mathbf{k}\in S_{N_{i(l)}}}\left|\hat{\varepsilon}^{u_j}_{\text{int},i(l)}(\mathbf{k})\right|^2} \le 2qn_V\,\Delta t^2\sqrt{L_1L_3}
\end{aligned} \tag{9.49}$$

$$\times \left[2\pi \underset{t \in [t_{n_{j-1}}, t_{n_j}]}{\text{Max}} \left| P_{N_{i(l)}} \dot{\mathbf{H}}_{\text{int}} \left(\mathbf{y}_{N_{i(l)}}(t), \mathbf{z}^N_{N_{i(l)}}(t) \right) \right|_{L^2(\Omega)^n} \right.$$

$$+ \underset{t \in [t_{n_{j-1}}, t_{n_j}]}{\text{Max}} \left| P_{N_{i(l)}} \Delta \dot{H}_{\text{int},2} \left(\mathbf{y}_{N_{i(l)}}(t), \mathbf{z}^N_{N_{i(l)}}(t) \right) \right|_{L^2(\Omega)}$$

$$\left. + \underset{t \in [t_{n_{j-1}}, t_{n_j}]}{\text{Max}} \left| \frac{\partial}{\partial x_2} P_{N_{i(l)}} (\nabla \cdot \dot{\mathbf{H}}_{\text{int}}) \left(\mathbf{y}_{N_{i(l)}}(t), \mathbf{z}^N_{N_{i(l)}}(t) \right) \right|_{L^2(\Omega)} \right]$$

for $j = 1, 3$.

Now we consider the error $\varepsilon^{\mathbf{u}}_{i(l)} = (\varepsilon^{u_j}_{i(l)})_j$, $j = 1, 2, 3$, injected in the computation of $\mathbf{u}$ at time $t_{n_{j-1}+l}$ by the quasistatic approximation of the scales $\mathbf{z}_i$, for $i = i(l), \ldots, m-1$. As in the homogeneous case (see Section 9.3), the scales $\mathbf{z}_i$ are kept constant on time intervals

$$I_{i,r} = \left(t_{n_{j-1}+(2r-1)q+i_1-i-1}, t_{n_{j-1}+(2r-1)q+i-i_1} \right)$$

for $r = 1, \ldots, n_V$. The quasistatic approximation on such intervals consists in neglecting the term $\int_{I_{i,r}} \hat{\dot{\mathbf{z}}}_i(\mathbf{k}, x_2, t)\, dt$. Let us write

$$\hat{\varepsilon}^{\mathbf{u}}_{i(l),r}(\mathbf{k}, x_2) = \int_{I_{i(l),r}} \hat{\dot{\mathbf{z}}}^N_{N_{i(l)}}(\mathbf{k}, x_2, t)\, dt.$$

Taking the $L^2(-1, +1)$ norm of this quantity, we obtain the following upper bound:

$$\begin{aligned} \hat{\varepsilon}^{\mathbf{u}}_{i(l),r}(\mathbf{k}) &= \left| \hat{\varepsilon}^{\mathbf{u}}_{i(l),r}(\mathbf{k}, x_2) \right|_{L^2(-1,+1)^n} \le \hat{\varepsilon}^{\mathbf{u}}_{i(l)}(\mathbf{k}) \\ &= \{2[i(l) - i_1] + 1\}\, \Delta t \underset{t \in [t_{n_{j-1}}, t_{n_j}]}{\text{Max}} \left| \hat{\dot{\mathbf{z}}}^N_{N_{i(l)}}(\mathbf{k}, x_2, t) \right|_{L^2(-1,+1)^n} \end{aligned} \tag{9.50}$$

for $\mathbf{k} \in S_N \backslash S_{N_{i(l)}}$.

With the Parseval identity (8.5), we have

$$\begin{aligned} \left| \varepsilon^{\mathbf{u}}_{i(l)} \right|_{L^2(\Omega)^n} &= \sqrt{L_1 L_3} \times \sqrt{\sum_{\mathbf{k} \in S_N \backslash S_{N_{i(l)}}} \left| \hat{\varepsilon}^{\mathbf{u}}_{i(l)}(\mathbf{k}) \right|^2} \\ &\le \{2[i(l) - i_1] + 1\} \Delta t \sqrt{L_1 L_3} \times \underset{t \in [t_{n_{j-1}}, t_{n_j}]}{\text{Max}} \left| \dot{\mathbf{z}}^N_{N_{i(l)}}(t) \right|_{L^2(\Omega)^n}. \end{aligned} \tag{9.51}$$

In Figure 9.5, we have represented $|\dot{\mathbf{z}}^N_{N_i}(t)|_{L^2(\Omega)^n}$ at a specified time, as a function of N_i. As we can see, this quantity decreases when N_i increases. So, from (9.51), we deduce that the error $|\varepsilon^{\mathbf{u}}_{i(l)}|_{L^2(\Omega)^n}$ decreases when $i(l)$ increases.

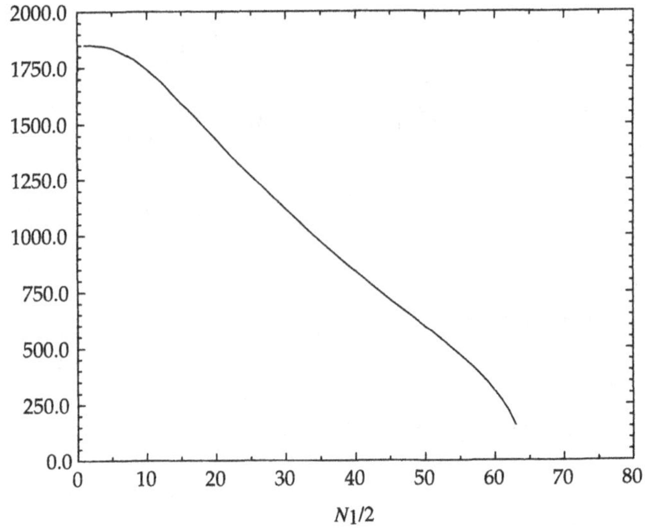

Figure 9.5. $|\dot{\mathbf{z}}_{N_1}^N|_0$ as a function of the wavenumber $k_{N_1} = N_1/2$.

In the same manner, we have observed, on numerical simulations, that the quantities

$$\left|P_{N_{i(l)}} \Delta \dot{H}_{\text{int},2}\left(\mathbf{y}_{N_{i(l)}}(t), \mathbf{z}_{N_{i(l)}}^N(t)\right)\right|_{L^2(\Omega)},$$

$$\left|\frac{\partial}{\partial x_2} P_{N_{i(l)}}(\nabla \cdot \dot{\mathbf{H}}_{\text{int}})\left(\mathbf{y}_{N_{i(l)}}(t), \mathbf{z}_{N_{i(l)}}^N(t)\right)\right|_{L^2(\Omega)},$$

and

$$\left|P_{N_{i(l)}} \dot{\mathbf{H}}_{\text{int}}\left(\mathbf{y}_{N_{i(l)}}(t), \mathbf{z}_{N_{i(l)}}^N(t)\right)\right|_{L^2(\Omega)^n}$$

decrease when N_i increases. Then, from (9.43) and (9.49) we can deduce that the errors $|\varepsilon_{\text{int},i(l)}^{u_j}|_{L^2(\Omega)}$, $j = 1, 2, 3$, decrease when $i(l)$ increases.

Now, as in the homogeneous case, we use the previous estimates to define the parameters $i_1(j)$, $i_2(j)$, and $n_V(j)$. We choose $i_1(j)$, as the lowest value $i_1 \in [1, m]$ satisfying

$$K_1 \left|\varepsilon_{i_1}^{\mathbf{u}}\right|_{L^2(\Omega)^n} \leq \xi(j), \tag{9.52}$$

with K_1 a given (tolerance) parameter ($K_1 \simeq 1$ or 2). Then we choose $i_2(j)$ as the lowest value $i_2 \in [i_1(j) + 2, m]$ satisfying

$$K_2 \left|\varepsilon_{i_2}^{\mathbf{u}}\right|_{L^2(\Omega)^n} \leq \xi(j), \tag{9.53}$$

where K_2 is a given (tolerance) parameter ($K_2 \simeq 1$ or 2). In fact, in (9.52) and (9.53), we replace $|\varepsilon_i^{\mathbf{u}}|_{L^2(\Omega)^n}$ by the majoration given by (9.51). The value of $i_1(j)$ is adjusted so that $i_1(j) = \mathrm{Max}(i_1(j), i_2(j) - \Delta_{\mathrm{level}} + 1)$, with Δ_{level} a given parameter. As for the number of V-cycles, it is estimated by imposing

$$|\varepsilon^{\mathbf{u}}_{\mathrm{int},i_1(j)}|_{L^2(\Omega)^n} \leq \xi(j), \tag{9.54}$$

where $\varepsilon^{\mathbf{u}}_{\mathrm{int},i(l)} = (\varepsilon^{u_j}_{\mathrm{int},i(l)})_j,\ j = 1, \dots, 3$.

In order to simplify the estimates of i_1, i_2 and n_V, we can observe that

$$\begin{aligned}
&\left| D^2_{\mathbf{k}} \hat{H}_{\mathrm{int},2,\mathbf{k}} \left(\mathbf{y}_{N_{i(l)}}(t), \mathbf{z}^N_{N_{i(l)}}(t)\right)\right|_{L^2(-1,+1)} \\
&\leq |\underline{\mathbf{k}}|^2 \left| \hat{H}_{\mathrm{int},2,\mathbf{k}} \left(\mathbf{y}_{N_{i(l)}}(t), \mathbf{z}^N_{N_{i(l)}}(t)\right)\right|_{L^2(-1,+1)} \\
&\quad + \left| \frac{\partial^2}{\partial x_2^2} \hat{H}_{\mathrm{int},2,\mathbf{k}} \left(\mathbf{y}_{N_{i(l)}}(t), \mathbf{z}^N_{N_{i(l)}}(t)\right)\right|_{L^2(-1,+1)} \\
&\leq |\underline{\mathbf{k}}|^2 \left| \hat{\mathbf{H}}_{\mathrm{int},\mathbf{k}} \left(\mathbf{y}_{N_{i(l)}}(t), \mathbf{z}^N_{N_{i(l)}}(t)\right)\right|_{L^2(-1,+1)^n} \\
&\quad + \left| \frac{\partial^2}{\partial x_2^2} \hat{\mathbf{H}}_{\mathrm{int},\mathbf{k}} \left(\mathbf{y}_{N_{i(l)}}(t), \mathbf{z}^N_{N_{i(l)}}(t)\right)\right|_{L^2(-1,+1)^n} .
\end{aligned} \tag{9.55}$$

In the same manner, we have

$$\begin{aligned}
&\left| \frac{\partial}{\partial x_2} D_{\mathbf{k}} \cdot \hat{\mathbf{H}}_{\mathrm{int},\mathbf{k}} \left(\mathbf{y}_{N_{i(l)}}(t), \mathbf{z}^N_{N_{i(l)}}(t)\right)\right|_{L^2(-1,+1)} \\
&\leq \sum_{j=1,3} \frac{2\pi}{L_j} k_j \left| \frac{\partial}{\partial x_2} \hat{H}_{\mathrm{int},j,\mathbf{k}} \left(\mathbf{y}_{N_{i(l)}}(t), \mathbf{z}^N_{N_{i(l)}}(t)\right)\right|_{L^2(-1,+1)} \\
&\quad + \left| \frac{\partial^2}{\partial x_2^2} \hat{H}_{\mathrm{int},2,\mathbf{k}} \left(\mathbf{y}_{N_{i(l)}}(t), \mathbf{z}^N_{N_{i(l)}}(t)\right)\right|_{L^2(-1,+1)} .
\end{aligned}$$

However,

$$\begin{aligned}
&\left| \frac{\partial}{\partial x_2} \hat{H}_{\mathrm{int},j,\mathbf{k}} \left(\mathbf{y}_{N_{i(l)}}(t), \mathbf{z}^N_{N_{i(l)}}(t)\right)\right|^2_{L^2(-1,+1)} \\
&= \left(\frac{\partial}{\partial x_2} \hat{H}_{\mathrm{int},j,\mathbf{k}} \left(\mathbf{y}_{N_{i(l)}}(t), \mathbf{z}^N_{N_{i(l)}}(t)\right), \frac{\partial}{\partial x_2} \hat{H}_{\mathrm{int},j,\mathbf{k}} \left(\mathbf{y}_{N_{i(l)}}(t), \mathbf{z}^N_{N_{i(l)}}(t)\right)\right)_{L^2(-1,+1)} \\
&= -\left(\hat{H}_{\mathrm{int},j,\mathbf{k}} \left(\mathbf{y}_{N_{i(l)}}(t), \mathbf{z}^N_{N_{i(l)}}(t)\right), \frac{\partial^2}{\partial x_2^2} \hat{H}_{\mathrm{int},j,\mathbf{k}} \left(\mathbf{y}_{N_{i(l)}}(t), \mathbf{z}^N_{N_{i(l)}}(t)\right)\right)_{L^2(-1,+1)} ,
\end{aligned}$$

with $j = 1, 3$. From the Cauchy–Schwarz inequality, it follows that

$$\begin{aligned}
&\left| \frac{\partial}{\partial x_2} \hat{\hat{H}}_{\mathrm{int},j,\mathbf{k}} \left(\mathbf{y}_{N_{i(l)}}(t), \mathbf{z}^N_{N_{i(l)}}(t)\right) \right|_{L^2(-1,+1)} \\
&\quad \le \left(\left| \hat{\hat{H}}_{\mathrm{int},j,\mathbf{k}} \left(\mathbf{y}_{N_{i(l)}}(t), \mathbf{z}^N_{N_{i(l)}}(t)\right) \right|_{L^2(-1,+1)} \right. \\
&\qquad \left. \times \left| \frac{\partial^2}{\partial x_2^2} \hat{\hat{H}}_{\mathrm{int},j,\mathbf{k}} \left(\mathbf{y}_{N_{i(l)}}(t), \mathbf{z}^N_{N_{i(l)}}(t)\right) \right|_{L^2(-1,+1)} \right)^{1/2}.
\end{aligned}$$

Finally, we obtain

$$\begin{aligned}
&\left| \frac{\partial}{\partial x_2} D_{\mathbf{k}} \cdot \hat{\hat{\mathbf{H}}}_{\mathrm{int},\mathbf{k}} \left(\mathbf{y}_{N_{i(l)}}(t), \mathbf{z}^N_{N_{i(l)}}(t)\right) \right|_{L^2(-1,+1)} \\
&\quad \le 2|\mathbf{k}| \times \left(\left| \hat{\hat{\mathbf{H}}}_{\mathrm{int},\mathbf{k}} \left(\mathbf{y}_{N_{i(l)}}(t), \mathbf{z}^N_{N_{i(l)}}(t)\right) \right|_{L^2(-1,+1)^n} \right. \\
&\qquad \left. \times \left| \frac{\partial^2}{\partial x_2^2} \hat{\hat{\mathbf{H}}}_{\mathrm{int},\mathbf{k}} \left(\mathbf{y}_{N_{i(l)}}(t), \mathbf{z}^N_{N_{i(l)}}(t)\right) \right|_{L^2(-1,+1)^n} \right)^{1/2} \\
&\qquad + \left| \frac{\partial^2}{\partial x_2^2} \hat{\hat{\mathbf{H}}}_{\mathrm{int},\mathbf{k}} \left(\mathbf{y}_{N_{i(l)}}(t), \mathbf{z}^N_{N_{i(l)}}(t)\right) \right|_{L^2(-1,+1)^n}.
\end{aligned} \tag{9.56}$$

The majorations (9.55) and (9.56) can be injected in the estimates (9.43) and (9.49) on $\varepsilon^{u_j}_{\mathrm{int},i(l)}$, $j = 1, 2, 3$. So we just need to compute $|\dot{\mathbf{z}}^N_{N_i}(t)|_{L^2(\Omega)^n}$, $|P_{N_{i(l)}} \dot{\mathbf{H}}_{\mathrm{int}}(\mathbf{y}_{N_{i(l)}}, \mathbf{z}^N_{N_{i(l)}})|_{L^2(\Omega)^n}$, and $|(\partial^2/\partial x_2^2)(P_{N_{i(l)}} \dot{\mathbf{H}}_{\mathrm{int}}(\mathbf{y}_{N_{i(l)}}, \mathbf{z}^N_{N_{i(l)}}))|_{L^2(\Omega)^n}$ in order to estimate the parameters i_1, i_2, and n_V at the beginning of each cycle.

In summary, the parameters $i_1(j)$ (associated with the lowest coarse level) and $i_2(j)$ (associated with the highest coarse level) are computed using (9.52) and (9.53). For the error $|\varepsilon^{\mathbf{u}}_i|_{L^2(\Omega)^n}$ we use the estimate given by (9.51). For the number of V-cycles in the cycle j, $n_V(j)$, we use (9.54). For the error $|\varepsilon^{\mathbf{u}}_{\mathrm{int},i}|_{L^2(\Omega)^n}$, we use the estimates (9.43) and (9.49), in which we have injected the majorations (9.55) and (9.56).

9.5.2. The DML Method Applied in the Nonhomogeneous Direction

In this section, we propose a dynamic multilevel algorithm for the nonhomogeneous direction. We denote by M the total number of modes retained in the direction x_2 normal to the walls. As in the periodic case, we choose a sequence of m cutoff levels $\{M_i\}_{i=1,m}$, satisfying

$$4 \le M_1 < M_2 < \cdots < M_i < M_{i+1} < \cdots < M_{m-1} \le M_m = M,$$

so that $\Delta_i = M_{i+1} - M_i$ remains approximately constant for $i = 1, \ldots, m-1$. For the cutoff levels N_i, the use of a one-dimensional fast Chebychev transform (FCT) at M_i modes for computing the nonlinear terms impose restrictions on the values of the cutoff levels. Indeed, the levels M_i have to be of the form $2^p 3^q 5^r 7^s$, $p, q, r, s \in \mathbb{N}$ with $p \geq 1$.

As previously, the algorithm consists in a succession of cycles based on multilevel V-cycles. A cycle j, $j \geq 1$, corresponds to a time interval $(t_{n_{j-1}}, t_{n_j})$. At time $t_{n_{j-1}}$ the following numerical approximation of $\mathbf{u}_N(t_{n_{j-1}})$ has been obtained (see (6.56) and (6.57)):

$$\tilde{u}_{2,M}^{n_{j-1}}(\mathbf{k}, x_2) = \tilde{u}_{2,M_{i_1(j-1)}}^{n_{j-1}}(\mathbf{k}, x_2) + \sum_{i=i_1(j-1)}^{i_2(j-1)-1} \tilde{u}_{2,i}^{\prime n_{j-1}}(\mathbf{k}, x_2) + \tilde{u}_{2,M_{i_2(j-1)}}^{\prime M, n_{j-1}}, \tag{9.57}$$

with $\tilde{u}_{2,M_{i_1(j-1)}}(\mathbf{k}) \in W_{M_{i_1}}$, $\tilde{u}'_{2,i}(\mathbf{k}) \in W_{M_{i+1}} \backslash W_{M_i}$, $\tilde{u}_{2,M_{i_2(j-1)}}^{\prime M} \in W_M \backslash W_{M_{i_2}}$, and

$$\tilde{u}_{j,M}^{n_{j-1}}(\mathbf{k}, x_2) = \tilde{u}_{j,M_{i_1(j-1)}}^{n_{j-1}}(\mathbf{k}, x_2) + \sum_{i=i_1(j-1)}^{i_2(j-1)-1} \tilde{u}_{j,i}^{\prime n_{j-1}}(\mathbf{k}, x_2) + \tilde{u}_{j,M_{i_2(j-1)}}^{\prime M, n_{j-1}} \tag{9.58}$$

for $j = 1, 3$, with $\tilde{u}_{j,M_{i_1(j-1)}}(\mathbf{k}) \in V_{M_{i_1}}$, $\tilde{u}'_{j,i}(\mathbf{k}) \in V_{M_{i+1}} \backslash V_{M_i}$, $\tilde{u}_{j,M_{i_2(j-1)}}^{\prime M} \in V_M \backslash V_{M_{i_2}}$. We have written $\mathbf{u}'_i = \mathbf{u}_{M_i}^{\prime M_{i+1}}$. We recall (see (6.58) and (6.59)) that (9.57) and (9.58) induce a similar decomposition on Δu_2 and ω_2:

$$\begin{aligned} &\Delta \tilde{u}_{2,M}(\mathbf{k}, x_2) \\ &\quad = \Delta \tilde{u}_{2,M_{i_1(j-1)}}(\mathbf{k}, x_2) + \sum_{i=i_1(j-1)}^{i_2(j-1)-1} \Delta \tilde{u}'_{2,i}(\mathbf{k}, x_2) + \Delta \tilde{u}_{2,M_{i_2(j-1)}}^{\prime M} \end{aligned} \tag{9.59}$$

and

$$\tilde{\omega}_{2,M}(\mathbf{k}, x_2) = \tilde{\omega}_{2,M_{i_1(j-1)}}(\mathbf{k}, x_2) + \sum_{i=i_1(j-1)}^{i_2(j-1)-1} \tilde{\omega}'_{2,i}(\mathbf{k}, x_2) + \tilde{\omega}_{2,M_{i_2(j-1)}}^{\prime M}. \tag{9.60}$$

Let ξ_r be a given nondimensional parameter corresponding to a relative error on the total kinetic energy. The DML method on the time interval $(t_{n_{j-1}}, t_{n_j})$ can then be summarized as follows:

1. *Computation of the absolute error* $\xi(j) = \xi_r |\mathbf{u}_N^{n_{j-1}}|_0$.
2. *Computation of* $i_1(j)$ associated with the lowest coarse level $M_{i_1(j)}$.
3. *Computation of* $i_2(j)$ associated with the highest coarse level $M_{i_2(j)}$.

4. *The value of* $i_1(j)$ *is adjusted so that* $i_1(j) = \max(i_1(j), i_2(j) - \Delta_{\text{level}} + 1)$, where Δ_{level} is a given parameter. So we impose that $q(j) = i_2(j) - i_1(j)$ is bounded by Δ_{level}.
5. *Computation of* $n_V(j)$ (the number of V-cycles in the cycle j).
6. *Computation of the scale components* $\{\tilde{u}_{2,M_{i_1}}^{n_{j-1}+l}, \tilde{u}_{2,i}^{\prime m_{j-1}+l}\}_{i=i_1,i(l)-1}$ *and* $\{\tilde{\omega}_{2,M_{i_1}}^{n_{j-1}+l}, \tilde{\omega}_{2,i}^{\prime m_{j-1}+l}\}_{i=i_1,i(l)-1}$ by integrating the following equations:

$$
\begin{aligned}
&\left(\frac{\partial}{\partial t} D_{\mathbf{k}}^2 \tilde{u}_{2,M_{i_1}}(\mathbf{k}) + \sum_{i=i_1}^{i(l)-1} \frac{\partial}{\partial t} D_{\mathbf{k}}^2 \tilde{u}'_{2,i}(\mathbf{k}), \psi_{l_{x_2}}\right) \qquad (9.61)_l \\
&\quad - \nu \left(D_{\mathbf{k}}^4 \tilde{u}_{2,M_{i_1}}(\mathbf{k}) + \sum_{i=i_1}^{i(l)-1} D_{\mathbf{k}}^4 \tilde{u}'_{2,i}(\mathbf{k}), \psi_{l_{x_2}} \right) \\
&\quad + \left(\hat{h}_{\mathbf{k}} \left(\mathbf{u}_{M_{i_1}} + \sum_{i=i_1}^{i(l)-1} \mathbf{u}'_i, \mathbf{u}_{M_{i_1}} + \sum_{i=i_1}^{i(l)-1} \mathbf{u}'_i \right), \psi_{l_{x_2}} \right) \\
&= -\left(\sum_{i=i(l)}^{i_2-1} \frac{\partial}{\partial t} D_{\mathbf{k}}^2 \tilde{u}'_{2,i}(\mathbf{k}) + \frac{\partial}{\partial t} D_{\mathbf{k}}^2 \tilde{u}_{2,M_{i_2}}^{\prime M}(\mathbf{k}), \psi_{l_{x_2}} \right) \\
&\quad + \nu \left(\sum_{i=i(l)}^{i_2-1} D_{\mathbf{k}}^4 \tilde{u}'_{2,i}(\mathbf{k}) + D_{\mathbf{k}}^4 \tilde{u}_{2,M_{i_2}}^{\prime M}(\mathbf{k}), \psi_{l_{x_2}} \right) \\
&\quad - \left(\hat{h}_{\text{int},\mathbf{k}} \left(\mathbf{u}_{M_{i_1}} + \sum_{i=i_1}^{i(l)-1} \mathbf{u}'_i, \sum_{i=i(l)}^{i_2-1} \mathbf{u}'_i + \mathbf{u}_{M_{i_2}}^{\prime M} \right), \psi_{l_{x_2}} \right)
\end{aligned}
$$

for $l_{x_2} = 0, \ldots, M_{i(l)-4}$, and

$$
\begin{aligned}
&\left(\frac{\partial}{\partial t} \tilde{\omega}_{2,M_{i_1}}(\mathbf{k}) + \sum_{i=i_1}^{i(l)-1} \frac{\partial}{\partial t} \tilde{\omega}'_{2,i}(\mathbf{k}), \phi_{l_{x_2}}\right) \qquad (9.62)_l \\
&\quad - \nu \left(D_{\mathbf{k}}^2 \tilde{\omega}_{2,M_{i_1}}(\mathbf{k}) + \sum_{i=i_1}^{i(l)-1} D_{\mathbf{k}}^2 \tilde{\omega}'_{2,i}(\mathbf{k}), \phi_{l_{x_2}} \right) \\
&\quad + \left(i\underline{k}^{\perp} \cdot \hat{\mathbf{H}}_{\mathbf{k}} \left(\mathbf{u}_{M_{i_1}} + \sum_{i=i_1}^{i(l)-1} \mathbf{u}'_i, \mathbf{u}_{M_{i_1}} + \sum_{i=i_1}^{i(l)-1} \mathbf{u}'_i \right), \phi_{l_{x_2}} \right) \\
&= -\left(\sum_{i=i(l)}^{i_2-1} \frac{\partial}{\partial t} \tilde{\omega}'_{2,i}(\mathbf{k}) + \frac{\partial}{\partial t} \tilde{\omega}_{2,M_{i_2}}^{\prime M}(\mathbf{k}), \phi_{l_{x_2}} \right)
\end{aligned}
$$

$$+\nu\left(\sum_{i=i(l)}^{i_2-1} D_{\mathbf{k}}^2\tilde{\omega}'_{2,i}(\mathbf{k}) + D_{\mathbf{k}}^2\tilde{\omega}'^M_{2,M_{i_2}}(\mathbf{k}), \phi_{l_{x_2}}\right)$$

$$-\left(i\underline{\mathbf{k}}^\perp \cdot \hat{\mathbf{H}}_{\text{int},\mathbf{k}}\left(\mathbf{u}_{M_{i_1}} + \sum_{i=i_1}^{i(l)-1}\mathbf{u}'_i, \sum_{i=i(l)}^{i_2-1}\mathbf{u}'_i + \mathbf{u}'^M_{M_{i_2}}\right), \phi_{l_{x_2}}\right)$$

for $l_{x_2} = 0, \ldots, M_{i(l)-2}$, over the time interval $(t_{n_{j-1}+(l-1)}, t_{n_{j-1}+l})$, where $n_{j-1}+l$ denotes the current iteration and $M_{i(l)}$ the current cutoff level, $l \in [1, 2qn_V]$. We recall that the nonlinear terms are treated with a third-order explicit Runge–Kutta scheme. The quasistatic approximation is performed on the nonlinear interaction terms at each sub-time-step of Runge–Kutta.

Knowing $\{\tilde{u}_{2,M_{i_1}}^{n_{j-1}+l}, \tilde{u}_{2,i}'^{n_{j-1}+l}\}_{i=i_1,i(l)-1}$ and $\{\tilde{\omega}_{2,M_{i_1}}^{n_{j-1}+l}, \tilde{\omega}_{2,i}'^{n_{j-1}+l}\}_{i=i_1,i(l)-1}$, we deduce $\{\tilde{u}_{1,M_{i_1}}^{n_{j-1}+l}, \tilde{u}_{1,i}'^{n_{j-1}+l}\}_{i=i_1,i(l)-1}$ and $\{\tilde{u}_{3,M_{i_1}}^{n_{j-1}+l}, \tilde{u}_{3,i}'^{n_{j-1}+l}\}_{i=i_1,i(l)-1}$ using the incompressibility constraint

$$\begin{aligned} ik_1\tilde{u}_1^{n_{j-1}+l}(\mathbf{k}, l_{x_2}) + ik_3\tilde{u}_3^{n_{j-1}+l}(\mathbf{k}, l_{x_2}) &= -\tilde{u}_2^{(1)n_{j-1}+l}(\mathbf{k}, l_{x_2}), \\ ik_3\tilde{u}_1^{n_{j-1}+l}(\mathbf{k}, l_{x_2}) - ik_1\tilde{u}_3^{n_{j-1}+l}(\mathbf{k}, l_{x_2}) &= \tilde{\omega}_2^{n_{j-1}+l}(\mathbf{k}, l_{x_2}) \end{aligned} \tag{9.63}$$

for $l_{x_2} = 0, \ldots, M_{i(l)-2}$ (see (6.63)). However, for $k_1 = k_3 = 0$ we cannot use (9.63). To compute $\tilde{u}_j^{n_{j-1}+l}(\mathbf{k} = \mathbf{0}, l_{x_2})$, $j = 1, 3$, we use the following equation:

$$\frac{\partial \tilde{u}_j}{\partial t}(\mathbf{0}) - \nu\frac{\partial^2 \tilde{u}_j}{\partial x_2^2}(\mathbf{0}) + \hat{H}_{j,\mathbf{0}}\left(\mathbf{u}_{M_{i_1}} + \sum_{i=i_1}^{i(l)-1}\mathbf{u}'_i, \mathbf{u}_{M_{i_1}} + \sum_{i=i_1}^{i(l)-1}\mathbf{u}'_i\right)$$
$$+\hat{H}_{\text{int},j,\mathbf{0}}\left(\mathbf{u}_{M_{i_1}} + \sum_{i=i_1}^{i(l)-1}\mathbf{u}'_i, \sum_{i=i(l)}^{i_2-1}\mathbf{u}'_i + \mathbf{u}'^M_{M_{i_2}}\right) + K_p\delta_{1j} = 0$$
$$\text{for } j = 1, 3.$$

A quasistatic approximation on the time interval $(t_{n_{j-1}+(l-1)}, t_{n_{j-1}+l})$ is applied to the small scales:

$$\mathbf{u}_i'^{n_j+l} = \mathbf{u}_i'^{n_j+l-1} \qquad \text{for } i = i(l), \ldots, m-1, \tag{9.64}$$

and

$$u_{\omega_2,i}'^{n_j+l} = u_{\omega_2,i}'^{n_j+l-1} \qquad \text{for } i = i(l), \ldots, m-1. \tag{9.65}$$

Note that when $l = (2r+1)q$, with $r = 0, \ldots, n_V - 1$, then $i(l) = i_1$ so that the intermediate scales $\sum_{i=i_1}^{i(l)-1}\tilde{u}'_{2,i}$ $(\sum_{i=i_1}^{i(l)-1}\tilde{\omega}'_{2,i})$ vanish in $(9.61)_l$ (in

(9.62)$_l$). The cutoff level corresponding to the time iteration $n_j + (2r+1)q$ is N_{i_1}, and only the large scales $\mathbf{y}_{N_{i_1}}$ and $y_{\omega_2, N_{i_1}}$ are updated.

7. *At time* $t_{n_j} = t_{n_{j-1}} + 2q(j)n_V(j)$ the approximate velocity field $\mathbf{u}_N^{n_j}$ is obtained according to (9.57) where $j-1$ is replaced by j. We then update j in $j+1$ and return to step 1.

This algorithm is repeated until the final time T of the integration is reached.

As in Section 9.5.1, error estimates are required to derive tests in order to compute the parameters $i_1(j)$, $i_2(j)$, and $n_V(j)$ of the algorithm. This work is in progress; details will be presented elsewhere.

10
Numerical Results

In this chapter we present the results of several numerical simulations and tests performed using the DML methods presented in Chapter 9. Two problems are considered: the homogeneous turbulence modeled by periodic flows, and nonhomogeneous turbulence. In the latter case, turbulent flows in an infinite three-dimensional channel are studied. For the homogeneous case, two-dimensional and three-dimensional simulations are reported and respectively described in the first and second sections. The two-dimensional fully periodic simulations correspond to resolutions of 128 and 256 modes in each spatial direction, while in the three-dimensional case 128^3 DNS and DML simulations at Reynolds number Re_λ of the order of 100 are described and analyzed in detail. For the channel flow problem, the simulation, presented in Section 10.3, corresponds to a Reynolds number $Re_* = 180$ with a resolution of $128 \times 129 \times 128$ modes. In this simulation we obtain, at reduced cost, results in good agreement with those presented in Kim, Moin, and Moser (1987).

10.1. Two-Dimensional Homogeneous Turbulent Flows

Several two-dimensional flows have been computed. The turbulence has been maintained by applying an external volume force acting on the large scales. These simulations have been presented in Dubois (1993), Debussche, Dubois, and Temam (1995), and Dubois, Jauberteau, and Temam (1995). The results obtained with the DML and pseudospectral Galerkin methods have been compared. Here, we present a numerical simulation described in Debussche, Dubois, and Temam (1995).

10.1.1. The Computational Parameters

The computational domain is $\Omega = (0, 2\pi)^2$. The external force is time-independent. In the spectral space, only a few low-frequency components of this force have nonzero coefficients, so that $\mathbf{f}(\mathbf{x}, t)$ is only active in the low-frequency components of the velocity field. These coefficients are defined as follows:

$$\hat{\mathbf{f}}(\mathbf{k}) = (\hat{f}_1(\mathbf{k}), \hat{f}_2(\mathbf{k})) \tag{10.1}$$

with

$$|\hat{f}_j(\mathbf{k})| = \begin{cases} c_8 & \text{if } |\mathbf{k}| = 3, \\ 0 & \text{otherwise,} \end{cases} \tag{10.2}$$

where $j = 1$ or 2, and c_8 is a constant chosen so that the L^2 norm of $\mathbf{f}$, $|\mathbf{f}|_0$, is equal to a given number. Finally, the Fourier coefficients of $\mathbf{f}$ are given by

$$\hat{f}_j(\mathbf{k}) = |\hat{f}_j(\mathbf{k})|\, e^{i\theta(\mathbf{k})}, \tag{10.3}$$

where the phases $\theta(\mathbf{k}) \in [0, 2\pi]$ are randomly generated by the computer. We choose the constant c_8 so that $|\mathbf{f}|_0 = 0.225$.

The initial condition is chosen so that its energy spectrum function fits a given shape. We first consider the spectrum of the initial vorticity $\omega_0^N = \nabla \times \mathbf{u}_0^N$, where

$$\mathbf{u}_0^N(\mathbf{x}) = \sum_{\mathbf{k}\in[1-N/2,N/2]\times[0,N/2]} \hat{\mathbf{u}}_0(\mathbf{k})\, e^{i\mathbf{k}\cdot\mathbf{x}}, \tag{10.4}$$

$\mathbf{x} = (x_1, x_2) \in \Omega = (0, 2\pi)^2$. We define ω_0^N by its Fourier coefficients

$$\hat{\omega}_0(\mathbf{k}) = |\hat{\omega}_0(\mathbf{k})|\, e^{i\theta(\mathbf{k})} \tag{10.5}$$

where $\theta(\mathbf{k}) \in [0, 2\pi]$ is generated, as for the external force, by a random function. The amplitude of $\hat{\omega}_0(\mathbf{k})$ satisfies

$$|\hat{\omega}_0(\mathbf{k})| = \begin{cases} \dfrac{c_9}{[k + (\sqrt{\nu}k)^5]^{1/2}} & \text{if } k = |\mathbf{k}| < k_\alpha, \\ 0, & \text{otherwise.} \end{cases} \tag{10.6}$$

The constant c_9 is chosen so that $|\omega_0^N|_\infty$ is equal to a prescribed value, and k_α is a given parameter. We choose k_α equal to 60, the kinematic viscosity $\nu = 10^{-3}$, and c_9 so that $|\omega_0^N|_\infty = 2.0$. If $|\hat{\omega}_0(\mathbf{k})| \simeq c_9 k^\beta$, the energy spectrum function of $\mathbf{u}_0$, namely (see Section 2.2.1)

$$E_0(k) = \frac{1}{2} \sum_{\mathbf{k}\in S_{k,1/2}} |\hat{\mathbf{u}}_0(\mathbf{k})|^2, \tag{10.7}$$

behaves like $c_9 k^{2\beta-1}$. So, in the case presented here, $\beta = -0.5$ implies that $E_0(k) \simeq k^{-2}$ (see Figure 10.1).

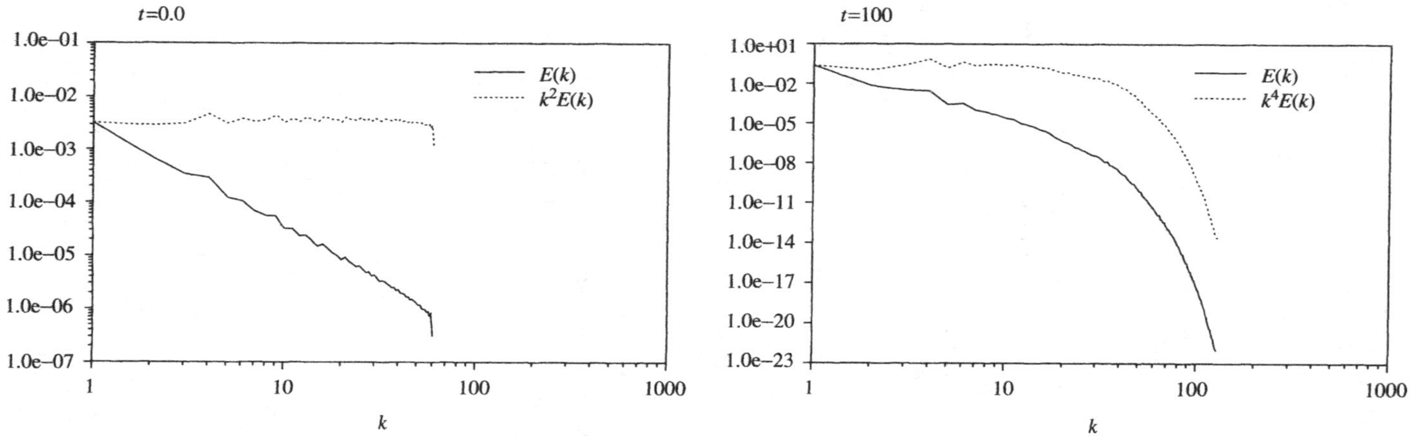

Figure 10.1. Energy spectrum function of the initial condition ($t = 0$) and at the final time of the computation ($t = 100$).

As has been noted in Section 4.1.1, in order to accurately compute all the scales of motion, the number of modes N retained in each direction must satisfy

$$k_N > k_d = \left[\frac{1}{\eta}\right], \tag{10.8}$$

where $[\cdot]$ denotes the integer part; this can be rewritten as

$$N \geq \mathrm{Re}_L^{1/2}. \tag{10.9}$$

For the simulation under consideration, the total number of modes retained, N^n, is taken equal to 256^2, so that the grid is fine enough to resolve the scales below the dissipative ones (DNS). The integral scale Reynolds number Re_L is equal to 784, and the integral scale $L = 1.72$. The dissipative wavenumber k_d varies between 15 and 25 during the time integration, and its average value is $k_d \simeq 20$, so that $\eta k_N \simeq 6.4$. The time step Δt is chosen according to accuracy and stability considerations. It is small enough to insure the numerical stability of the scheme, that is, the following Courant–Friedrichs–Lewy (CFL) condition is required:

$$k_N \,\Delta t \, U_\infty < \alpha \;(= \sqrt{3}). \tag{10.10}$$

We have set Δt to 10^{-3}. Since the L^∞ (supremum) norm U_∞ of $\mathbf{u}_N$ has reached a maximum value of 1.15, the CFL number $k_N \,\Delta t\; U_\infty$ remains below 0.15; hence, (10.10) is satisfied. Moreover, this value of Δt allows us to compute all the scales with enough accuracy. Indeed, the scheme is of the third order in time, and the smallest scales are of the order of 10^{-10} in the $|\cdot|_0$ norm (see Figure 10.2).

10.1.2. Description and Analysis of the Numerical Results

The numerical results obtained with this two-dimensional simulation were described in detail in Debussche, Dubois, and Temam (1995). We recall here the main parts of these results.

In Figure 10.1, the slope of the energy spectrum function appears to be smaller than -3 and close to -4. The decay of the energy spectrum function is then faster than for a fully developed turbulent flow (see Section 2.2.4). These results are in agreement with those obtained by Orszag (1977) and by Brachet et al. (1988), which show that a k^{-3} energy spectrum function can only be obtained when the Reynolds number is of the order of 25,000. With the characteristic

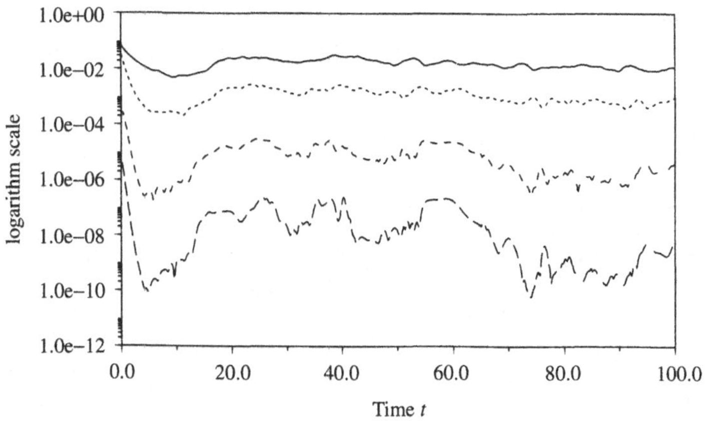

Figure 10.2. Time evolution of $|\mathbf{z}_{N_1}^N|_0$ for $N_1 = 32$ (solid line), 64 (dotted line), 128 (dashed line), and 196 (long-dashed line).

length scale L and the root mean square velocity u we can define, as in the three-dimensional case (see Section 2.2.4), a characteristic time unit

$$\tau_e = \frac{L}{u}, \tag{10.11}$$

called the eddy-turnover time. For this simulation, τ_e is approximately equal to 3.82. The flow was computed over 10^5 iterations, that is, on the time interval $[0, 100]$, which corresponds to 26 eddy-turnover times. This simulation was performed on a Cray 2. Two codes, using respectively the DML method described in Section 9.2 and the pseudospectral Galerkin method (DNS) described in Section 3.1, have been developed and were used simultaneously so that their results could be compared.

In order to compare the DML and DNS simulations, the accuracy parameter ξ of the DML method is first set to 10^{-9}. In such a case, all the scales of the flow can be recovered. Indeed, $k_{N_{i_1}}$ is larger than the Kolmogorov wavenumber k_d, and it lies far inside the dissipation range; DML is then close to DNS. This can be seen when we compare the time evolution of the first component u_1 of the velocity field at two separated points of the computational domain Ω, namely $\mathbf{x}^1 = (2\pi/3, 2\pi/3)$ and $\mathbf{x}^2 = (\pi/3, \pi/3)$. In Table 10.1 (Table 10.2) we compare $u_1(\mathbf{x}^1)$ $(u_1(\mathbf{x}^2))$ obtained with DNS (first column) and DML (second column) at different times. These results show that the DNS and DML trajectories remain close to each other during the whole computation. The distance between these trajectories is of the order of ξ. In Figure 10.3, we can see that the lowest coarse level N_{i_1} is always larger than 128.

Table 10.1. *Comparison of the first component u_1 of the velocity at the point $\mathbf{x}^1 = (2\pi/3, 2\pi/3)$ and at different times t between 0 and 100*

	u_1		
t	DNS	DML	Difference
0	$-0.9650082490 \times 10^{-2}$	$-0.9650082490 \times 10^{-2}$	0.000
25	0.2980770146	0.2980770146	$<1.0 \times 10^{-10}$
50	0.3592114126	0.3592114126	$<1.0 \times 10^{-10}$
75	0.3819454889	0.3819454893	1.0×10^{-9}
100	-0.1233746011	-0.1233746004	1.0×10^{-9}

Table 10.2. *Comparison of the first component u_1 of the velocity at the point $\mathbf{x}^2 = (\pi/3, \pi/3)$ and at different times t between 0 and 100*

	u_1		
t	DNS	DML	Difference
0	$0.4501442245 \times 10^{-1}$	$0.4501442245 \times 10^{-1}$	0.000
25	0.3879730195	0.3879730197	2.0×10^{-10}
50	0.2458823192	0.2458823192	$<1.0 \times 10^{-10}$
75	-0.3332440190	-0.3332440188	2.0×10^{-10}
100	0.3196664945	0.3196664945	$<1.0 \times 10^{-10}$

So the associated wavenumber $k_{N_{i_1}} = 64$ is always larger than the Kolmogorov wavenumber $k_d \simeq 20$. Finally, we want to mention that the DML method used for this test, with $\xi = 10^{-9}$, requires twice less CPU time than the DNS. These results are consistent with the analysis of the multilevel method performed in Section 9.3 and with the implementation of the DML algorithm described in Section 9.4.

With the same DNS simulation, another numerical test has been performed with the DML method. The accuracy parameter ξ is taken equal to 10^{-2} in order to decrease the coarse level N_{i_1} so that $k_{N_{i_1}} = 12$, that is, $\eta k_{N_{i_1}} \simeq 0.5$ (see Figure 10.4); hence parts of the inertial range, as well as the whole dissipation range, are modeled by the DML strategy. In such a case, DML is close to an LES in the sense that only the largest scales $k < k_{N_{i_1}} = 12$ are resolved at each time iteration. The nonlinear interaction term between the small and large scales is modeled via a quasistatic approximation. However, in the DML approach, the small scales are resolved (with less accuracy than the large ones), and they are not resolved in LES methods. In Figure 10.5, the energy spectrum function of

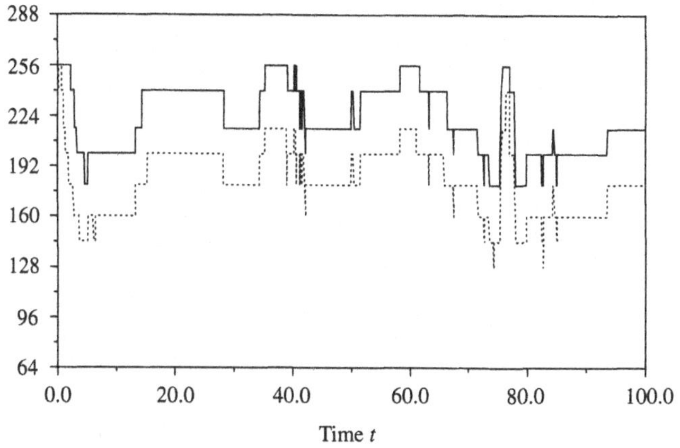

Figure 10.3. Time evolution of the cutoff levels $N_{i_1(j)}$ (dotted line) and $N_{i_2(j)}$ (solid line).

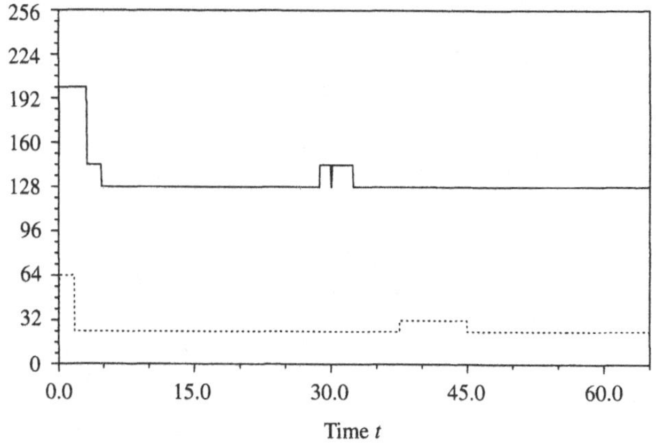

Figure 10.4. Time evolution of the cutoff levels $N_{i_1(j)}$ (dotted line) and $N_{i_2(j)}$ (solid line).

the solution computed with the DML method is represented. Note that there is no energy pile-up at high wavenumbers and the slope of the spectrum is well preserved. If we compare the vorticity structures obtained with the DNS and DML methods, we note (see Debussche, Dubois, and Temam (1995)) that the DML method allows us to recover the large structures of the flow. The difference in the energy norm $|\cdot|_0$ between DNS and DML is of the order of ξ $(= 10^{-2})$. The upper level N_{i_2} of the multilevel procedure is approximately equal to 128 during the whole computation, so that $\eta k_{N_{i_2}} \simeq 3.2$.

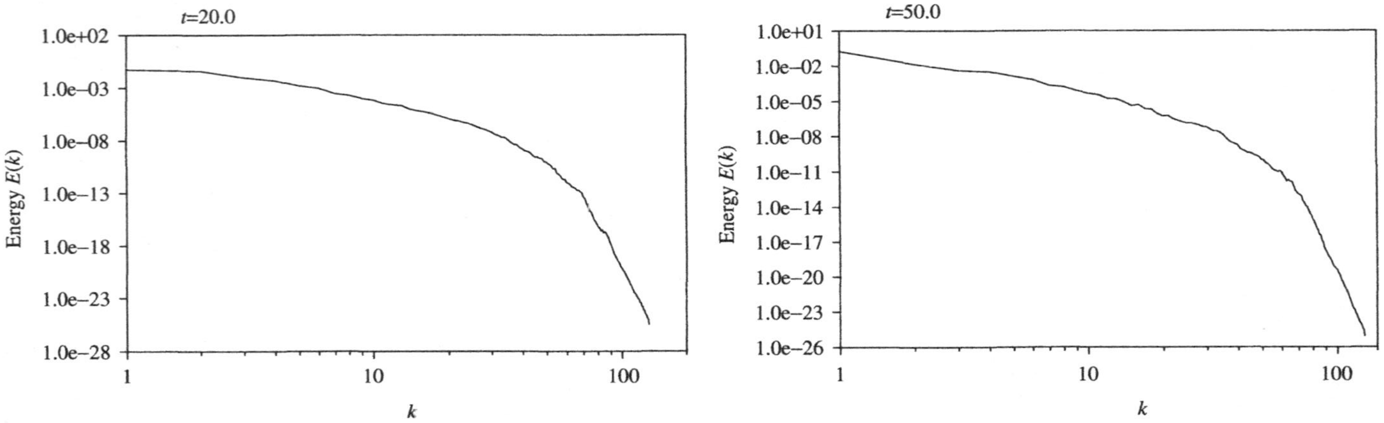

Figure 10.5. Energy spectrum function at $t = 20$ and at $t = 50$.

Finally, an additional numerical test has been performed in order to test the efficiency of the DML method. We have integrated a low-resolution ($k_N < k_d$) Galerkin simulation over 50,000 time iterations, that is, 13 eddy-turnover times, with a total number of degrees of freedom $N^2 \simeq \mathrm{Re}_L$, that is, $N = 32$ instead of the 256 previously, so that $k_N = 16$ and $\eta k_N = 0.8$. We have noticed, by looking at the vorticity structures obtained with this simulation (see Debussche, Dubois, and Temam (1995)), that the large scales of the flow cannot be computed with such a value of N: the vorticity seems to correspond to an unstructured field. Figure 10.6 shows the energy spectrum function obtained for this test. The dissipation of enstrophy is not preserved, and the slope of the spectrum is not recovered. Indeed, by taking N too small, we do not allow the appearance of the small scales or their dissipation mechanism occurring in the viscous range. The dynamics of the flow is drastically modified. However, this computation is stable, since no quantity is artificially increased until an overflow occurs. A new (nonphysical) equilibrium is reached. Then we have set $N = 48$, so that $k_N = 24$, $\eta k_N = 1.2$. In this case, the Galerkin simulation can be considered as a DNS. We have noted that the large scales are reasonably well captured. However, the DML method with $N = 256$ and the accuracy parameter $\xi = 10^{-2}$ ($k_{N_{i_1}} = 12$) provides better qualitative results and requires less than half of the 2730-seconds, CPU time required for the 48^2 modes DNS simulation.

Another simulation of the same type, but with a larger Reynolds number, $\mathrm{Re}_L = 6{,}328$ instead of 784, has been run; Re_L was increased by changing the kinematic viscosity ν to 2.5×10^{-4}. The resolution is higher, namely, $N = 512$ (512^2 degrees of freedom) instead of $N = 256$. The behavior of the DML method is similar to the previous cases, and the simulation is presented in Dubois (1993).

10.2. Three-Dimensional Homogeneous Turbulent Flows

In this section, we present the results of three-dimensional simulations of homogeneous isotropic turbulent flows. In Dubois and Jauberteau (1998), several three-dimensional simulations have been described. They correspond to a Reynolds number Re_λ in the range of 50 to 150 and to resolutions of 64^3, 96^3, 128^3, and 256^3 modes. These simulations and the results obtained are described in detail in Dubois and Jauberteau (1998). Different simulations with $\mathrm{Re}_\lambda \simeq 100$ and with a spatial resolution of 128^3 are reported and described here. The computational domain retained here is $\Omega = (0, 2\pi)^3$.

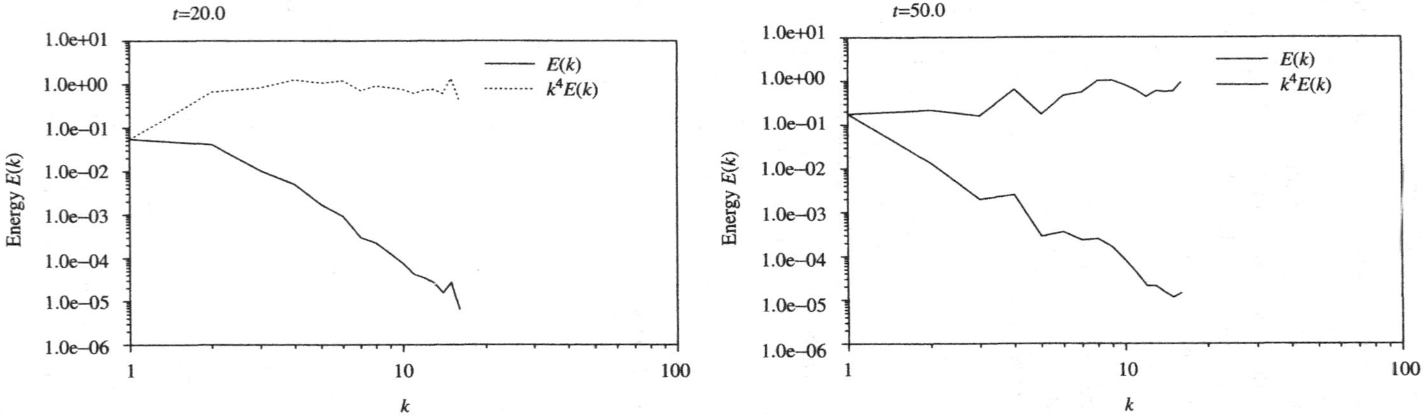

Figure 10.6. Energy spectrum function at $t = 20$ and at $t = 50$.

10.2.1. The Computational Parameters

The Initial Condition

The energy spectrum function of the initial condition,

$$E_0(k) = \frac{1}{2} \sum_{\mathbf{k} \in S_{k,1/2}} |\hat{\mathbf{u}}_0(\mathbf{k})|^2 \qquad \text{for} \quad k = 1, \ldots, k_N, \tag{10.12}$$

has an imposed shape

$$E_0(k) = c_{10}\, k^{-5/3}\, e^{-(k/k_0)^2}, \tag{10.13}$$

where k_0 and c_{10} are given parameters. The phases of the Fourier components $\hat{\mathbf{u}}_0(\mathbf{k})$ are randomly generated. With such an energy spectrum function for the initial velocity field, we expect the small scales to be close to their equilibrium state.

In the following, we show that estimates of the parameters k_0, c_0 as well as the viscosity can be obtained by specifying the Reynolds number as well as bounds for the eddy-turnover time τ_e and the Kolmogorov length scale η. The Reynolds number Re_λ at the initial time is taken equal to $\mathrm{Re}_\lambda(t = 0) = 100$. This value is in agreement with the numerical tests performed by Kerr (1985), Vincent and Ménéguzzi (1991), and Jiménez et al. (1993).

The parameter c_{10} can be expressed as a function of the viscosity and of the parameter k_0. From the definitions of the characteristic velocity (2.75) and of the Taylor microscale Reynolds number (2.76) and from (2.73), we have, at time $t = 0$,

$$\mathrm{Re}_\lambda = \sqrt{\frac{5}{3}}\, \frac{\left\langle |\mathbf{u}_0^N|^2 \right\rangle_\Omega}{\nu \left\langle |\boldsymbol{\omega}_0^N|^2 \right\rangle_\Omega^{1/2}}, \tag{10.14}$$

where the operator $\langle \cdot \rangle_\Omega$ denotes the spatial average (see Section 2.2.1, Example 2.1). By recalling (2.47), we have

$$E_0 = \frac{1}{2} \left\langle |\mathbf{u}_0^N|^2 \right\rangle_\Omega = \frac{1}{2} \sum_{\mathbf{k} \in S_N} |\hat{\mathbf{u}}_0(\mathbf{k})|^2 = c_{10} \sum_{k=1}^{k_N} k^{-5/3} e^{-k^2/k_0^2}.$$

Similarly (2.55) can be rewritten as

$$\left\langle |\boldsymbol{\omega}_0^N|^2 \right\rangle_\Omega = \sum_{\mathbf{k} \in S_N} |\mathbf{k}|^2\, |\hat{\mathbf{u}}_0(\mathbf{k})|^2 = \sum_{k=1}^{k_N} \epsilon_0(k).$$

Therefore, from (10.14), we derive the following expression for the parameter c_{10}:

$$c_{10} = \frac{1}{2}\sqrt{\frac{3}{5}}\nu \mathrm{Re}_\lambda \left(\sum_{k=1}^{k_N} \epsilon_0(k) \right)^{1/2} \left(\sum_{k=1}^{k_N} k^{-5/3} e^{-k^2/k_0^2} \right)^{-1} . \tag{10.15}$$

From (2.46) and (2.48) we obtain a relation between the energy and enstrophy spectrum functions, namely,

$$2\left(k - \tfrac{1}{2}\right)^2 E_0(k) \le \epsilon_0(k) < 2\left(k + \tfrac{1}{2}\right)^2 E_0(k) \qquad \text{for any} \quad k > 0.$$

Therefore, lower and upper bounds for the mean-square vorticity follow:

$$\begin{aligned} 2c_{10} \sum_{k=1}^{k_N} \left(k - \frac{1}{2}\right)^2 k^{-5/3}\, e^{-k^2/k_0^2} &\le \left\langle |\omega_0^N|^2 \right\rangle_\Omega \\ &< 2c_{10} \sum_{k=1}^{k_N} \left(k + \frac{1}{2}\right)^2 k^{-5/3}\, e^{-k^2/k_0^2}. \end{aligned} \tag{10.16}$$

By inserting (10.16) into (10.15), we deduce that

$$\Phi_-(k_0, \nu) \le c_{10} < \Phi_+(k_0, \nu), \tag{10.17}$$

where

$$\begin{aligned} &\Phi_\pm(k_0, \nu) \\ &= \frac{3}{10}\, \nu^2\, \mathrm{Re}_\lambda^2 \left(\sum_{k=1}^{k_N} \left(k \pm \frac{1}{2}\right)^2 k^{-5/3} e^{-k^2/k_0^2} \right) \left(\sum_{k=1}^{k_N} k^{-5/3} e^{-k^2/k_0^2} \right)^{-2} . \end{aligned}$$

An estimate of the parameter k_0 in (10.13) can be obtained by imposing, in agreement with Jiménez et al. (1993) concerning the resolution requirement for DNS, that $\eta k_N = \frac{3}{2}$ at $t = 0$. From the definition (2.109) of the Kolmogorov length scale η, we deduce that

$$k_N = \frac{3}{2}\left(\frac{\epsilon}{\nu^3}\right)^{1/4} .$$

Recalling that $\epsilon = \nu \langle |\omega_N|^2 \rangle_\Omega$, we have

$$k_N^2 = \frac{9}{4\nu} \left\langle |\omega_N|^2 \right\rangle_\Omega^{1/2} .$$

Therefore, from (10.16) we obtain

$$\frac{9c_{10}^{1/2}}{2\sqrt{2}\nu}\left(\sum_{k=1}^{k_N}\left(k-\frac{1}{2}\right)^2 k^{-5/3}e^{-(k/k_0)^2}\right)^{1/2} \le k_N^2 < \frac{9c_{10}^{1/2}}{2\sqrt{2}\nu}\left(\sum_{k=1}^{k_N}\left(k+\frac{1}{2}\right)^2 k^{-5/3}e^{-(k/k_0)^2}\right)^{1/2}.$$

By using (10.17), the above inequalities are rewritten as

$$\frac{9}{4}\sqrt{\frac{3}{5}}\,\mathrm{Re}_\lambda \frac{\sum_{k=1}^{k_N}\left(k-\frac{1}{2}\right)^2 k^{-5/3}e^{-k^2/k_0^2}}{\sum_{k=1}^{k_N}k^{-5/3}e^{-k^2/k_0^2}} \le k_N^2 < \frac{9}{4}\sqrt{\frac{3}{5}}\,\mathrm{Re}_\lambda \frac{\sum_{k=1}^{k_N}\left(k+\frac{1}{2}\right)^2 k^{-5/3}e^{-k^2/k_0^2}}{\sum_{k=1}^{k_N}k^{-5/3}e^{-k^2/k_0^2}},$$

which can be reformulated as follows:

$$\begin{aligned}\sum_{k=1}^{k_N}\left[\frac{4}{9}k_N^2-\sqrt{\frac{3}{5}}\,\mathrm{Re}_\lambda\left(k+\frac{1}{2}\right)^2\right]k^{-5/3}e^{-k^2/k_0^2} < 0,\\ \sum_{k=1}^{k_N}\left[\frac{4}{9}k_N^2-\sqrt{\frac{3}{5}}\,\mathrm{Re}_\lambda\left(k-\frac{1}{2}\right)^2\right]k^{-5/3}e^{-k^2/k_0^2} \ge 0.\end{aligned} \tag{10.18}$$

For $\mathrm{Re}_\lambda = 100$ and $N = 128$, the first inequality in (10.18) is valid for $k_0 > k_0^{\min} = 20$, while the second inequality is valid for $k_0 \le k_0^{\max} = 25$. As $(k_0^{\min}, k_0^{\max}]$ provides an interval of allowable values for k_0 consistent with $\eta k_N = 1.5$, we have set $k_0 = 25$.

In order to derive an estimated value for the viscosity, we consider the eddy-turnover time defined in Section 2.2.4 as $\tau_e = L/u$. Recalling the definition (2.103) of the integral length scale, we have

$$\tau_e = \frac{3\pi}{4u}\frac{\sum_{k=1}^{k_N}k^{-1}E_0(k)}{\sum_{k=1}^{k_N}E_0(k)}.$$

Now, recalling (2.75) and (2.47), we can express the eddy-turnover time as a function of the energy spectrum function, namely

$$\tau_e = \frac{3\pi}{4}\sqrt{\frac{3}{2}}\frac{\sum_{k=1}^{k_N}k^{-1}E_0(k)}{\left(\sum_{k=1}^{k_N}E_0(k)\right)^{3/2}},$$

which can be rewritten as follows:

$$\tau_e = \frac{3\pi}{4}\sqrt{\frac{3}{2}}\, c_{10}^{-1/2} \frac{\sum_{k=1}^{k_N} k^{-8/3} e^{-k^2/k_0^2}}{\left(\sum_{k=1}^{k_N} k^{-5/3} e^{-k^2/k_0^2}\right)^{3/2}}.$$

Using (10.17), we derive the following lower bound for τ_e:

$$\tau_e \geq \frac{3\sqrt{5}\,\pi}{4\,\nu \mathrm{Re}_\lambda} \left(\sum_{k=1}^{k_N} k^{-8/3} e^{-k^2/k_0^2} \right) \times \left(\sum_{k,l=1}^{k_N} \left(k + \frac{1}{2}\right)^2 (kl)^{-5/3} e^{-(k^2+l^2)/k_0^2} \right)^{-1/2}. \tag{10.19}$$

In order to converge rapidly to a statistically steady state, we require the eddy-turnover time to be small. At time $t = 0$, we have imposed that $\tau_e \leq 3\sqrt{3}\pi/4$. From (10.19) we derive a lower bound for the value of the viscosity:

$$\nu \geq \sqrt{\frac{5}{3}} \mathrm{Re}_\lambda^{-1} \left(\sum_{k=1}^{k_N} k^{-8/3} e^{-k^2/k_0^2} \right) \times \left(\sum_{k,l=1}^{k_N} \left(k + \frac{1}{2}\right)^2 (kl)^{-5/3} e^{-(k^2+l^2)/k_0^2} \right)^{-1/2}. \tag{10.20}$$

With $\mathrm{Re}_\lambda(t = 0) = 100$, the lower bound given by (10.20) decreases rapidly from 8.4×10^{-3} to 1.86×10^{-3} when k_0 is increased, $k_0 \in [1, 20]$. Then, a plateau is reached and a much slower decrease is noticed. As the value of the parameter k_0 was set to 25, we have chosen $\nu = 2 \times 10^{-3}$, so that we have $\tau_e(t = 0) \simeq 3.3$. The transient time period needed to reach a statistical equilibrium for the large scales can be reduced by decreasing τ_e. However, in such a case, we have noticed that the numerical stability constraint becomes stronger, so that the time step Δt must be taken smaller.

With the values $\mathrm{Re}_\lambda(t = 0) = 100$ and $\nu = 2.0 \times 10^{-3}$, we can verify that both the lower bound $\Phi_-(k_0, \nu)$ and the upper bound $\Phi_+(k_0, \nu)$ for c_{10} given by (10.17) are positive functions and that they increase to 0.46 and 0.52, respectively, when k_0 increases from 1 to 100. Moreover, we have

$$|\Phi_+(k_0, \nu) - \Phi_-(k_0, \nu)| < 0.06 \qquad \text{for} \quad k_0 > 0.$$

For $k_0 = 25$, we have $0.15 < c_{10} < 0.19$.

The different parameters used for the DNS and DML simulations studied in this section are summarized in Table 10.3.

Table 10.3. *Computational parameters for the DNS and DML simulations*

Mesh	k_N	$\mathrm{Re}_\lambda(t=0)$	k_0	ν	Δt
128^3	64	100	25	2.0×10^{-3}	1.0×10^{-2}

The External Force

In order to reach a statistically steady state, the turbulence has been maintained by forcing the large scales in a deterministic way. As in the two-dimensional case (Section 10.1), the external force is time-independent. Therefore, no randomness is artificially introduced into the system. The Fourier coefficients $\hat{f}_j(\mathbf{k})$ are defined by

$$\hat{f}_j(\mathbf{k}) = |\hat{f}_j(\mathbf{k})|\, e^{i\theta(\mathbf{k})} \qquad \text{for} \quad j = 1, 2, 3, \tag{10.21}$$

where $\theta(\mathbf{k}) \in [0, 2\pi]$ is randomly generated and $\hat{f}_j(\mathbf{k}) \neq 0$ only if $|\mathbf{k}| = 1$. Hence, the external force acts only on the lowest modes of the velocity field (large structures). A similar external force was used by Vincent and Ménéguzzi (1991). As the energy-containing eddies are unstable for small viscosity (see Voke and Collins (1983)), we can hope to reduce the transient period.

In Jiménez et al. (1993), the flow is forced in such a way that the numerical resolution ηk_N remains nearly constant during the whole simulation. Therefore, the amplitude of the energy injection rate oscillates in time. As pointed out by Jiménez et al., this forcing procedure avoids an increase (or decrease) of the Kolmogorov wavenumber k_d. However, with the choice of initial condition and parameters described in Section 10.2.1, small fluctuations of k_d were noticed on our simulations (see below).

The amplitude of the external volume force is chosen so that, at initial time, the energy injection rate $\langle \mathbf{g}_N \cdot \mathbf{u}_0^N \rangle_\Omega$, where $\mathbf{g}_N$ is the projection of $\mathbf{f}_N$ onto the space of free divergence functions (see Chapter 3), and the rate of dissipation of the kinetic energy $\epsilon = \nu \langle |\omega_0^N|^2 \rangle_\Omega$ are in balance, that is,

$$\left\langle \mathbf{g}_N \cdot \mathbf{u}_0^N \right\rangle_\Omega \simeq \epsilon. \tag{10.22}$$

Indeed, if we consider the energy equation, we have (see Section 2.2.3)

$$\frac{dE}{dt} = \left\langle \mathbf{g}_N \cdot \mathbf{u}_0^N \right\rangle_\Omega - \epsilon. \tag{10.23}$$

So (10.22) implies that $dE/dt \simeq 0$, where E is the total kinetic energy. Hence, we expect to avoid strong variations of the kinetic energy during the transient period, that is, during the first eddy-turnover times.

10.2.2. Comparison and Analysis of the DNS and DML Simulations

In order to compare the DML method presented in Section 9.4 with DNS results, simulations with the same set of data (initial condition, forcing terms, and viscosity) have been computed with the DNS and DML codes.

The Parameters for the DML Simulations

We first present the DML simulations, that is, we list the parameters (ξ_r, K_1, K_2) required by the DML algorithm described in Section 9.4. We also describe the cutoff levels, number of coarse grids, and number of QS iterations for the nonlinear interaction terms, which are computed by the DML algorithm during the simulations.

Three different DML simulations with the parameter ξ_r, representing a relative error in the norm $|\cdot|_0$ on the velocity field, respectively set to 5.0×10^{-3}, 7.5×10^{-3}, and 1.0×10^{-2}, have been performed. In all cases, the parameters (K_1, K_2) have been set to $K_1 = 2$ and $K_2 = 3$.

The sequence of coarse levels provided for the DML algorithm is

$$S_c = \{24, 32, 48, 64, 80, 96, 108, 128\}.$$

These levels are possible choices for the cutoff levels corresponding to the jth cycle, $N_{i_1(j)}$ and $N_{i_2(j)}$, which belongs to S_c. Therefore, the sequence of intermediate levels is fully determined, namely

$$N_{i_1(j)},\ N_{i_1(j)+1}, \dots, N_{i_2(j)-1},\ N_{i_2(j)} \qquad \text{with}$$
$$N_{i_1(j)+k} \in S_c,\ k = 0, \dots, i_2(j) - i_1(j).$$

For the different values of the parameter ξ_r the average values of the coarse levels $N_{i_1(j)}$ and $N_{i_2(j)}$ satisfy

for $\quad \xi_r = 5.0 \times 10^{-3}$:

$$64 \le \overline{N_{i_1(j)}}^j \le 80,$$
$$108 \le \overline{N_{i_2(j)}}^j \le 128,$$

for $\quad \xi_r = 7.5 \times 10^{-3}$ and $\xi_r = 1.0 \times 10^{-2}$:

$$48 \le \overline{N_{i_1(j)}}^j \le 64,$$
$$96 \le \overline{N_{i_2(j)}}^j \le 108,$$

where $^{-j}$ denotes the average with respect to j.

Table 10.4. *Computational parameters of the DML simulations*

ζ_r	K_1	K_2	$\overline{\eta k_{N_{i_1(j)}}}^j$	$\overline{\eta k_{N_{i_2(j)}}}^j$	$\overline{n_V(j)}^j$	$\overline{q(j)}^j$	CPU time/ iteration (s)
5×10^{-3}	2	3	0.88	1.45	3	4	2.60
7.5×10^{-3}	2	3	0.74	1.30	2	4	2.11
10^{-2}	2	3	0.65	1.24	3	4	1.66

In Figure 10.7, the time evolution of the coarse levels $N_{i_1(j)}, N_{i_2(j)}$ is represented for the three DML simulations. Note that the values of these levels decrease while the control parameter ξ_r is increased, which is consistent with the analysis performed in Section 9.3.

In Table 10.4, the values of the parameters chosen for the DML simulations are summarized as well as the mean values of the cutoff levels, of the number of coarse levels $q(j)$, and of the number $n_V(j)$ of V-cycles performed on the time interval $(t_{n_{j-1}}, t_{n_j})$. Recalling that a whole cycle corresponds to $2q(j)n_V(j)$ time iterations, the average number of iterations per cycle is of the order of 16 or 24, depending on the value of ξ_r. This represents the number of time iterations during which the nonlinear interaction terms corresponding to every coarse level $N_{i_1(j)+k}$, $k = 0, \ldots, p(j)$, are kept constant.

Description of the DNS and DML Simulations

The DNS and DML simulations presented in this section have been conducted over the time interval $[0, T]$ with $T = 40\tau_e$, τ_e being the eddy-turnover time (see Section 2.2.4). The time step Δt was set to 1.0×10^{-2} in order to insure numerical stability (CFL condition). The CFL number $k_N \, \Delta t \, U_\infty$ oscillates around 1.15 during the computations. With $N = 128$ modes in each spatial direction, the DNS code requires 4.025 s per time step on one processor of a Fujitsu VPP300. Therefore, the whole simulation represents 17 h, that is, 25 min per eddy-turnover time. The performance of the code was evaluated on a Cray C90; at this resolution, approximately 450 megaflops per processor were obtained, and the CPU time per time step is equal to 6.27 s. The DNS and DML codes exhibit similar performance for vectorization and multitasking (see Section 3.1).

As a result of having a forcing term independent of time (constant), the energy injection rate $\langle \mathbf{g}_N \cdot \mathbf{u}_N \rangle_\Omega$ has large time fluctuations. Therefore, global quantities such as the kinetic energy E, the energy dissipation rate ϵ, the length scales (L, λ, η), and the Reynolds number vary in time. Let $\phi(t)$, $t \in [0, T]$,

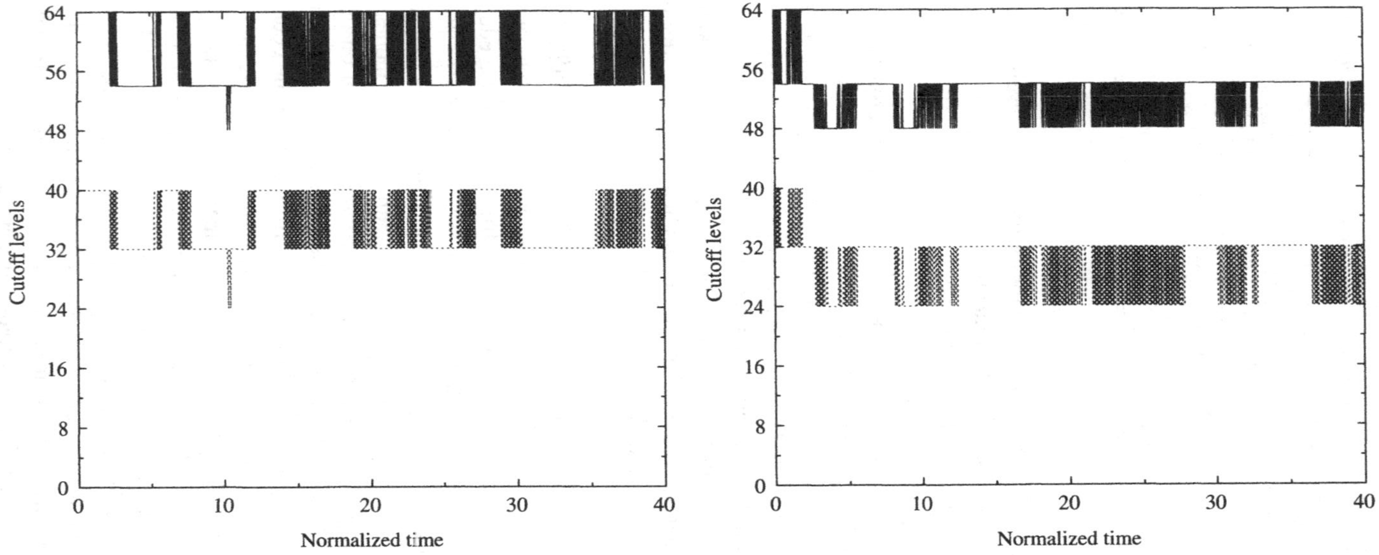
Cutoff levels
64
56
48
40
32
24
16
8
0
0
10
20
30
40
Normalized time
Cutoff levels
64
56
48
40
32
24
16
8
0
0
10
20
30
40
Normalized time

(a)

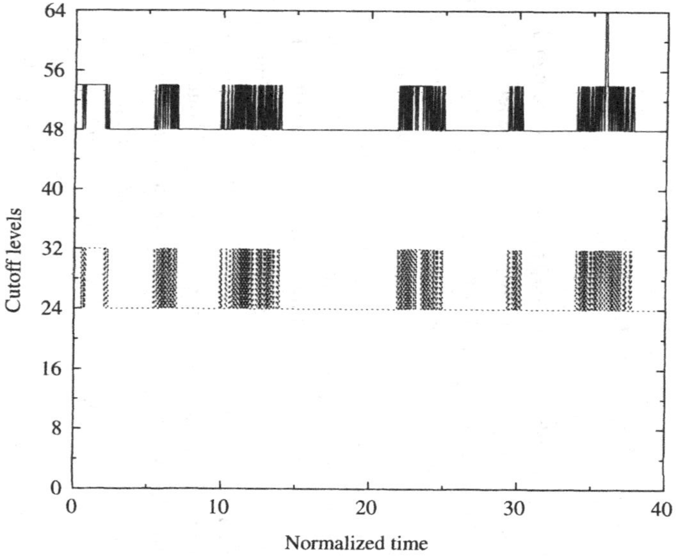

(b)

Figure 10.7. Time evolution of the cutoff levels $k_{N_{i_1(j)}}$ (dotted line) and $k_{N_{i_2(j)}}$ (solid line) for the DML simulations with values of the parameter ξ_r given by (a) 5.0×10^{-3} (left curve), 7.5×10^{-3} (right curve); (b) 1.0×10^{-2}. The time has been normalized by the eddy-turnover time τ_e.

Table 10.5. *Characteristic values of the DNS simulations*

Mesh	Re_λ	E	ϵ	τ_e	L	λ	η	ηk_N
128^3	99.61	0.2497	2.121×10^{-2}	3.72	1.514	0.488	2.483×10^{-2}	1.59
96^3	101.45	0.2586	2.187×10^{-2}	3.69	1.531	0.489	2.463×10^{-2}	1.18
64^3	100.07	0.2478	2.065×10^{-2}	3.78	1.535	0.491	2.494×10^{-2}	0.79

be any of these quantities characterizing a homogeneous turbulent flow (see Chapter 2). We define its time average:

$$\overline{\phi}(t) = \frac{1}{t} \int_0^t \phi(s)\, ds, \qquad 0 < t < T\,.$$

The 128^3 DNS simulation has reach a statistically steady state in the sense that on the time interval $[2T/3, T]$ the time-averaged kinetic energy $\overline{E}(t)$ fluctuates from its mean value E with amplitudes of the order of 0.51 percent; the energy dissipation rate $\overline{\epsilon}(t)$, of the order of 0.65 percent. Amplitudes smaller than 0.55 percent are observed for the Reynolds number as well as the length scales, that is, the integral scale L, the Taylor microscale λ, and the Kolmogorov scale η. The mean values of these quantities are reported in Table 10.5.

The Taylor microscale Reynolds number Re_λ is of the order of 100. For the 128^3 simulation, the Kolmogorov scale η and the largest wavenumber k_N satisfy $\eta k_N \simeq 1.6$, which, according to Jiménez et al. (1993), insures a fine enough accuracy to resolve the small scales of the motion. The nonlinear terms in our simulations were computed by a pseudospectral method. The aliasing errors were not removed for the 128^3 and 96^3 simulations reported in Table 10.5. The same simulations have been run on a shorter time interval, $[0, 20\tau_e]$, with a fully dealiased computation of the nonlinear terms via the $\frac{3}{2}$ rule (see e.g. Canuto et al. (1988)). No noticeable differences were noted from the aliased simulations for the quantities listed in Table 10.5. As is mentioned below, the effects of the aliasing errors on the statistical properties of the flow are stronger than on global spatial quantities, like the ones listed here, and larger differences appear in the high-order moments of the velocity and its derivatives. For the lowest resolution (the 64^3 simulation), the aliasing errors were systematically removed. The aliasing errors have the strongest effects when the numerical resolution is not sufficient to resolve all the scale of motions, which is the case for the 64^3 simulation, as $\eta k_N = 0.8 < 1.0$.

For the DML simulations with $\xi_r = 7.5 \times 10^{-2}$ and $\xi_r = 1.0 \times 10^{-2}$, the lowest cutoff level was $\overline{N_{i_1(j)}}^j \simeq 50$ and the largest one $\overline{N_{i_2(j)}}^j \simeq 96$. In order to estimate the effective resolution of the small scales by the DML algorithm,

Table 10.6. *Characteristic values of the DML simulations*

ξ_r	Re_λ	E	ϵ	τ_e	L	λ	η	ηk_N
5.0×10^{-3}	100.00	0.2528	2.156×10^{-2}	3.73	1.526	0.487	2.474×10^{-2}	1.58
7.5×10^{-3}	99.06	0.2485	2.119×10^{-2}	3.75	1.525	0.487	2.483×10^{-2}	1.59
1.0×10^{-2}	98.62	0.2491	2.147×10^{-2}	3.75	1.527	0.484	2.475×10^{-2}	1.58

64^3 and 96^3 simulations have been performed with the DNS code; the CPU time per iteration is equal to 1.85 s for $N = 96$ and to 1.5 s for $N = 64$. As for the 128^3 DNS, a statistically steady state has been reached for these simulations, and the observed amplitudes, with respect to the mean values reported in Table 10.5, are of the order of 0.75 percent for $\overline{E}(t)$, 0.9 percent for $\overline{\epsilon}(t)$, and smaller than 0.63 percent for the other quantities. These variations are slightly larger than the 128^3 DNS ones for the kinetic energy as well as the energy dissipation rate. In the case $N = 96$, the mean kinetic energy and energy dissipation rate are 3 percent larger than the 128^3 DNS values. The time evolution of $\overline{E}(t)$ and $\overline{\epsilon}(t)$, represented in Figure 10.8, shows that the 128^3 and 96^3 simulations reach slightly different statistically steady states. Curves corresponding to both simulations remain very close to each other on the interval $[0, 7\tau_e]$. A noticeable separation appears for $t \geq 7\tau_e$. The same behavior has been seen for the 96^3 dealiased simulation. Note also that a similar deviation appears in the evolution of the time-averaged energy dissipation rate $\overline{\epsilon}(t)$. The curve corresponding to the 64^3 simulation are closer to the 128^3 one. However, $\overline{\epsilon}(t)$ is lower than the 128^3 DNS one, and, as will be discussed later, this solution has nonphysical behavior.

In Table 10.6, the mean values of the characteristic quantities previously analyzed are reported for the three DML simulations. Amplitudes larger than those reported for the 128^3 DNS are obtained. For increasing values of ξ_r, they are respectively of the order of 1.64, 1.15, and 0.62 percent for the time-averaged kinetic energy, and of the order of 1.23, 0.47, and 0.43 percent for $\overline{\epsilon}(t)$. For the other listed quantities, the amplitudes are smaller than 0.82 percent. By comparing the values reported in Table 10.5 and Table 10.6, we notice that the DML statistical steady states are very close to the 128^3 DNS one. The difference between the DML and the 128^3 DNS kinetic energy is of the order of 1.25 percent for $\xi_r = 5.0 \times 10^{-3}$ and of the order 0.5 percent for the other values of ξ_r. The time evolution of the kinetic energy $\overline{E}(t)$ and of the energy dissipation rate $\overline{\epsilon}(t)$ for the 128^3 DNS and DML simulations is represented in Figure 10.9. All simulations appear to tend to the same statistical steady state. By comparing Figures 10.8 and 10.9, we note that the time-averaged energy

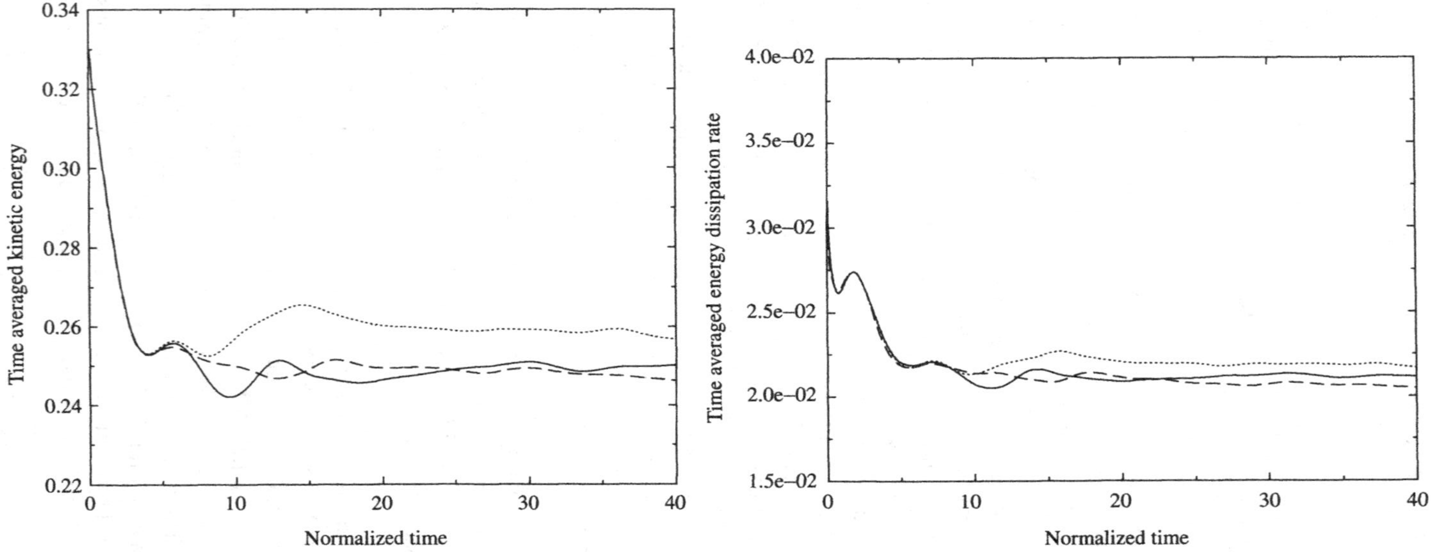

Figure 10.8. Time-averaged kinetic energy $\overline{E}(t)$ (left) and energy dissipation rate $\overline{\epsilon}(t)$ (right) plotted versus the normalized time t/τ_γ. The curves correspond to the 128^3 simulation (solid line), the 96^3 simulation (dotted line), and the 64^3 simulation (dashed line).

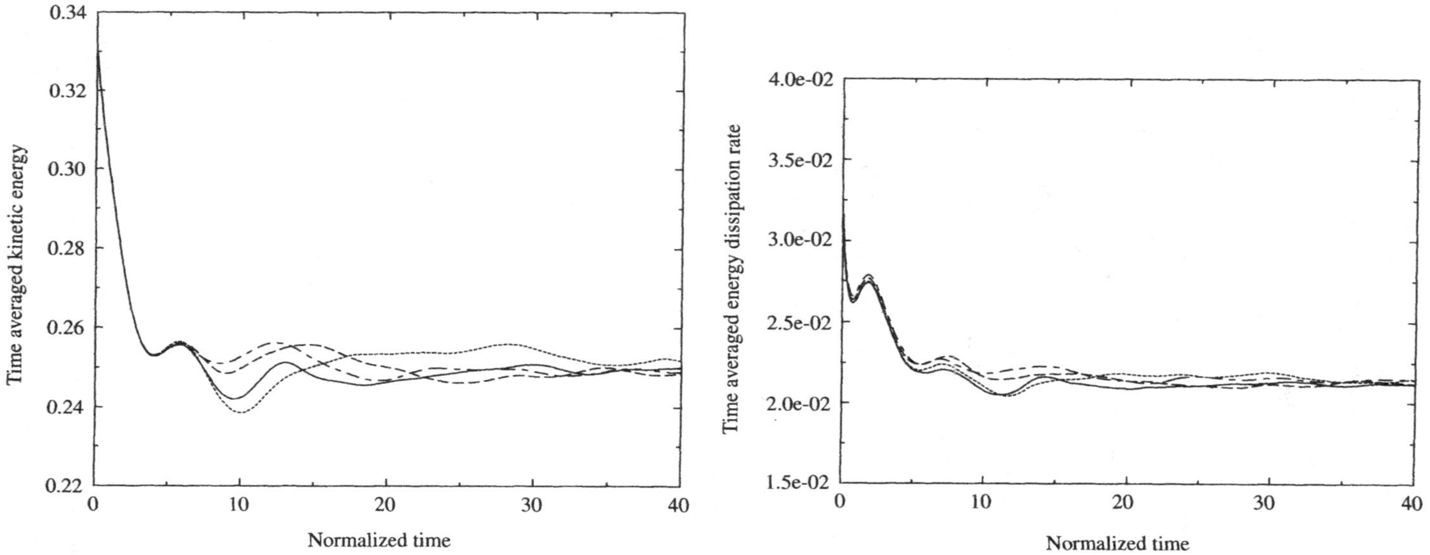

Figure 10.9. Time-averaged kinetic energy $\overline{E}(t)$ (left) and energy dissipation rate $\overline{\epsilon}(t)$ (right) plotted versus the normalized time t/τ_e. The curves correspond to the 128^3 simulation (solid line) and to the DML simulations with $\xi_r = 5.0 \times 10^{-3}$ (dotted line), $\xi_r = 7.5 \times 10^{-3}$ (dashed line), and $\xi_r = 1.0 \times 10^{-2}$ (dotted–dashed line).

dissipation rate and kinetic energy corresponding to the DML simulation are very close to the 128^3 DNS simulation, closer than the two other DNS simulations at lower resolution. Note that the DML curve corresponding to $\xi_r = 5.0 \times 10^{-3}$ gets closer to the other DML ones and to the 128^3 DNS one for $t > 33\tau_e$. All curves have very similar behavior, and no separation, as was previously reported for the 96^3 simulation, is observed here.

Hence, the statistical steady states reached with the DML algorithm are very close to the 128^3 DNS one. This indicates that the DML method is stable for long-time integration. There is no amplification of the errors injected at each time iteration (see Section 9.3). The differences between the DML and DNS instantaneous values of the kinetic energy $E(t)$ are larger than ξ_r^2, which corresponds to an upper bound of the relative errors tolerated at each time iteration (see Sections 9.3 and 9.4). By comparing the time evolutions of $E(t)$ for the 128^3 DNS and the DML simulation with $\xi_r = 1.0 \times 10^{-2}$, represented on Figure 10.10, it appears that the absolute distance between the DNS and DML values remains smaller than ξ_r^2 over approximately 5 eddy-turnover times. After this transient period (for $t > 5\tau_e$), the curves have different oscillating behaviors, but their means are very close, according to the previous discussion. The same behavior can be observed for the energy injection rate as well as for the

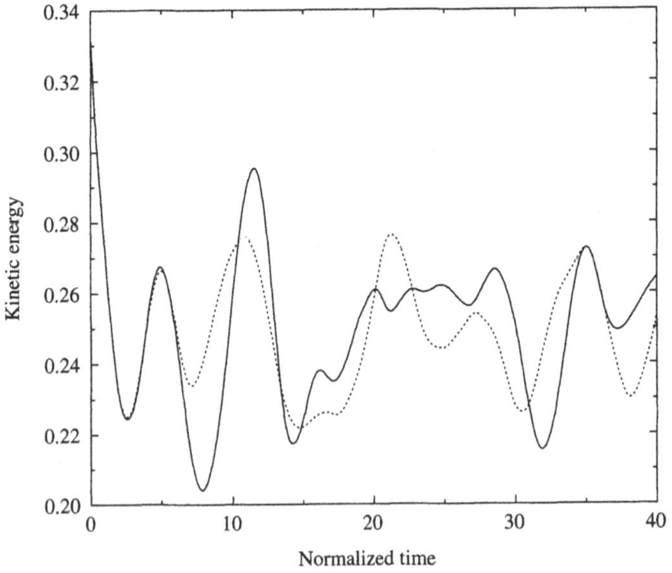

Figure 10.10. Instantaneous kinetic energy $E(t)$ plotted versus t/τ_e. The curves correspond to the 128^3 simulation (solid line) and to the DML simulation with $\xi_r = 1.0 \times 10^{-2}$ (dotted line).

energy dissipation rate. This illustrates a well-known property of nonlinear systems. Indeed, even with very small differences in the initial data, two solutions will separate and follow different dynamics after a short transient period of time.

As in Vincent and Méneguzzi (1991), we have computed the isotropy coefficient

$$I(t) = \frac{\left(\sum_{\mathbf{k}\in S_N} |\hat{u}_{p,1}(\mathbf{k}, t)|^2\right)^{1/2}}{\left(\sum_{\mathbf{k}\in S_N} |\hat{u}_{p,2}(\mathbf{k}, t)|^2\right)^{1/2}},$$

where $\hat{u}_{p,j}(\mathbf{k}, t)$, $j = 1, 2$, are the components of $\hat{\mathbf{u}}(\mathbf{k}, t)$ in the spectral subspace orthogonal to the wavenumber $\mathbf{k}$. The instantaneous values $I(t)$, for $t \in [0, T]$, fluctuates between 0.8 and 1.2 for all the simulations reported here. The time-averaged values of I are of the order of 1.0 ± 6 percent for the DNS and the DML simulations. Our simulated turbulent flows are then statistically isotropic.

Energy and Enstrophy Spectrum Functions

In the case of homogeneous turbulence, when the flow is assumed to be isotropic, the two-point correlation tensor and its spectral representation can be expressed in terms of the energy spectrum function $E(k)$ (see Chapter 2 and the references therein). Therefore, this spectrum function is often used to characterize such turbulent flows. The energy spectrum function depends on the energy transfer terms and on the energy dissipation spectrum; it then depends on how the energy is transferred and dissipated through the different scales of motion. In order to estimate the effect, on the representation of turbulence, of the DML modeling of the small scales and of the nonlinear interaction terms in the large-scale equation, we compare here the shape of the different energy spectra for the DNS and DML simulations.

The energy spectrum functions $E(k, t)$ have been time-averaged; as previously, the averaging has been performed over the time interval $[0, T]$. By analyzing these spectra, we can determine whether there are significant truncation errors. Indeed, when the number of modes N retained in each spatial direction is not sufficiently large, an accumulation of energy appears at high wavenumbers. Nevertheless, if the dissipation range is sufficiently extended, this spurious tail is not significant. By inspecting the high wavenumbers of the DNS spectra (Figure 10.11), we notice that the size of the upturn is acceptable for the 128^3 simulation and it is of the same order as the ones appearing in Vincent and Méneguzzi (1991) and in Jiménez et al. (1993). A larger upturn appears on the lower resolution (96^3); however, in this case $\eta k_N = 1.18$ (see Table 10.5),

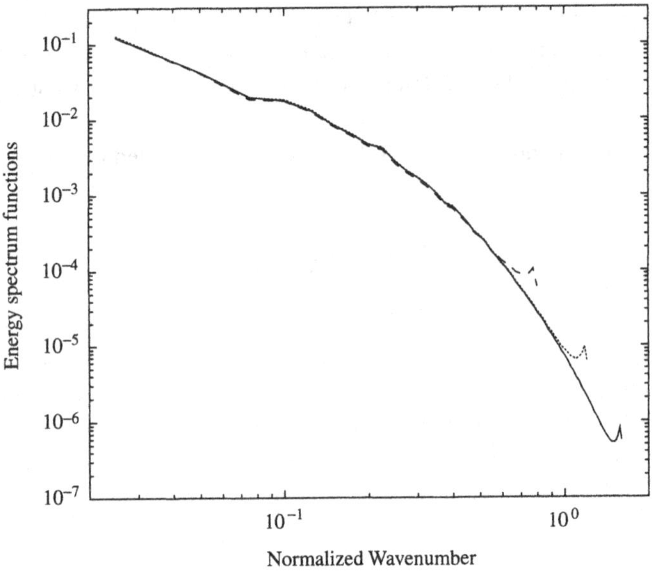

Figure 10.11. Time-averaged energy spectrum functions $\overline{E(k)}(40\tau_e)$ plotted versus the normalized wavenumber ηk; η is the Kolmogorov scale. The curves correspond to the 128^3 simulation (solid line), the 96^3 simulation (dotted line), and the 64^3 simulation (dashed line).

which is less than the resolution requirement $\eta k_N = 1.5$ (see Jiménez et al. (1993)). No amplifications were noted during the long-time integration of this simulation. However, the high-order moments may be slightly affected by this low resolution. An upturn of the same order of magnitude has been observed for the energy spectrum function corresponding to the dealiased 96^3 simulation. For the 64^3 simulation, a larger deviation from the other energy spectrum functions can be seen at large wavenumbers, that is, for $\eta k \in [0.6, 0.8)$. As the size of the dissipation range is not large enough, the energy transferred from the large scales cannot be fully dissipated and it tends to accumulate near the cutoff wavenumber $k_N = 32$. A deviation from the other spectrum functions can be seen even at small wavenumbers, for $k \leq 10$. A nonphysical situation occurs: the energy that is not dissipated at large wavenumbers moves backward and contaminates the large scales. This phenomenon will be amplified for lower resolutions, that is, $N \leq 64$, leading to a decrease of the energy at low wavenumbers and to a deviation from the $-\frac{5}{3}$ Kolmogorov spectrum.

On comparing the DNS and DML spectra (Figure 10.11 and Figure 10.12) at high wavenumbers, we note that the upturn appearing on the DNS spectra is not present on the DML ones. The DML method has a dissipative effect in the

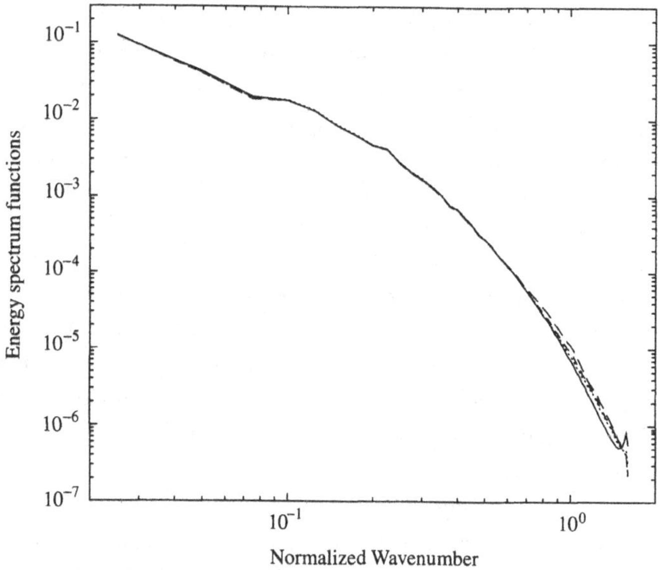

Figure 10.12. Time-averaged energy spectrum functions $\overline{E(k)}(t = 40\tau_e)$ plotted versus the normalized wavenumber ηk; η is the Kolmogorov scale. The curves correspond to the 128^3 simulation (solid line) and to the DML simulations with $\xi_r = 5.0 \times 10^{-3}$ (dotted line), $\xi_r = 7.5 \times 10^{-3}$ (dashed line), and $\xi_r = 1.0 \times 10^{-2}$ (dotted–dashed line).

neighborhood of the largest wavenumber, that is, for $k \simeq k_N$. The shape and the slope of the energy spectrum function are preserved for wavenumbers k smaller than the lower cutoff level $k_{N_{i_1(j)}}$ (see Table 10.4). In fact, the 128^3 DNS and the DML spectra are almost indistinguishable for $k \leq k_{N_{i_1(j)}}$. Hence, the DML method provides an accurate modeling of the nonlinear interaction term in the large-scales equation, especially at small wavenumbers. At wavenumbers in the range $k \in [k_{N_{i_1(j)}}, k_{N_{i_2(j)}}]$ the energy is slightly increased as compared to the 128^3 DNS spectrum. However, the shape of the spectrum is preserved that is, the energy contained in the small scales decays at a sensibly similar rate. This reflects the ability of the DML algorithm to mimic the energy transfer from the large scales to the small ones. The analysis of the statistical properties of the simulated flows, presented below, confirms this point. Hence, the DML method provides accurate modeling of the nonlinear interaction term in the large-scales equation, especially at small wavenumbers.

As was mentioned in Chapter 2 (Section 2.2.4), Smith and Reynolds (1991) and Manley (1992) suggested the following analytical form for the energy spectrum function for an isotropic homogeneous turbulent flow:

$$E(k) = C_K \epsilon^{2/3} k^{-5/3} e^{\alpha_1 \eta k} \qquad \text{for} \quad k = 1, \ldots, k_N,$$

where C_K is called the Kolmogorov constant and $\alpha_1 = [2C_K\Gamma(\frac{4}{3})]^{3/4}$. By fitting the spectra corresponding to our different simulations with the above form, the constant C_K can be estimated via a least-squares method. The values corresponding to the DNS spectra are respectively 1.734, 1.756, and 1.758 for the resolutions $N = 128, 96$, and 64. A noticeable increase appears when N is decreased. The values of C_K corresponding to the DML simulations are of the same order as the 128^3 DNS one, namely, $C_K = 1.730$ for $\xi_r = 5.0 \times 10^{-3}$, 7.5×10^{-3}, and $C_K = 1.725$ for $\xi_r = 1.0 \times 10^{-2}$. In Dubois and Jauberteau (1998), the Kolmogorov constant was shown to decrease from 1.95 to 1.61 when the Reynolds number was increased from $\mathrm{Re}_\lambda = 61$ to $\mathrm{Re}_\lambda = 150$. Because experiments are done at much larger Reynolds numbers than DNS, the estimated values computed here are in agreement with the usual experimental values $C_K \simeq 1.5$ mentioned in Monin and Yaglom (1975).

The High-Order Moments of the Velocity and Its Derivatives

We now investigate the statistical properties, such as skewness and flatness factors (see Section 2.2.5), of the turbulent flows previously described. As in Jiménez et al. (1993), we study the high-order moments of the velocity component u_1 and its longitudinal and transverse derivatives, namely $\partial u_1/\partial x_1$ and $\partial u_1/\partial x_2$, corresponding to the DNS and DML simulations under consideration. We recall that the mth moment of the velocity component u_1 is defined by

$$\mu_m(u_1(t)) = \langle u_1^m(t) \rangle_\Omega ,$$

when an ergodic hypothesis is made and the statistical average (see Chapter 2, Section 2.1) is replaced by the spatial average. Such a moment is then time-dependent and is widely fluctuating in time. As usual, we consider nondimensional moments defined by

$$\frac{\mu_m(u_1(t))}{\mu_2(u_1(t))^{m/2}}.$$

In order to compare the different simulations reported here, we have averaged these moments with respect to time and defined

$$\overline{\mu_m}(u_1, t) = \frac{1}{t} \int_0^t \mu_m(u_1(s))\, ds.$$

The first moment $\overline{\mu_1}(u_1, t)$ corresponds to the mean value of u_1 over the domain $\Omega \times (0, t)$. The odd moments provide information on the symmetry of the probability distribution function (pdf) of u_1; all odd moments μ_{m+1} of pdfs

Table 10.7. *High-order moments of the velocity component u_1 and its longitudinal and transverse derivatives*

		u_1		$\partial u_1/\partial x_1$				$\partial u_1/\partial x_2$	
		F_4	F_6	$-F_3$	F_4	$-F_5$	F_6	F_4	F_6
DNS	Mesh								
	128^3	2.71	11.31	0.467	5.14	8.72	76.16	7.36	185.1
	96^3	2.75	11.83	0.462	4.98	7.97	67.60	7.07	165.0
	64^3	2.75	11.57	0.411	4.58	6.52	52.27	6.30	120.1
DML	ξ_r								
	5.0×10^{-3}	2.73	11.50	0.478	5.18	8.98	77.91	7.47	196.0
	7.5×10^{-3}	2.77	11.85	0.468	5.07	8.48	73.50	7.24	181.3
	1.0×10^{-2}	2.75	11.66	0.459	5.13	8.65	78.41	7.05	170.6

that are symmetric with respect to their mean value vanish. We associate with $\overline{\mu_m}(u_1, t)$ nondimensional moments defined by

$$F_m(u_1, t) = \frac{\overline{\mu_m}(u_1, t)}{\overline{\mu_2}(u_1, t)^{m/2}}.$$

The third nondimensionalized moment is called the *skewness* factor, and the fourth one is called the *flatness* factor.

In Table 10.7, we have listed the mean values over the time interval $[2T/3, T]$ of $F_m(u_1, t)$ for $m = 2$ and 4 and of $F_m(\partial u_1/\partial x_1, t)$ and $F_m(\partial u_1/\partial x_2, t)$ for $3 \leq m \leq 6$, corresponding to the DNS and DML simulations presented in the preceding subsection. For the DNS simulations, the amplitudes of these moments with respect to their mean on the interval $[2T/3, T]$ are less than 2.15 percent except for the sixth moment of $\partial u_1/\partial x_2$ corresponding to the 128^3 DNS, which shows a variation of about 4.5 percent. Concerning the DML simulations all the amplitudes are smaller than 2.25 percent. It takes a longer time for this statistical quantities to reach a statistical steady state; they are indeed much more oscillating in time than the kinetic energy and the energy dissipation rate.

The distribution of homogeneous flows is often compared to a Gaussian one; as was mentioned in Section 2.2.5, the fourth (the sixth) moment of a Gaussian distribution is equal to 3.0 (15.0). For our simulations, the velocity skewness factor $-F_3(u_1, t)$ has a zero mean. The flatness factor is of the order of 2.75, while the sixth moment $F_6(u_1, t)$ is of the order of 11.5 and thus is smaller than the sixth moment of a Gaussian distribution. Similar values for these moments have been reported in Jiménez et al. (1993).

The values corresponding to the high-order moments of the velocity derivatives are slightly smaller than, but of the same order as, the ones reported in Jiménez et al. (1993) at a similar Reynolds number. The differences with the latter results are due to the facts that the aliasing errors have not been removed in our DNS simulations at resolution $\eta k_N > 1$ and that in our case the integrations have been conducted on a much larger time interval. By accumulating the statistics over the time interval $[5\tau_e, 10\tau_e]$ for the 128^3 simulation, we obtain

$$-F_3(\partial u_1/\partial x_1, t) = 0.52, \qquad F_4(\partial u_1/\partial x_1, t) = 5.35,$$
$$-F_5(\partial u_1/\partial x_1, t) = 10.0, \qquad F_6(\partial u_1/\partial x_1, t) = 85.0,$$

in agreement with the values given in Jiménéz et al. (1993). As can be seen on Figure 10.13, the third and fifth order moments decrease for $t > 10\tau_e$ and reach a plateau corresponding to a slightly smaller mean value. The same time behavior was observed for the dealiased simulations. However, for the resolutions $N = 128$ and $N = 96$ we have noticed that the presence of aliasing errors tends to reduce by a small percentage the high-order moments of the velocity derivatives. For instance, the moments corresponding to the dealiased 96^3 DNS are of the same order as the values corresponding to the aliased 128^3 DNS, while the results of the aliased 96^3 DNS are smaller than those of the 128^3 DNS (see Table 10.7).

By first comparing the DNS simulations, we note that the moments of the velocity derivatives are sensitive to the numerical resolution. Indeed, they decrease when the resolution k_N is decreased. The values corresponding to the DML simulation with $\xi_r = 5.0 \times 10^3$ are of the same order as the 128^3 DNS ones. The two other DML simulations give moments slightly smaller but of the order of the 128^3 DNS ones and larger than the 96^3 DNS ones. We recall that the cutoff levels for these simulations are such that the scales corresponding to wavenumbers $k \in [25, 48]$ are not computed exactly but are approximated via the DML algorithm based on a quasistatic (QS) approximation. Therefore, the statistic properties of these scales, as well as of the smallest ones (i.e. $48 \leq k \leq 64$), are preserved with the DML algorithm. The DML algorithm does not act as a filter.

In Figure 10.14, we have represented the time evolution of the instantaneous longitudinal velocity derivative skewness and flatness factors, namely

$$\frac{\mu_{2j+1}\left(\frac{\partial u_1}{\partial x_1}(t)\right)}{\mu_2\left(\frac{\partial u_1}{\partial x_1}(t)\right)^{(2j+1)/2}}, \qquad j = 1, 2,$$

for the 128^3 DNS and the DML simulations with $\xi_r = 1.0 \times 10^{-2}$. Note that they are strongly oscillating functions of time. These curves are different; however,

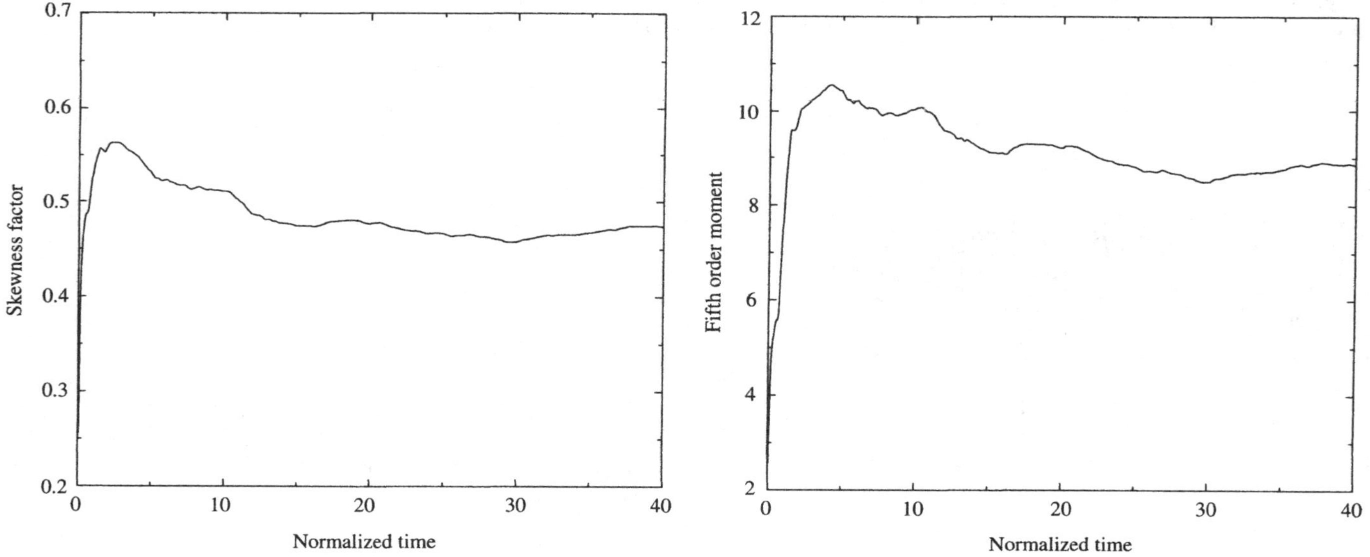

Figure 10.13. Third (left) and fifth (right) nondimensionalized moments $F_{2j+1}(\partial u_1/\partial x_1, t)$, $j = 1, 2$, of the longitudinal derivative of the velocity component u_1, plotted as a function of $t/\tau\psi_e$These curves correspond to the 128^3 simulation.

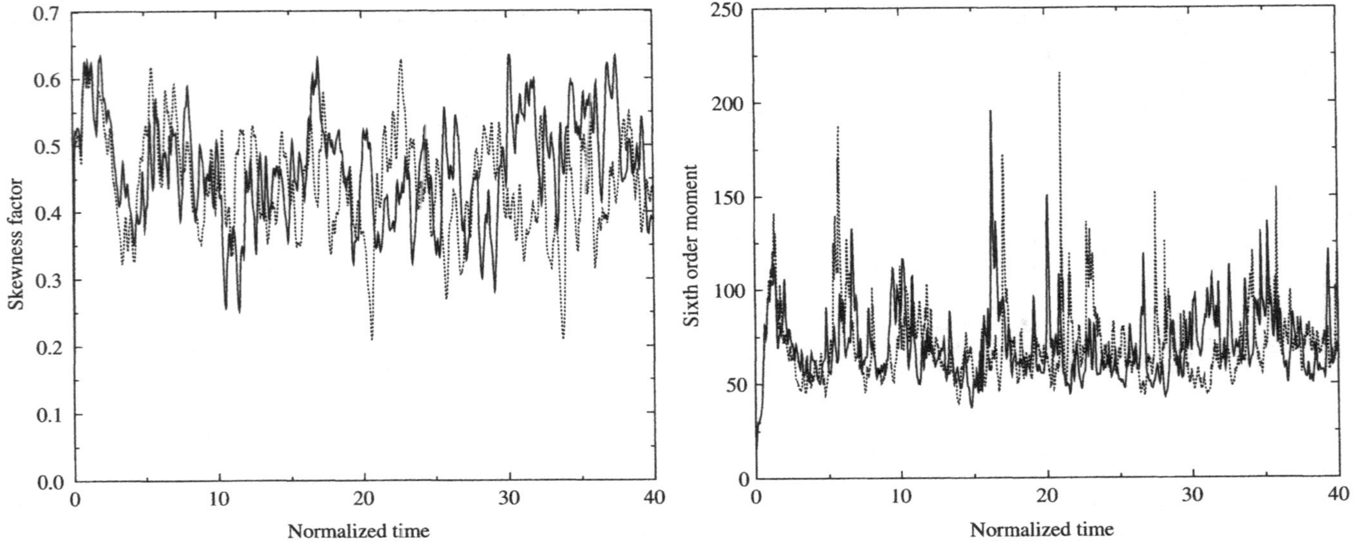

Figure 10.14. Instantaneous third (left) and sixth (right) nondimensionalized moments $\mu_{3j}(\partial u_1(t)/\partial x_1)/\mu_2^{3j/2}(\partial u_1(t)/\partial x_1)$, $j = 1, 2$, of the longitudinal derivative of the velocity component u_1, plotted as a function of t/τ_e. These curves correspond to the 128^3 simulation (solid line) and the DML simulation with $\xi_r = 1.0 \times 10^{-2}$ (dotted line).

their mean values are very close to each other. This indicates that the DML method computes small scales with a dynamics similar to the small scales of the 128^3 DNS.

The Probability Density Function

In Figures 10.15 and 10.16, we have represented the probability density function (pdf) of the velocity component u_1 for the DNS and DML simulations. First, we note that all distribution functions have a similar shape and only minor differences appear. They all match in the interval $[-1, 1]$. As was previously mentioned, this pdf is known to be approximately Gaussian. Our distribution functions are indeed close to a Gaussian distribution, represented in Figure 10.15, and they are in agreement with the results shown in Vincent and Ménéguzzi (1991).

While the distribution of u_1 appears not to be sensitive to the resolution, noticeable differences appear on the distributions of the longitudinal and transverse derivatives represented in Figure 10.17. The extremal values of these pdfs decrease when N is decreased, so that the 64^3 simulation gives a distribution less intermittent than the other ones. As we compare simulations with the same

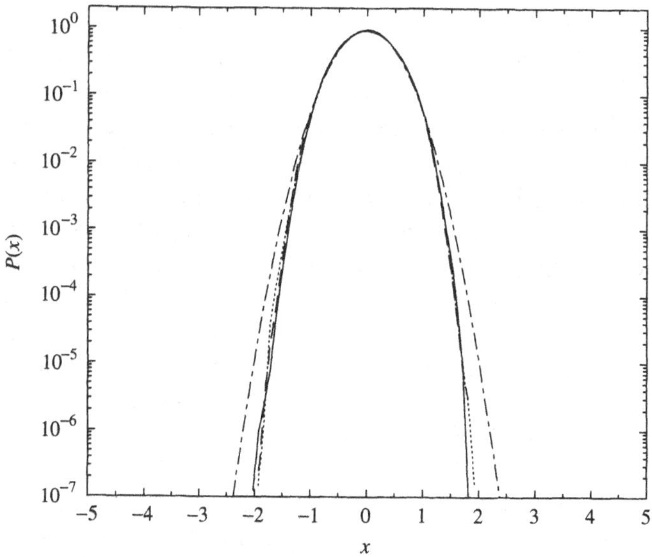

Figure 10.15. Probability distribution functions of the DNS simulations shown together with a Gaussian distribution. The curves correspond to the 128^3 DNS (solid line), the 96^3 DNS (dotted line), the 64^3 DNS (dashed line), and the Gaussian distribution (dotted–dashed line).

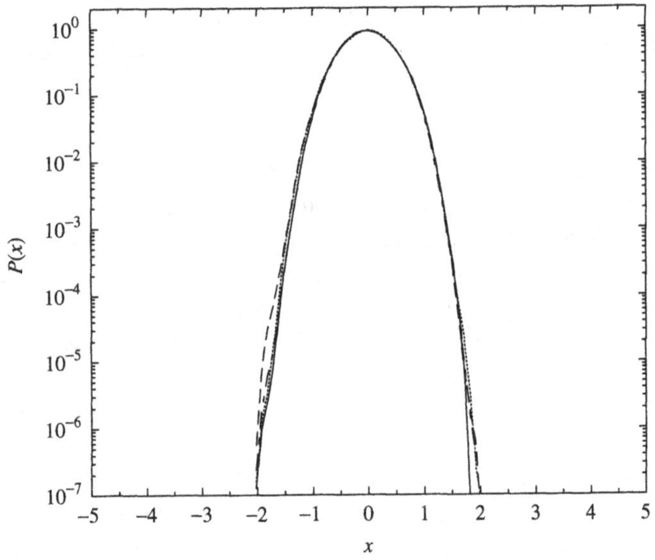

Figure 10.16. Probability distribution functions of the DML simulations shown together with the 128^3 DNS. The curves correspond to the 128^3 DNS simulation (solid line) and to the DML simulations with $\xi_r = 5.0 \times 10^{-3}$ (dotted line), $\xi_r = 7.5 \times 10^{-3}$ (dashed line), and $\xi_r = 1.0 \times 10^{-2}$ (dotted–dashed line).

Reynolds number and same set of data and with parameters differing only in the resolution, the distributions have not been normalized as usual. The distribution of $\partial u_1/\partial x_2$ takes larger extremal values, inducing larger even moments for $\partial u_1/\partial x_2$ than for $\partial u_1/\partial x_1$ (see Table 10.7). As a consequence, it appears that the distribution of $\partial u_1/\partial x_2$ is more sensitive to the resolution than that of $\partial u_1/\partial x_1$.

By comparing the distributions corresponding to the 128^3 DNS and the DML simulations on Figure 10.18, we note that they match perfectly at small values, that is, near the average of the signal, and that minor differences appear on the wings of the distribution, where the values are larger. The extremal values, even for the transverse derivative of u_1, which has a wider distribution, are of the same order. The shape of the distributions is then preserved by the DML algorithm. This is consistent with the analysis previously presented of the high-order moments of the velocity derivatives.

The distributions of the velocity derivatives presented here are in agreement with those reported in Vincent and Ménéguzzi (1991) and in Jiménez et al. (1993). The same properties are noted: the distribution of $\partial u_1/\partial x_2$ is more symmetric than the distribution of $\partial u_1/\partial x_1$. This implies that the odd-order moments associated with $\partial u_1/\partial x_2$ are lower than those associated with $\partial u_1/\partial x_1$.

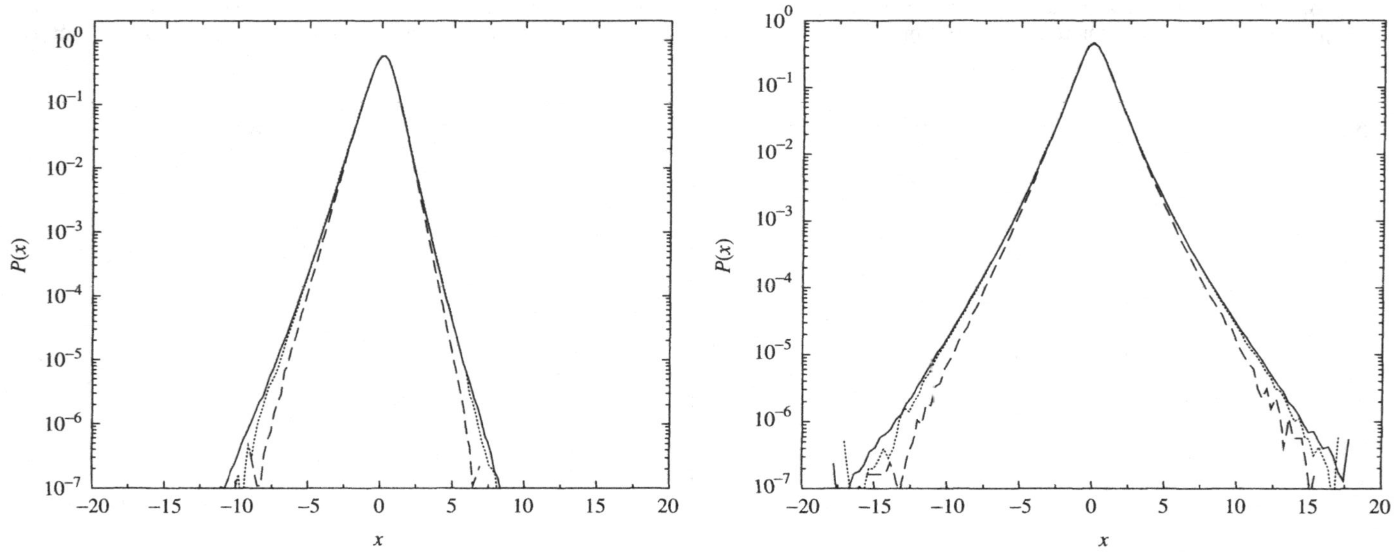

Figure 10.17. Probability distribution functions of the longitudinal derivative $\partial u_1/\partial x_1$ and of the tranverse derivative $\partial u_1/\partial x_2$ for the DNS simulations. The curves correspond to the 128^3 DNS (solid line), the 96^3 DNS (dotted line), the 64^3 DNS (dashed line), and the Gaussian distribution (dotted–dashed line).

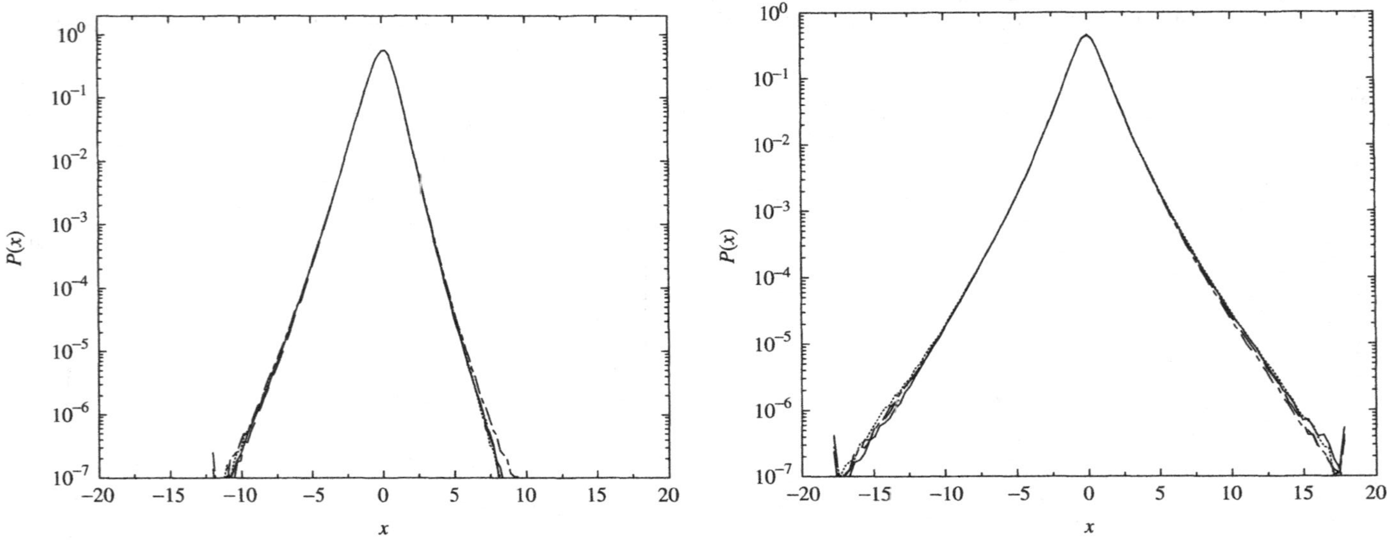

Figure 10.18. Probability distribution functions of the longitudinal derivative $\partial u_1/\partial x_1$ and of the tranverse derivative $\partial u_1/\partial x_2$ for the DML and the 128^3 DNS simulations. The curves correspond to the 128^3 DNS simulation (solid line) and to the DML simulations with $\xi_r = 5.0 \times 10^{-3}$ (dotted line), $\xi_r = 7.5 \times 10^{-3}$ (dashed line), and $\xi_r = 1.0 \times 10^{-2}$ (dotted–dashed line).

Moreover, the distribution of $\partial u_1/\partial x_2$ has larger wings than the distribution of $\partial u_1/\partial x_1$.

Summary

In this section, we have estimated the effect of the resolution, that is, of k_N, on the statistics of forced three-dimensional homogeneous turbulent flows. Long-time integrations, over 40 eddy-turnover times, of the Navier–Stokes equations at $\mathrm{Re}_\lambda = 100$ and at different resolution, $\eta k_N \in [0.79, 1.59]$, have been done so that statistically steady states have been reached and analyzed. Simulations with the DML algorithm, described in the preceding chapter, and with different values of the parameter ξ_r have been conducted with the same data and parameters as the DNS simulations. We have shown that with cutoff levels in the range $[0.65k_d, 1.3k_d]$ the DML simulations yield very similar statistical properties to the DNS simulation at resolution $k_N = 1.59k_d$. The DNS with resolution $k_N = 0.79k_d$ fails to reproduce these properties, while the DNS with $k_N = 1.18k_d$ has lower high-order moments for the velocity derivatives than the DML simulations, and it reaches a statistical state slightly different; namely, the kinetic energy and the energy dissipation rate are increased as compared to other simulations. It should be observed that the CPU time required by the DML simulations for $\xi_r = 1.0 \times 10^{-2}$ (7.5×10^{-3}) is about 2.5 (2) times faster than the DNS simulation with $k_N = 1.59k_d$. We have therefore shown the efficiency of the DML methodology and its ability to estimate this range of scales at lower costs. While computing different small scales than the standard Galerkin algorithm used for DNS, the DML method preserves the statistical properties, such as the intermittency, of the small scales. However, in order to be used for practical applications, the DML method has to be optimized. For instance, other (more flexible) strategies than the one based on V-cycles should be tested and compared. The study of improved (more efficient) DML algorithms is in progress, and the results will be reported in future works.

10.3. Turbulent Channel Flows

In this section we present results on the numerical simulation of the flow in a three-dimensional channel. The computational domain is $\Omega = (0, L_1) \times (-h, +h) \times (0, L_3)$ with $h = 1$, $L_1 = 4\pi$ (streamwise direction), and $L_3 = 2\pi$ (spanwise direction). The boundary conditions are periodic in the streamwise and spanwise directions. For the direction normal to the walls, we retained a (no-slip) homogeneous Dirichlet boundary condition. As indicated in Section 3.2.1, the flow is sustained by a mean pressure gradient K_p. Here we have chosen $K_p = -1$. The channel lengths L_1 and L_3 must be chosen in order to insure that the turbulence fluctuations are uncorrelated at a separation of one half period in the two homogeneous directions. In the following, as in Section 3.2.1, we denote by $\langle\cdot\rangle$ the average operator in the homogeneous directions x_1 and x_3. The decrease of the two-point correlations $R_{u_i u_i} = \langle u_i'(x_j) u_i'(x_j+\delta x_j)\rangle$, $i = 1, 2, 3$, in the streamwise ($j = 1$) and spanwise ($j = 3$) directions shows the adequacy of the computational domain. The values of the channel lengths have been chosen in order to satisfy the above requirement in Kim, Moin, and Moser (1987) for a simulation with the same Reynolds number and resolution as the one presented here.

The viscosity is chosen equal to $\nu = 1/180 = 5.5 \times 10^{-3}$. The Reynolds number based on the friction velocity $u_* = 1$ and introduced in (2.142), $\mathrm{Re}_* = hu_*/\nu$, is equal to 180. The number of grid points retained in each direction for this simulation is $128 \times 129 \times 128$, that is, $N = 128$ and $M = 128$. So the grid spacings in the streamwise and spanwise directions are uniform and equal to $\Delta x_1 = L_1/N = 0.1$ and $\Delta x_3 = L_3/N = 0.05$. In the normal direction, a nonuniform mesh is used, based on the Chebyshev–Gauss–Lobatto collocation points $\cos(k\pi/M)$, $i = 0, \ldots, M$. The first mesh point away from the upper wall is at $x_2 = \cos(\pi/M) = 0.9997$. The grid resolution is sufficiently fine to resolve the turbulent flow. Indeed, in Figure 8.7 (Figure 8.8) we can see that there is no energy pile-up at high wavenumbers on the one-dimensional energy spectra, in the streamwise (the spanwise) direction.

This simulation has been conducted from the initial time $t = 0$ until the final time of the computation, $T = 30$. This represents 30 nondimensional time units tu_*/h. The gradient $(dU/dx_2)(\pm h)$ oscillates around the value $-K_p h/\nu = 180$, showing that a statistically steady state is reached. The time step is taken equal to $\Delta t = 10^{-3}$ to insure numerical stability. Indeed, the stability constraint (see Section 4.1.2) is

$$\frac{2}{h\pi^2}\,\Delta t\, M^2 U_\infty \simeq 1.1.$$

The CPU time required for the direct numerical simulation is 144 CPU hours on one processor of a Fujitsu VPP300, and the memory used is 500 Mego-Octets.

The nonlinear term is computed in the rotational form $\boldsymbol{\omega} \times \mathbf{u}$ (see (3.19)) to preserve the conservation of energy in the absence of viscosity. Moreover, we use the $\frac{3}{2}$ rule, in the two periodic directions, to remove the aliasing errors appearing in the computation of the nonlinear terms by pseudospectral methods (see Gottlieb and Orszag (1977), Canuto et al. (1988)).

The cutoff levels used in the DML method applied in the homogeneous directions x_1 and x_3 (see Section 9.5.1) are $N_{i_1} = 64$ and $N_{i_2} = N = 128$. These levels are constant during the whole computation. Five coarse levels

$N_i \in [N_{i_1}, N_{i_2}]$ have be retained for the V-cycle strategy, namely,

$$N_i \in S_c = \{64, 80, 96, 108, 128\}.$$

With these cutoff levels used in the two homogeneous directions, the DML code runs approximatively 2.5 times faster than the DNS one.

In the following, we present several quantities computed with the velocity and with the velocity and pressure fluctuations (see Section 2.3): $u'_i = u_i - \langle u_i \rangle$, $i = 1, 2, 3$, and $p' = p - \langle p \rangle$.

The time evolution of the L_2 norm $|\mathbf{u}_{NM}|_0$, H_1 norm $\|\mathbf{u}_{NM}\|$ and L^∞ (supremum) norm $|\mathbf{u}_{NM}|_\infty$ of the velocity field $\mathbf{u}_{NM}$ computed with the DNS and DML methods are represented on Figure 10.19. For each quantity, the curves corresponding to the DNS and DML simulations are very close to each other, even for large times.

Figure 10.20 shows the time evolution of the turbulent kinetic energy

$$\frac{1}{4} \int_{-1}^{+1} \left\langle \left|u'_i\right|^2 \right\rangle dx_2$$

for the DNS and DML simulations. The profiles of the two curves are similar. The turbulent kinetic energy computed with the DML method is slightly lower than the one computed with DNS, showing that DML is slightly dissipative. The time-averaged value of the turbulent kinetic energy is 2.04 for DNS and

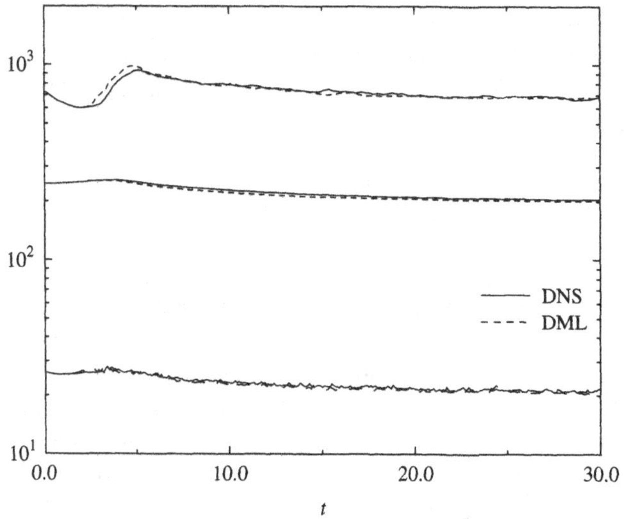

Figure 10.19. Time evolution of the L_2 norm (medium curve), H_1 norm (top curve), and L^∞ norm (bottom curve) of the velocity field, computed with DNS and DML methods.

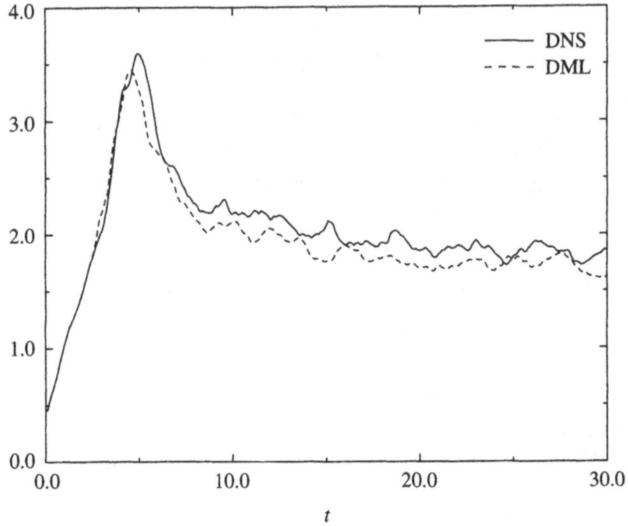

Figure 10.20. Time evolution of the turbulent kinetic energy, computed with DNS and DML methods.

1.93 for DML, so that the relative difference between the two simulations is of the order of 5.4 percent.

Now we consider the boundary conditions. In Figure 10.21 we have represented the time evolution of $\langle(\partial u_1/\partial x_2)(\pm 1)\rangle$, computed with the DNS and DML methods. The profiles and the values of the time evolution, computed with the two methods, are quite identical. Moreover, the curves are symmetrical with respect to the centerline of the channel. According to the theory, it is expected (see Section 2.2) that

$$\lim_{T\to+\infty}\frac{1}{T}\int_0^T\left\langle\frac{\partial u_1}{\partial x_2}(-1)\right\rangle dt = -\frac{hK_p}{\nu}.$$

In particular, when a steady state is reached, the quantity $\langle\partial u_1/\partial x_2(-1)\rangle$ oscillates around $-K_p h/\nu = 180$. As has been said previously, this follows because the average values are -186.6 and 182.5 for DNS, -182.7 and 186.6 for DML.

Now we consider the time evolution of the supremum of $u_i(\mathbf{x})$, for $x_2 = \pm 1$ and $i = 1, 2, 3$, for the DNS and DML simulations. These quantities have very close mean values. Indeed, at the final time of the computation the relative differences for the time-averaged values are 3.8 percent for $i = 1$, 1.9 percent for $i = 2$, and 4.8 percent for $i = 3$.

Figure 10.22 (Figure 10.23) represents one-dimensional energy spectra similar to those represented on Figures 8.7 (Figure 8.8), computed with the DNS

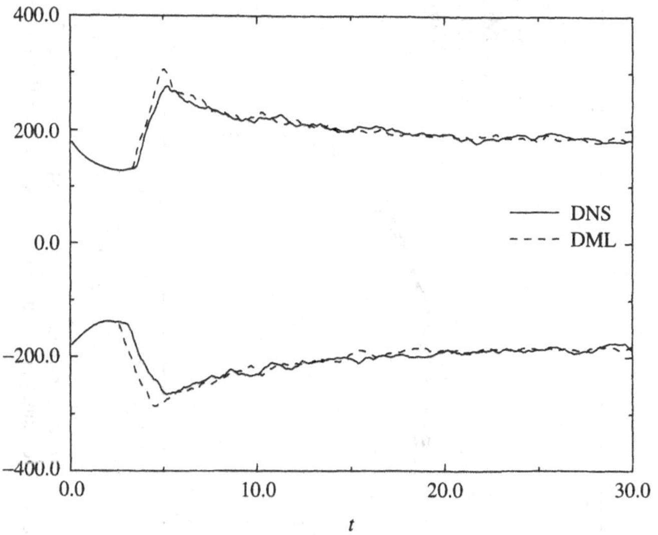

Figure 10.21. Time evolution, computed with DNS and DML methods, of $\langle \partial u_1/\partial x_2 (\pm 1)\rangle$. The top curve corresponds to $x_2 = -1$, and the bottom curve to $x_2 = +1$.

and DML methods. As we can see, the low parts of the spectra, associated with small wavenumbers (large scales), are in agreement. At large wavenumbers, DML spectra are higher than DNS spectra. The first slide in the DML spectra corresponds to the lowest cutoff level N_{i_1}. The other slides are associated with the other cutoff levels, between N_{i_1} and N_{i_2}.

According to the theory, the friction velocity u_* must be equal to $\sqrt{-K_p h} = 1$, since $h = 1$ and $K_p = -1$ (see Section 2.3). According to the definition (2.140) of the friction velocity, we have $u_* = \nu(dU/dx_2)(-h) = 1.01$ (i.e. an error of 1.4 percent of the theoretical value) for DNS and 1.04 for DML (error equal to 3.9 percent of the theoretical value). The Reynolds number based on the friction velocity $\mathrm{Re}_* = hu_*/\nu$ (see (2.142)) is equal to 181.8 for DNS and 187.2 for DML.

In Figure 10.24, we have represented the DNS mean-velocity profile non-dimensionalized by the friction velocity, $\langle u_1 \rangle / u_*$, in wall coordinates $x_2^+ = x_2 u_*/\nu$. The collapse of the mean-velocity profiles corresponding to the upper and lower halves of the channel indicates the adequacy of the grid used for the computation (see Kim, Moin, and Moser (1987)). We can observe that the profiles agree well with the law of the wall for small values of x_2^+. As for the log law, there is a small difference between the profiles of DNS and of the log law for large values of x_2^+. Our profiles are quite similar to those reported in Kim, Moin, and Moser (1987).

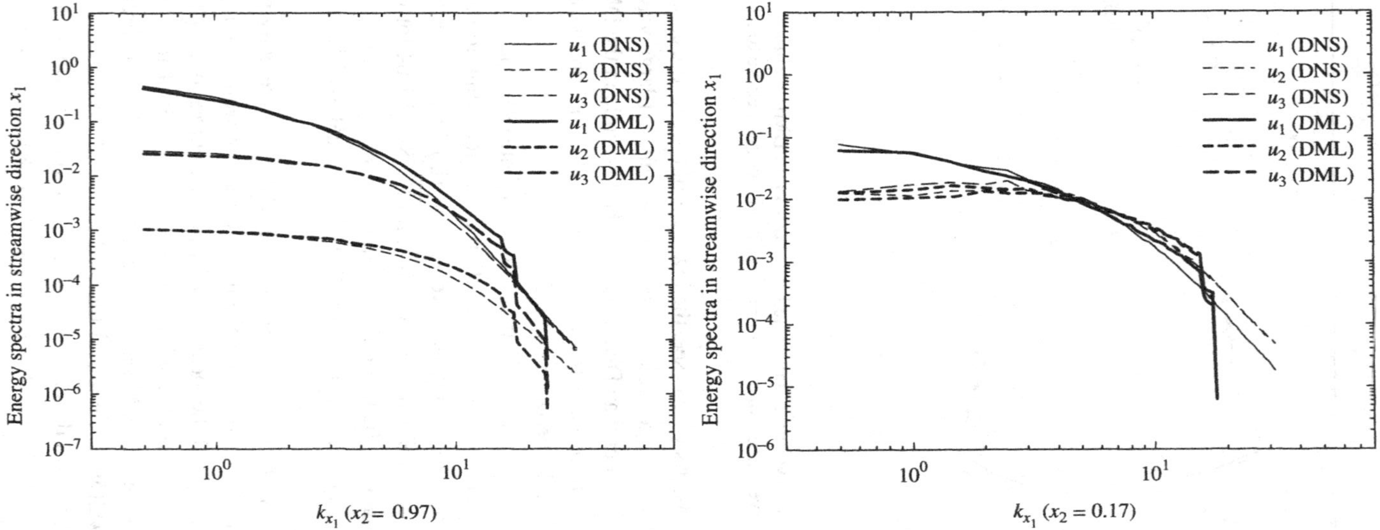

Figure 10.22. One-dimensional energy spectrum functions in the streamwise direction x_1 for each component of the velocity u_i, $i = 1, 2, 3$, and for two values of $x_2 : x_2 = 0.97$ (near the upper wall) and $x_2 = 0.17$ (near the centerline of the channel), computed with DNS and DML methods.

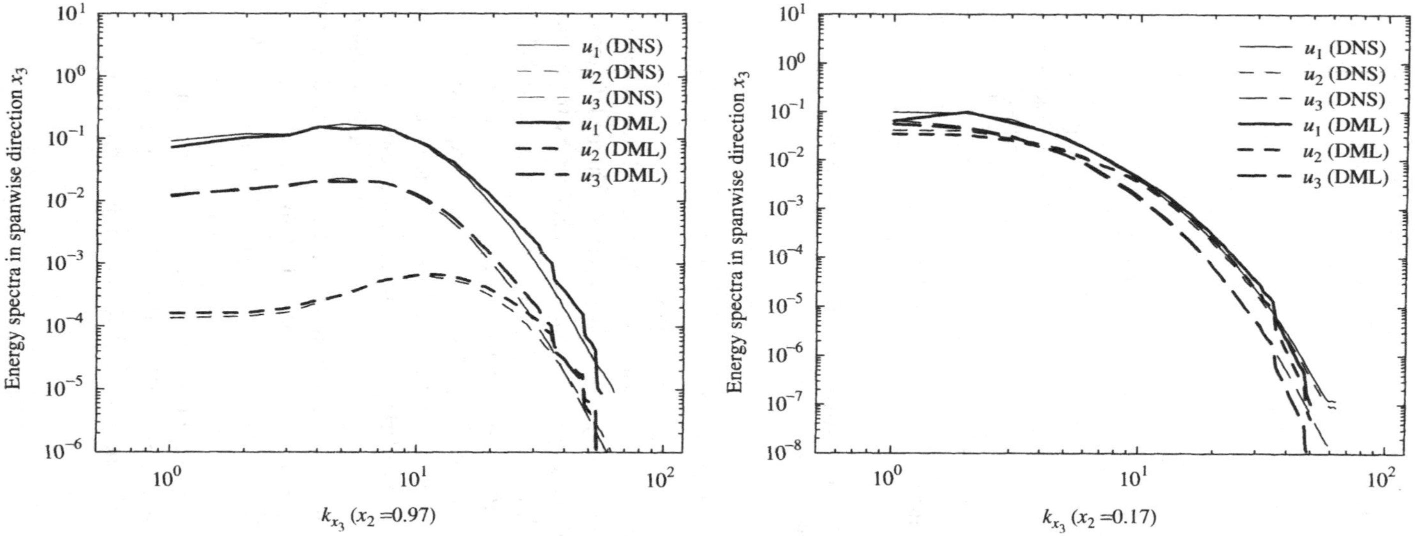

Figure 10.23. One-dimensional energy spectrum functions in the spanwise direction x_3 for each component of the velocity u_i, $i = 1, 2, 3$, and for two values of x_2: $x_2 = 0.97$ (near the upper wall) and $x_2 = 0.17$ (near the centerline of the channel), computed with DNS and DML methods.

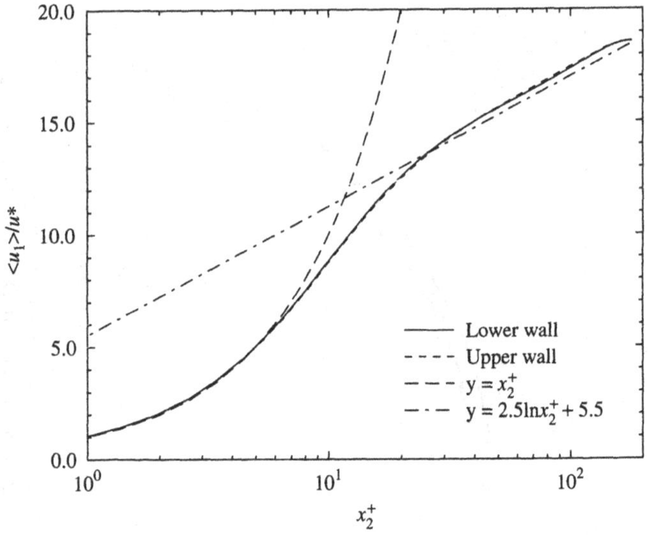

Figure 10.24. DNS mean-velocity profile, normalized by the friction velocity, $\langle u_1 \rangle / u_*$, in wall coordinates $x_2^+ = x_2 u_* / \nu$.

Now let us consider the DML results. In Figure 10.25, we compare the DNS and DML mean-velocity profiles, in wall coordinates and in global coordinates x_2/h. We first notice that with the DML method, the collapse of the upper and lower mean-velocity profiles is preserved. On the left, the mean-velocity profile, in wall coordinates, is nondimensionalized by the friction velocity as in Figure 10.24, and on the right, corresponding to the global coordinates, it is normalized with the bulk mean velocity U_m (see Section 4.1.2). The bulk mean velocity computed with DNS (with DML) is equal to 17.21 (16.87). The Reynolds number based on U_m, $\mathrm{Re}_m = hU_m/\nu$, is equal to 3098 for DNS and 3037 for DML. Considering the mean centerline velocity $U_c = \langle u_1 \rangle \, (x_2 = 0)$, we have $U_c = 19.75$ for DNS and 19.53 for DML. The Reynolds number based on U_c, $\mathrm{Re}_c = hU_c/\nu$, is equal to 3555 for DNS and to 3515 for DML. The ratio U_c/U_m is equal to 1.15 for DNS (1.16 for DML). Following Kim, Moin, and Moser (1987), Dean's correlation gives $U_c/U_m = 1.28 \, \mathrm{Re}_m^{-0.0116}$. With this law, we find that this ratio is equal to 1.17 for DNS and DML. This is in good agreement with the numerical results, since the ratio U_c/U_m is equal to 1.15 for DNS and 1.16 for DML. Considering Figure 10.25, we can see that the DNS and DML mean values are very similar throughout the width of the channel.

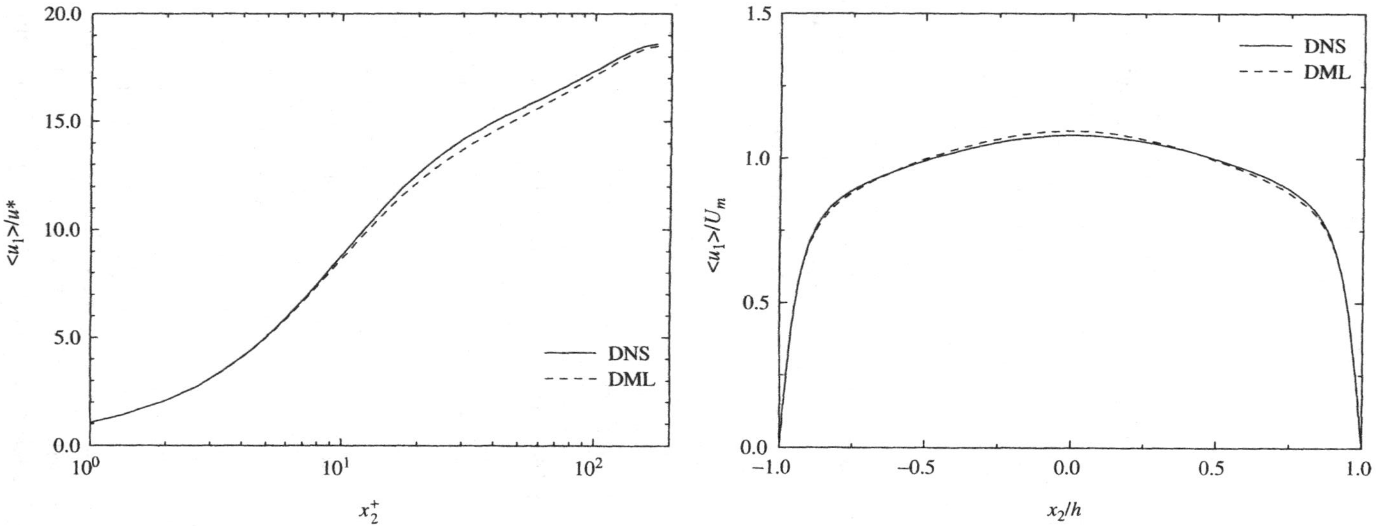

Figure 10.25. DNS and DML mean-velocity profiles: normalized by the friction velocity, $\langle u_1 \rangle / u_*$, in wall coordinates $x_2^+ = x_2 u_* / \nu$ (left), and normalized by the bulk mean velocity $\langle u_1 \rangle / U_m$, in global coordinates x_2 / h (right).

Now, we consider the root-mean-square velocity fluctuations (turbulence intensities) for u_i, $i = 1, 2, 3$, normalized by the friction velocity:

$$u_{i,\text{rms}} = \frac{\langle u_i'^2 \rangle^{1/2}}{u_*},$$

in global and in wall coordinates (Figure 10.26), for the DNS and DML simulations. The symmetry of the profiles with respect to the channel centerline indicates the adequacy of the grid computation, as has been noted previously. The DNS profiles are in agreement with those presented in Kim, Moin, and Moser (1987). If we compare the DNS and DML results, as we can see in Figure 10.26 the curves associated with the two methods are identical for u_2' and u_3'. A small difference appears only for u_1' (overestimation of the maximum values).

In Figure 10.27 the root-mean-square pressure fluctuation $p_{\text{rms}} = \langle p'^2 \rangle^{1/2}$, normalized by the wall shear velocity $\tau(-h) = u_*^2$ (see Section 2.3), is represented at the final time of the computation, in wall and in global coordinates. As previously, the profiles obtained with DNS and the profiles obtained by Kim, Moin, and Moser (1987) are similar. Now, if we compare the results obtain with DNS and DML, the curves are very close.

Figure 10.28 represents the Reynolds shear stress $\langle u_1' u_2' \rangle$, normalized with the friction velocity u_*, in global coordinates and in wall coordinates. We can note that the Reynolds shear stress is antisymmetric and vanishes at the center of the channel ($x_2 = 0$). This indicates that the flow reaches an equilibrium state (see Kim, Moin, and Moser (1987)). If we compare the DNS and DML results, we can see that the curves are identical, even near the wall.

The root-mean-square vorticity fluctuations $\langle \omega_i'^2 \rangle^{1/2}$, normalized by the mean shear $dU/x_2(-h) = u_*^2/\nu$ (see (2.140)) – that is, $\nu \langle \omega_i'^2 \rangle^{1/2} / u_*^2$, $i = 1, 2, 3$ – are shown in Figure 10.29, in global coordinates and in wall coordinates. Comparing the DNS results with those of Kim, Moin, and Moser (1987), we can see that the profiles are similar. The three components of the fluctuating vorticity are identical away from the wall. Near the wall, the streamwise vorticity ω_1' shows a local maximum for $x_2^+ = 20$ and a local minimum for $x_2^+ = 5$, before it reaches its maximum value at the wall. Following Kim, Moin, and Moser (1987), the center of the streamwise vortex is located on average at $x_2^+ = 20$ with radius $r^+ = 15$. Now if we compare the DNS and DML results, we see that the curves computed with DNS and those computed with DML are identical in the center of the channel. Near the wall, the wall coordinate representations show that the DNS and DML curves are identical up to the wall for ω_3'. However, a slight difference appears in an intermediate region lying between $x_2^+ = 5$ and

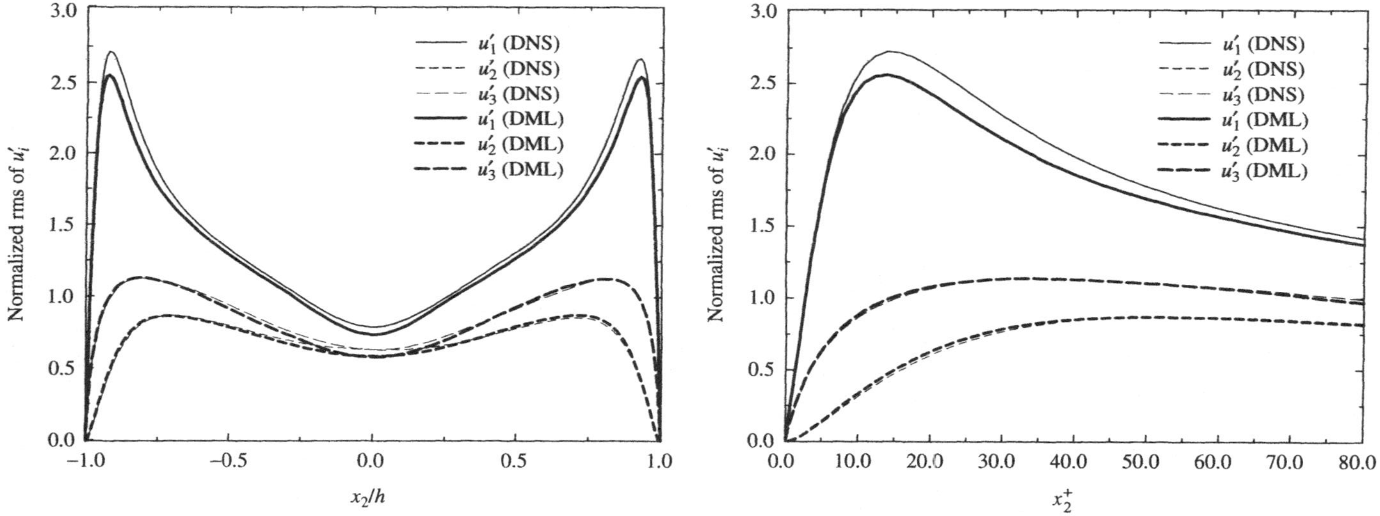

Figure 10.26. DNS and DML root-mean-square velocity fluctuations, normalized by the friction velocity, $\langle u_i'^2 \rangle^{1/2}/u_*$, $i = 1, 2, 3$, in global coordinates x_2/h (left) and in wall coordinates $x_2^+ = x_2 u_*/\nu$ (right).

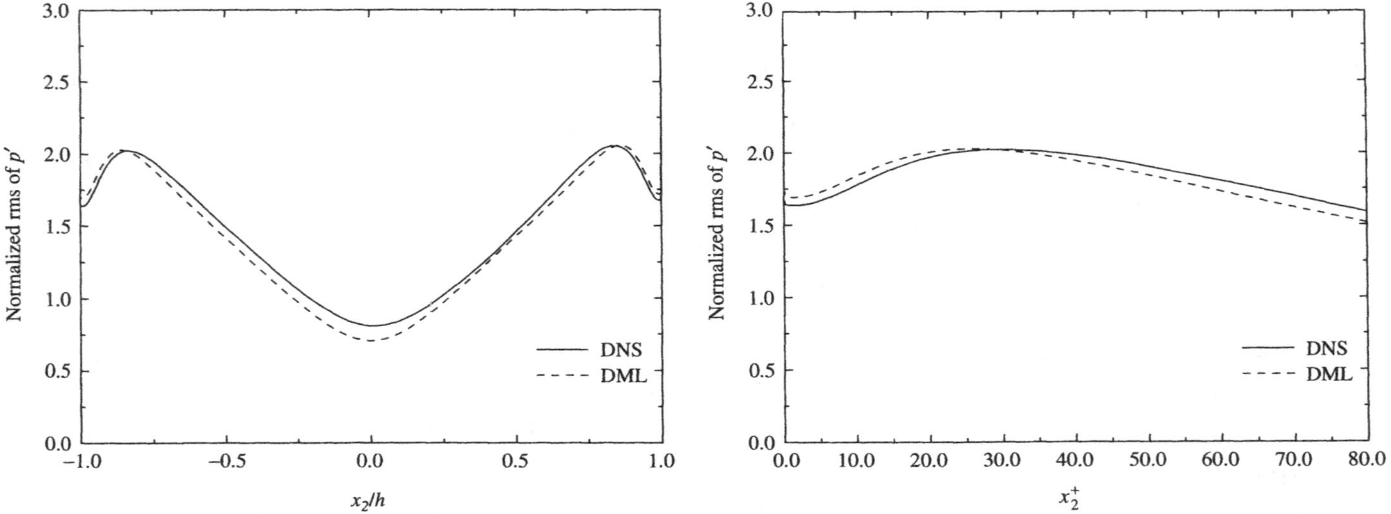

Figure 10.27. DNS and DML root-mean-square pressure fluctuations, normalized by the wall shear velocity, $\langle p'^2\rangle^{1/2}/u_*^2$, in global coordinates x_2/h (left) and in wall coordinates $x_2^+ = x_2 u_*/\nu$ (right).

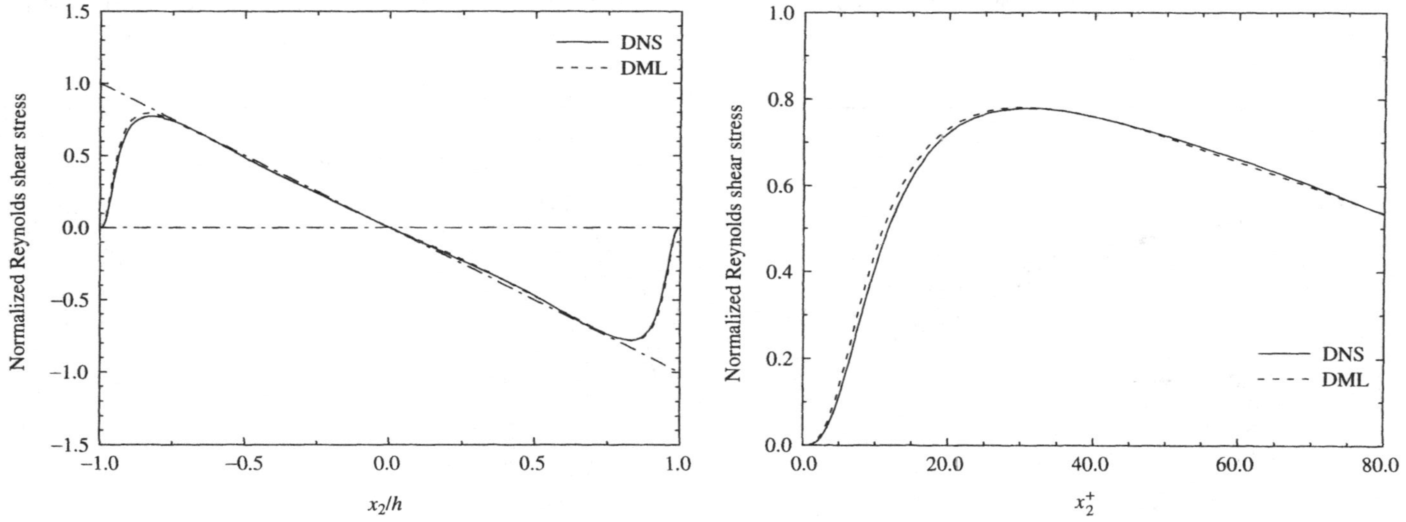

Figure 10.28. DNS and DML Reynolds shear stress, normalized with the wall shear velocity, $\langle u_1' u_2' \rangle / u_*^2$, in global coordinate x_2/h (left) and in wall coordinates $x_2^+ = x_2 u_* / \nu$ (right).

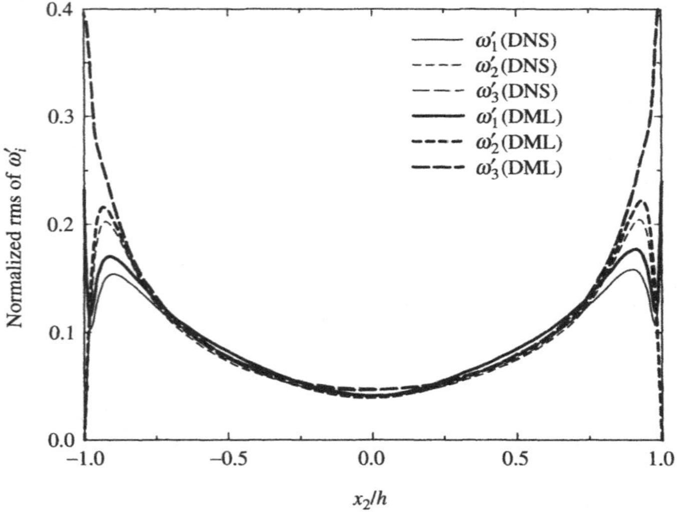

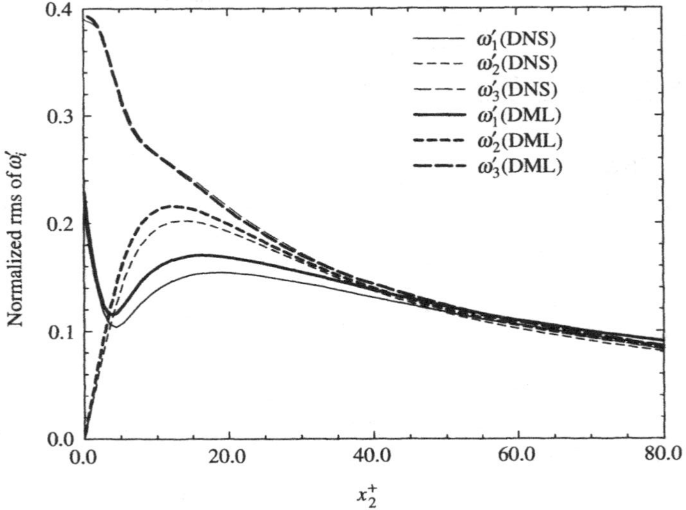

Figure 10.29. DNS and DML root-mean-square vorticity fluctuations, normalized by the mean shear, $\nu\langle\omega_i'^2\rangle^{1/2}/u_*^2$, $i = 1, 2, 3$, in global coordinates x_2/h (left) and in wall coordinates $x_2^+ = x_2 u_*/\nu$ (right).

$x_2^+ = 25$. In this region, the root-mean-square vorticity fluctuations are slightly overestimated with DML for ω_i', $i = 1, 2$. This difference decreases for ω_i' when i increases. However, the profiles are preserved. Moreover, the location of the local minimum $x_2^+ = 5$ of the streamwise vorticity is preserved with DML, and the local maximum is slightly translated towards the wall, that is, with DML the radius of the streamwise vortex is preserved but the location is shifted towards the wall.

To conclude this section, we study the skewness and flatness factors for the fluctuation quantities u_i', $i = 1, 2, 3$. Figure 10.30 (Figure 10.31) represents the skewness factors $F_3(u_i')$ (the flatness factors $F_4(u_i')$) in global and in wall coordinates. These quantities have been computed with the DNS and DML methods. As in the results presented by Kim, Moin, and Moser (1987), we notice that the skewness and flatness factors are significantly different from the values for a Gaussian distribution (see Section 2.2.5: 0 for the skewness and 3 for the flatness). Near the walls, the intermittency of the velocity fluctuations increases, in particular for the normal velocity u_2'. By comparison with the values reported in Kim, Moin, and Moser (1987), the skewness and flatness values computed here are similar. The comparison of the DNS and DML results obtained for the skewness factors (Figure 10.30) shows that the values in the channel are quite similar for the three velocity components. Near the wall, as we can see with the wall coordinate representation, the results are very close for u_1' and u_3', but for u_2' a difference appears when x_2^+ lies in the viscous sublayer. For this velocity component, the skewness factor estimated with DML is lower than with DNS. For the flatness factors (see Figure 10.31), the DNS and DML profiles are nearly the same. This similarity exists, also, near the wall, as we can see with the wall coordinate representation, except for the normal component u_2'. As for the skewness, the intermittency is underestimated, showing the dissipative action of DML on this component in the viscous sublayer.

Finally, we want to note that the DML method has been applied only in the two homogeneous directions x_1 and x_3 in the channel simulation previously described. The saving in CPU time obtained with the DML method versus DNS is a factor of 2.5.

These results are preliminary ones for this problem. They show the ability of the DML method to reproduce the mean velocity profile as well as the turbulence statistics. A detailed study of the DML method has to be done in order to derive an optimal algorithm for this problem. In the future, we should study the behavior of the DML method when the cutoff level is reduced. Moreover, a DML method will be used in the nonhomogeneous direction x_2 (see Section 9.5.2). To obtain the scale separation, the Galerkin basis described in Section 3.2.4 will be used (see Section 6.2.2).

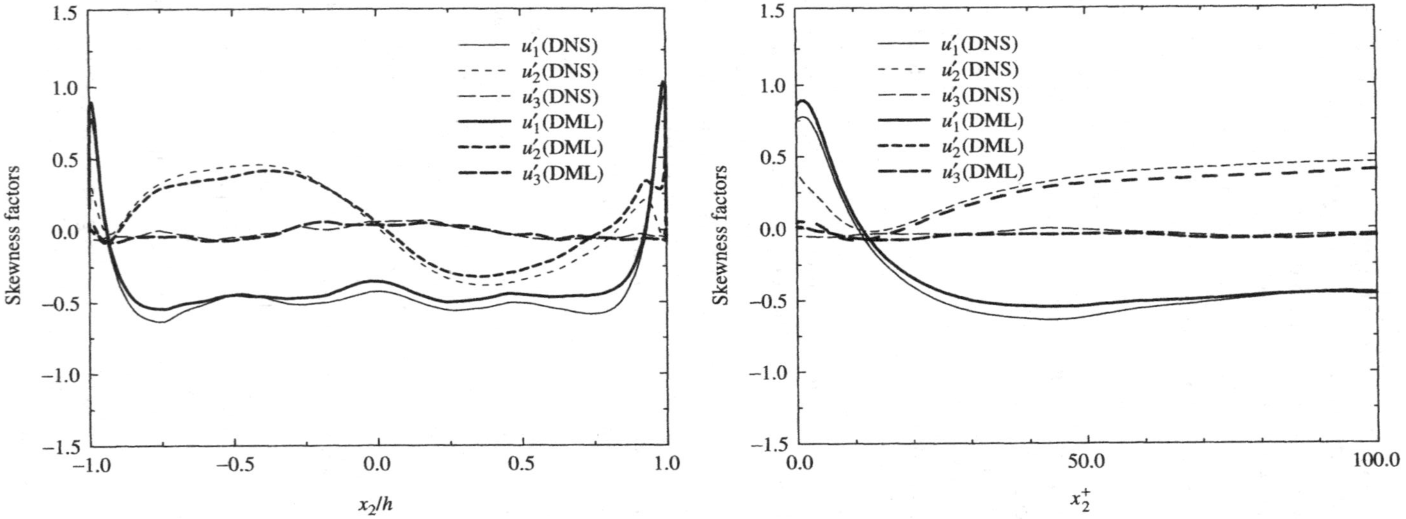

Figure 10.30. DNS and DML skewness factors $F_3.u_i'/, i = 1, 2, 3$, in global coordinates x_2/h (left) and in wall coordinates $x_2^+ = x_2 u_*/\nu$ (right).

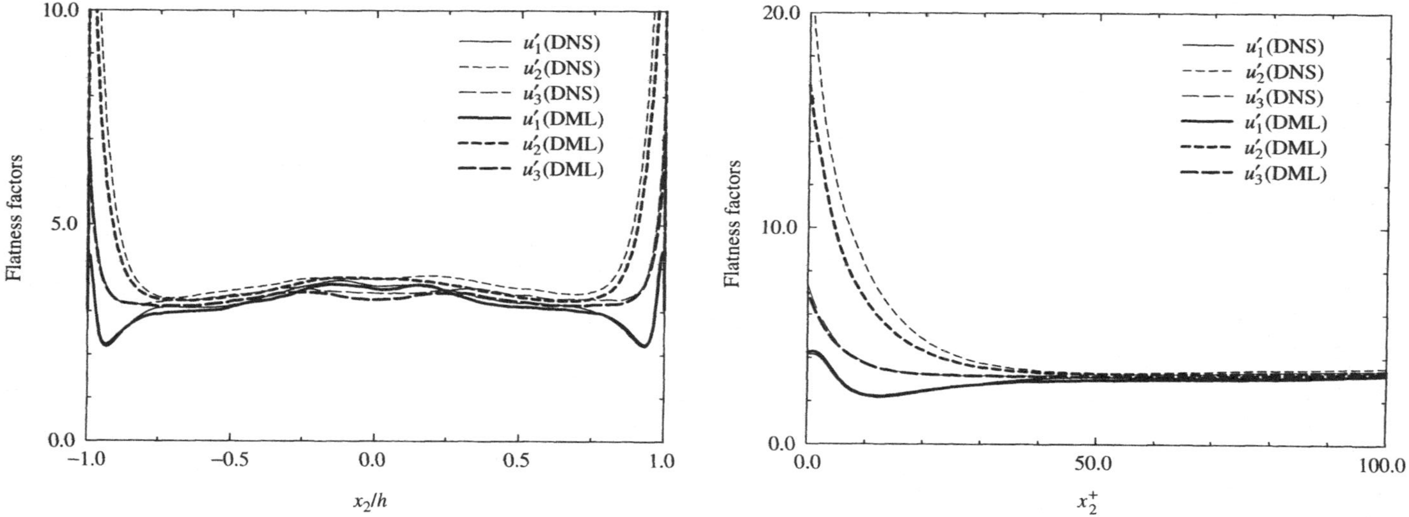

Figure 10.31. DNS and DML flatness factors $F_4.u'_i/$, $i = 1, 2, 3$, in global coordinates x_2/h (left) and in wall coordinates $x_2^+ = x_2 u_*/\nu$ (right).

First results for the one-dimensional Burgers equation, with homogeneous Dirichlet boundary conditions, have been obtained. These results show good behavior of the DML methodology. For the Navier–Stokes equations, the work is in progress and will be presented in Bouchon, Dubois, and Jauberteau (1998). Moreover, another Galerkin basis, the divergence-free Galerkin basis, described in Section 3.2.5, can be used in the DML methodology.

Conclusion

In this work, we have described the implementation of dynamic multilevel methods with application to the numerical simulation of turbulent flows. In this context, the dynamic multilevel (DML) methodology has been described and a retrospect of several DML methods has been presented. Two types of turbulent flows have been considered, namely homogeneous and nonhomogeneous turbulent flows; for the latter case, wall-bounded flows – more precisely, flows in an infinite channel – have been retained.

The DML methodology is based on scale separations with different numerical treatments for the small and large scales. The starting point with the DML methodology is the decomposition of the velocity field into several variables

$$\mathbf{u}_N = \mathbf{y}_{N_{i_1}} + \sum_{i=i_1}^{i_2-1} \mathbf{z}_{N_i}^{N_{i+1}} + \mathbf{z}_{N_{i_2}}^{N},$$

where $\mathbf{y}_{N_{i_1}}$ represents the large scales of the flow and $\mathbf{z}_{N_i}^{N_{i+1}}$ successive range of small scales. Essential in the DML methodology is the dynamical adaptative form of this decomposition; of course the judicious and practical choice of the cutting-wavenumbers $N_{i_1}, \ldots, N$, is crucial. These new flow variables satisfy a set of coupled equations very similar (and equivalent) to the initial equations. However, they enjoy different properties. The small scales contain less and less kinetic energy for increasing cutoff levels N_i. Also, their characteristic (eddy turnover) time decreases so that they have an increasing rotation rate. By studying the time variation properties of these different scales, we have shown that the small scales have a relatively small time variation over time intervals of a few time steps. The time variation rate is a decreasing function of the cutoff level. Based on these experimentally (and computationally) observed behaviors, we have developed a multilevel strategy treating in different ways the small and large eddies of the flow. The small scales are not updated at each time iteration of the temporal integration of the system of equations. The less energy they contain, the longer in time they are kept fixed. A V-cycle-like algorithm is used to determine the length of the time interval over which the scales $\mathbf{z}_{N_i}^{N_{i+1}}$ are frozen. The parameters, such as the levels N_{i_1} and N_{i_2}, are dynamically computed and are updated during the time integration. The tests used to define these parameters are based on a dynamic control of the errors introduced in the system by the simplified numerical treatment of the small scales as well as of the nonlinear interaction terms.

The DML method schematically summarized above has been implemented and tested in the case of the numerical simulations of forced homogeneous turbulent flows as well as turbulent channel flows, that is, nonhomogeneous turbulence. Here, the DML method has been applied in the context of DNS, that is, the smallest scales retained in the computation are of the order of the Kolmogorov scale η. In the homogeneous case, the effects of the resolution, that is, k_N, on the statistics of forced three-dimensional homogeneous turbulent flows have been estimated. Integrations of the Navier–Stokes equations at Reynolds number based on the Taylor microscale $\mathrm{Re}_\lambda = 100$ and at different resolutions ($\eta k_N \in [0.8, 1.6]$) have been conducted over long time integration; therefore, statistical steady states have been reached and analyzed. Simulations with the DML algorithm (Chapter 9) and with different values of the parameter ξ_r, providing an estimate of the order of accuracy (see Section 9.4), have been performed with the same data and parameters as the DNS simulations. We have studied the performances of the DML methodology and proved its ability to estimate this range of scales. While computing small scales different than the standard Galerkin algorithm used for DNS, the DML method preserves the statistical properties, such as intermittency, of the small scales. The multilevel approach, based on different approximation schemes for different ranges of scales, is successful in the context of direct simulation. The method is time-saving but requires a little more memory. The CPU time is about 2.5 times less than the similar DNS simulation and the memory required by the DML algorithm is about 10 to 20 percent larger than by DNS.

For the three-dimensional channel flow problem, a DML algorithm in the periodic directions has been proposed and tested; the flow variables $\mathbf{z}_{N_i}^{N_{i+1}}$ are obtained by scale decomposition in the directions parallel to the walls. The statistically steady states reached by the DNS and DML simulations were compared by analyzing different turbulent statistics like the mean flow properties, the root-mean square of the velocity, vorticity and pressure fluctuations, the one-dimensional spectrum functions, the Reynolds shear stresses, and the high-order moments of the velocity fluctuations. The results are preliminary ones for this problem; further developments will appear elsewhere. However, they show the ability for the DML method to reproduce accurately the mean velocity profile as well as the turbulence statistics mentioned above. A more detailed study of the DML method has to be done in order to derive an optimal algorithm for this problem. The problem of the scale separation in the nonhomogeneous direction has been discussed and an algorithm has been proposed; it has not been implemented yet. Galerkin bases derived from Legendre and Chebyshev polynomials are used. In such a context, the scale decomposition by cutoff into the spectrum of the velocity leads to small and large scales satisfying the boundary conditions at the walls. A coupling between the different scales still appears in the continuity equation. To overcome this difficulty, a divergence-free Galerkin basis with elements formed by linear combinations of the Legendre polynomials has been derived and described. Therefore, a Galerkin approximation of the Navier–Stokes equations in the primitive variable form can be achieved. The study of multilevel schemes in the normal direction is under progress and will be reported in future works.

In order to be useful in practical applications, the DML methodology has to be implemented and tested in the context of the large eddy simulation (LES), that is, by retaining in the computation as few modes (or grid points) as possible. Typically in LES, the mesh size is about ten to twenty times larger than in DNS. In most LES models, the subgrid stress (sgs) tensor is modeled in terms of gradients of the resolved fields. As a consequence, the sgs forces are dissipative; they remove energy from the large scales at every instant. The dissipative process occurring in turbulence is more complicated than this simple mechanism. However, most eddy-viscosity models perform reasonably well in actual LES as they provide the right amount of mean dissipation. Based on the DML methodology, more sophisticated models can be developed. Indeed, rather than modeling the nonlinear operator itself, the DML methodology can be used to estimate scales that are smaller, say, by two times, than the grid size retained for the resolved ones. These scales can then be used to evaluate the sgs tensor. A similar approach, using a different methodology, has been used in Domaradzki and Saiki (1997). The study of multilevel schemes in the context of LES is under progress and will be reported in forthcoming articles.

In this work, DML methods have been presented in the context of spectral methods. Other theoretical and numerical works have been done for different spatial discretizations, such as finite elements and finite differences. For the finite elements, which are attractive when considering complex geometries, the DML methodology uses hierarchical bases to decompose the solution in components of different sizes. Several algorithms have been proposed (see, for example, Marion and Temam (1990), Laminie, Pascal, and Temam (1993,

1994), Calgaro, Laminie, and Temam (1995) and the references therein). The incremental unknowns, introduced in Temam (1990b), lead to a multilevel method intended for the resolution of nonlinear problems when finite differences are used. The incremental unknowns carry out a decomposition of the solution into components with different sizes. The application of the incremental unknowns to the solution of linear elliptic problems was described by Chen and Temam (1991, 1993). In Chehab (1995) and Chehab and Temam (1995), the incremental unknowns are used to compute unstable solutions of nonlinear elliptic eigenvalue problems. Much work remains to be done in order to extend and apply these approaches in actual simulations of turbulent flows in complex geometries, but the principle of decomposing the unknowns according to the size of the eddies is very appealing and looks fully natural in such problems involving many scales. Differentiated treatments of small and large eddies have also been used in other areas of physics (such as meteorology), with a decomposition based on physical intuition rather than on spectral properties of the solutions.

References

B.K. Alpert and V. Rokhlin (1991), *A fast algorithm for the evaluation of Legendre expansions*, SIAM J. Sci. Stat. Comp. **12**, 158–179.

A.V. Babin and M.I. Vishik (1983), *Attractors of partial differential equations and estimate of their dimension*, Uspekhi Mat. Nauk **38**, 133–187 (in Russian). Russian Math. Surveys **38**, 151–213 (in English).

J. Bardina, J.H. Ferziger, and W.C. Reynolds (1983), *Improved turbulence models based on large-eddy simulation of homogeneous, incompressible, turbulent flows*, Rep. TF-19, Stanford University.

G.K. Batchelor (1969), *Computation of the energy spectrum in homogeneous two-dimensional turbulence*, High-speed computing in fluid dynamics, Phys. Fluids Suppl. II **12**, 233–239.

G.K. Batchelor (1971), *The theory of homogeneous turbulence*, 2nd ed., Cambridge University Press.

G.K. Batchelor and A.A. Townsend (1947), *Decay of vorticity in isotropic turbulence*, Proc. Roy. Soc. A **190**, 534.

H. Bercovici, P. Constantin, C. Foias, and O.P. Manley (1995), *Exponential decay of the power spectrum of turbulence*, J. Stat. Phys. **80** (3), 4, 579–602.

F. Bouchon, T. Dubois, and F. Jauberteau (1998), *Dynamic multilevel methods and nonhomogeneous turbulence*, Proceedings of the 16th International Conference on Numerical Methods in Fluid Dynamics (ICNMFD), C.H. Bruneau, ed., Lecture Notes in Physics, Springer-Verlag, Heidelberg, 1998.

J. Boussinesq (1877), *Théorie de l'écoulement tourbillonant*, Mém. Prés. Acad. Sci. Paris, 23–46.

M.E. Brachet, M. Ménéguzzi, H. Politano, and P. Sulem (1988), *The dynamics of freely decaying two-dimensional turbulence*, J. Fluid Mech. **194**, 333–349.

W.L. Briggs (1987), *A multigrid tutorial*, Society for Industrial and Applied Mathematics, Philadelphia.

J.B. Burie and M. Marion (1997), *Multi-level methods in space and time for the Navier–Stokes equations*, SIAM J. Numer. Anal. **34** (4), 1574–1599.

C. Calgaro, J. Laminie, and R. Temam (1995), *Dynamical multilevel schemes for the solution of evolution equations by hierarchical finite element discretization*, Appl. Numer. Math. **21**, 1–40.

C. Canuto, M.Y. Hussaini, A. Quarteroni, and T.A. Zang (1988), *Spectral methods in fluid dynamics*, Springer-Verlag, New York.

J.P. Chehab (1995), *A nonlinear adaptive multi-resolution method in finite differences with incremental unknowns*, Math. Model. Numer. Anal. (M2AN), **29** (4), 451–475.

J.P. Chehab and R. Temam (1995), *Incremental unknowns for solving nonlinear eigenvalue problems: new multiresolution methods*, Numer. Methods Partial Differential Eqs. **11** (3), 199–218.

M. Chen and R. Temam (1991), *Incremental unknowns for solving partial differential equations*, Numer. Math. **59**, 255–271.

M. Chen and R. Temam (1993), *Incremental unknowns in finite differences: condition number of the matrix*, SIAM J. Matrix Anal. Appl. (SIMAX) **14** (2), 432–455.

S. Chen, G.D. Doolen, R.H. Kraichnan, and Z.S. She (1993), *On statistical correlations between velocity increments and locally averaged dissipation in homogeneous turbulence*, Phys. Fluids A **5** (2), 458–463.

P. Constantin and C. Foias (1989), *Navier–Stokes equations*, University of Chicago Press, Chicago.

P. Constantin, C. Foias, and R. Temam (1985), *Attractors representing turbulent flows*, Mem. Amer. Math. Soc. **53** (114), 67 + vii pages.

P. Constantin, C. Foias, O.P. Manley, and R. Temam (1985), *Determining modes and fractal dimension of turbulent flows*, J. Fluid Mech. **150**, 427–440.

J.W. Deardorff (1970), *A numerical study of three-dimensional turbulent channel flow at large Reynolds numbers*, J. Fluid Mech. **41** (2), 453–480.

A. Debussche and T. Dubois (1994), *Approximation of exponential order of the attractor of a turbulent flow*, Physica D **72**, 372–389.

A. Debussche, T. Dubois, and R. Temam (1995), *The nonlinear Galerkin method: a multiscale method applied to the simulation of homogeneous turbulent flows*, Theoret. Comp. Fluid Dynamics **7** (4), 279–315.

A. Debussche and R. Temam (1991), *Inertial manifolds and the slow manifolds in meteorology*, Diff. Integral Eqs. **4** (5), 897–931.

A. Debussche and R. Temam (1994), *Convergent families of approximate inertial manifolds*, J. Math. Pures Appl. **73**, 485–522.

L. Dettori, D. Gottlieb, and R. Temam (1995), *Nonlinear Galerkin method: the two-level Chebyshev-collocation case*, in Proc. ICOSAHOM'95, Houston J. Math. 75–82.

Ch. Doering and X. Wang (1998), *Attractor dimension estimates for two-dimensional shear flows*, Physica D, to appear.

J.A. Domaradzki and E.M. Saiki (1997), *A subgrid-scale model based on the estimation of unresolved scales of turbulence*, Phys. Fluids **9** (7), 2148–2164.

T. Dubois (1993), *Simulation numérique d'écoulements homogènes et non homogènes par des méthodes multi-résolution*, Thèse, Université de Paris-Sud.

T. Dubois and F. Jauberteau (1998), *A dynamical multilevel model for the simulation of the small structures in three-dimensional homogeneous isotropic turbulence*, J. Sci. Comp., to appear.

T. Dubois and F. Jauberteau, *On spectral Galerkin approximations for the simulation of turbulent channel and pipe flows*, in preparation.

T. Dubois, F. Jauberteau, and R. Temam (1993), *Solution of the incompressible Navier–Stokes equations by the nonlinear Galerkin method*, J. Sci. Comp. **8** (2), 167–194.

T. Dubois, F. Jauberteau, and R. Temam (1995), *Dynamic multilevel methods in turbulence simulations*, in *Computational Fluid Dynamics Review* 1995, M. Hafez and K. Oshima, eds., Wiley, 679–694.

T. Dubois, F. Jauberteau, and Y. Zhou (1997), *Influences of subgrid scale dynamics on resolvable scale statistics in large-eddy simulations*, Physica D, **100**, 390–406.

T. Dubois and A. Miranville (1994), *Existence and uniqueness results for a velocity formulation of Navier–Stokes equations in a channel,* Applicable Anal. **55**, 103–138.

T. Dubois and R. Temam (1992), *The nonlinear Galerkin method applied to the simulation of turbulence in a channel flow*, in Proc. ICNMFD **13** (Rome), N. Napolitano and F. Sabetta, eds., Lecture Notes in Physics, Springer-Verlag.

T. Dubois and R. Temam (1993), *Separation of scales in turbulence using the nonlinear Galerkin method*, in *Advances in computational fluid dynamics*, W.G. Habashi and M. Hafez, eds., Gordon and Breach.

G. Erlebacher, M.Y. Hussaini, C.G. Speziale, and T.A. Zang (1992), *Toward the large-eddy simulation of compressible turbulent flows*, J. Fluid Mech. **238**, 155–185.

C. Foias, M.S. Jolly, I.G. Kevrekidis, G.R. Sell, and E.S. Titi (1988), *On the computation of inertial manifolds*, Phys. Lett. A **131**, 433–436.

C. Foias, O.P. Manley, and R. Temam (1987), *Sur l'interaction des petits et grands tourbillons dans les écoulements turbulents*, C.R. Acad. Sci. Paris Série I **305**, 497–500.

C. Foias, O.P. Manley, and R. Temam (1988), *Modelling of the interaction of small and large eddies in two-dimensional turbulent flows*, Math. Model. and Numer. Anal. (M2AN) **22** (1), 93–114.

C. Foias, G.R. Sell, and R. Temam (1985), *Variétés inertielles des équations différentielles dissipatives*, C.R. Acad. Sci. Paris Série I **301**, 139–142.

C. Foias, G.R. Sell, and R. Temam (1988), *Inertial manifolds for nonlinear evolutionary equations*, J. Diff. Eqs. **73**, 309–353.

C. Foias and R. Temam (1979), *Some analytic and geometric properties of the evolution Navier–Stokes equations*, J. Math. Pures Appl. **58**, 339–368.

C. Foias and R. Temam (1989), *Gevrey class regularity for the solutions of the Navier–Stokes equations*, J. Funct. Anal. **87**, 359–369.

D. Foster, D. Nelson, and M. Stephen (1977), *Large-distance and long-time properties of a randomly stirred fluid*, Phys. Rev. A **16**, 732.

M. Germano, U. Piomelli, P. Moin, and W.H. Cabot (1991), *A dynamic subgrid-scale eddy viscosity model*, Phys. Fluids A **3** (7), 1760–1765.

S. Ghosal and P. Moin (1995), *The basic equations for the large eddy simulation of turbulent flows in complex geometry*, J. Comp. Phys. **118**, 24–37.

S. Ghosal, T.S. Lund, P. Moin, and K. Akselvoll (1995), *A dynamic localization model for large-eddy simulation of turbulent flows*, J. Fluid Mech. **286**, 229–255.

I.I. Gikhman and A.V. Skorokhod (1969), *Introduction to the theory of random processes*, Saunders Mathematics Books.

D. Gottlieb and S.A. Orszag (1977), *Numerical analysis of spectral methods: theory and applications*, CBMS-NSF Regional Conf. Ser. Appl. Math. SIAM, Philadelphia.

W. Hackbusch and U. Trottenberg (1982), *Multigrid methods*, Springer-Verlag, Berlin.

J.R. Herring, S.A. Orszag, R.H. Kraichnan, and D.G. Fox (1974), *Decay of the two-dimensional homogeneous turbulence*, J. Fluid Mech. **66** (3), 417–444.

F. Jauberteau (1990), *Résolution numérique des équations de Navier–Stokes instationnaires par méthodes spectrales. Méthode de Galerkin non linéaire*, Thèse, Université de Paris-Sud.

F. Jauberteau, C. Rosier, and R. Temam (1989), *The nonlinear Galerkin method in computational fluid dynamics*, Appl. Numer. Math. **6**, 361–370.

F. Jauberteau, C. Rosier, and R. Temam (1990), *A nonlinear Galerkin method for the Navier–Stokes equations*, Comp. Meth. Appl. Mech. and Eng. **80**, 245–260.

J. Jiménez (1990), *Transition to turbulence in two-dimensional Poiseuille flow*, J. Fluid Mech. **218**, 265–297.

J. Jiménez (1994), *Resolution requirements for velocity gradients in turbulence*, Annual Research Briefs, Center for Turbulence Research, Stanford University.

J. Jiménez, A.A. Wray, P.G. Saffman, and R.S. Rogallo (1993), *The structure of intense vorticity in isotropic turbulence*, J. Fluid Mech. **255**, 65–90.

M.S. Jolly and C. Xiong (1995), *On computing the long-time solution of the Navier–Stokes equations*, Theoret. Comput. Fluid Dynamics **7** (4), 261–278.

D.A. Jones, L.G. Margolin, and E.S. Titi (1995), *On the effectiveness of the approximate inertial manifold: a computational study*, Theoret. Comput. Fluid Dynamics **7** (4), 243–260.

R.M. Kerr (1985), *Higher-order derivative correlations and the alignment of small-scale structures in isotropic numerical turbulence*, J. Fluid Mech. **153**, 31–58.

A.I. Khinchin (1949), *Mathematical foundation of statistical mechanics*, Dover, New York.

J. Kim, P. Moin, and R. Moser (1987), *Turbulence statistics in fully developed channel flow at low Reynolds number*, J. Fluid Mech. **177**, 133–166.

L. Kleiser and U. Schuman (1980), *Treatment of incompressibility and boundary conditions in 3D numerical spectral simulations of plane channel flows*, in Proc. 3rd GAMM Conf. on Numerical Methods in Fluid Mechanics, E.H. Hirschel, ed., Vieweg, Braunschweig, 165–173.

A.N. Kolmogorov (1941a), *The local structure of turbulence in incompressible viscous liquid*, Dokl. Akad. Nauk. SSSR **30**, 301–305.

A.N. Kolmogorov (1941b), *On degeneration of isotropic turbulence in an incompressible viscous liquid*, Dokl. Akad. Nauk. SSSR **31**, 538–541.

A.N. Kolmogorov (1941c), *Dissipation of energy in locally isotropic turbulence*, C.R. Acad. Sci. URSS **32**, 16.

R.H. Kraichnan (1967), *Inertial ranges in two-dimensional turbulence*, Phys. Fluids **10**, 1417–1423.

O.A. Ladyzhenskaya (1969), *The mathematical theory of viscous incompressible flow*, 2nd ed., Gordon and Breach, New York.

J. Laminie, F. Pascal, and R. Temam (1993), *Implementation of finite element nonlinear Galerkin methods using hierarchical bases*, J. Comp. Mech. **11**, 384–407.

J. Laminie, F. Pascal, and R. Temam (1994), *Implementation and numerical analysis of the nonlinear Galerkin methods with finite elements discretization*, Appl. Numer. Math. **15**, 219–246.

Y.W. Lee (1960), *Statistical theory of communication*, Wiley, New York.

C.E. Leith (1968), *Diffusion approximation for two-dimensional turbulence*, Phys. Fluids **11**, 671–673.

A. Leonard (1974), *On the energy cascade in large-eddy simulations of turbulent fluid flows*, Adv. Geophys. A **18**, 237–248.

J. Leray (1933), *Etude de diverses équations intégrales non linéaires et de quelques problèmes que pose l'hydrodynamique*, J. Math. Pures Appl. **12**, 1–82.

J. Leray (1934a), *Essai sur les mouvements plans d'un liquide visqueux que limitent des parois*, J. Math. Pures Appl. **13**, 331–418.

J. Leray (1934b), *Essai sur les mouvements d'un liquide visqueux emplissant l'espace*, Acta Math. **63**, 193–248.

M. Lesieur (1990), *Turbulence in fluids*, 2nd ed., Kluwer Academic, London.

D.C. Leslie (1973), *Developments in the theory of turbulence*, Clarendon Press, Oxford.

D.C. Leslie and G.L. Quarini (1979), *The application of turbulence theory to the formulation of subgrid modeling procedures*, J. Fluid Mech. **91** (1), 65–91.

D.K. Lilly (1971), *Numerical simulation of developing and decaying two-dimensional turbulence*, J. Fluid Mech. **45** (2), 395–415.

D.K. Lilly (1992), *A proposed modification of the Germano subgrid-scale closure method*, Phys. Fluids A **4** (3), 633–635.

J.L. Lions (1969), *Quelques méthodes de résolution des problèmes aux limites non linéaires*, Dunod, Paris.

J.L. Lions, R. Temam, and S. Wang (1996), *Splitting up methods and numerical analysis of some multiscale problems*, Int. J. Comp. Fluid Dynamics **5** (2), 157–202.

V.X. Liu (1993), *A sharp lower bound for the Hausdorff dimension of the global attractor of the 2D Navier–Stokes equations*, Comm. Math. Phys. **158**, 327–339.

J.L. Lumley (1970), *Stochastic tools in turbulence*, Academic Press, London.

Y. Maday and B. Métivet (1987), *Chebyshev spectral approximation of Navier–Stokes equations in a two-dimensional domain*, Math. Modelling and Numer. Anal. **21** (1), 93–123.

O.P. Manley (1992), *The dissipation range spectrum*, Phys. Fluids A **4** (6), 1320–1321.

M. Marion and R. Temam (1989), *Nonlinear Galerkin methods*, SIAM J. Numer. Anal. **26**, 1139–1157.

M. Marion and R. Temam (1990), *Nonlinear Galerkin methods: the finite elements case*, Numer. Math. **57**, 205–226.

P. Moin and J. Kim (1982), *Numerical investigation of turbulent channel flow*, J. Fluid Mech. **118**, 341–377.

A.S. Monin and A.M. Yaglom (1975), *Statistical fluid mechanics: mechanics of turbulence*, **2**, MIT Press, Cambridge.

R.D. Moser (1994), *Kolmogorov inertial range spectra for inhomogeneous turbulence*, Phys. Fluids **6** (2), 794–801.

R.D. Moser, P. Moin, and A. Leonard (1983), *A spectral numerical method for the Navier–Stokes equations with applications to Taylor–Couette flow*, J. Comput. Phys. **52**, 524–544.

S.A. Orszag (1970), *Analytical theories of turbulence*, J. Fluid Mech. **41** (2), 363–386.

S.A. Orszag (1973), *Lectures on the statistical theory of turbulence*, in Proc. Summer School in Theoretical Physics, Les Houches.

S.A. Orszag (1977), *Turbulence and transition: a progress report*, in Proc. 5th Int. Conf. on Numerical Methods in Fluid Dynamics, Springer-Verlag, Heidelberg, Lecture Notes in Physics **59**, 32.

S.A. Orszag and G.S. Patterson (1972), *Numerical simulation of three-dimensional homogeneous isotropic turbulence*, Phys. Rev. Lett. **28**, 76.

A.T. Patera and S.A. Orszag (1980), *Transition and turbulence in plane channel flows*, in Proc. Seventh Int. Conf. on Numerical Methods in Fluid Dynamics, W.C. Reynolds and R.W. MacCormack, eds., Springer-Verlag, Berlin/Heidelberg/New York, 329–335.

S.B. Pope (1985), *Pdf methods for turbulent reactive flows*, Prog. Energy Combust. Sci. **11**, 119–192.

O. Reynolds (1895), *On the dynamical theory of incompressible viscous fluids and the determination of the criterion*, Phil. Trans. Roy. Soc. London A **186**, 123–164.

H.A. Rose (1977), *Eddy diffusivity, eddy noise and subgrid-scale modelling*, J. Fluid Mech. **81**, 719–734.

B.L. Rozhdestvensky and I.N. Simakin (1984), *Secondary flows in a plane channel: their relationship and comparison with turbulent flows*, J. Fluid Mech. **147**, 261–289.

Z.S. She, E. Jackson, and S.A. Orszag (1988), *Scale-dependent intermittency and coherence in turbulence*, J. Sci. Comp. **3** (4), 407–434.

Z.S. She, E. Jackson, and S.A. Orszag (1991), *Structure and dynamics of homogeneous turbulence: models and simulations*, Proc. Roy. Soc. London A **434**, 101–124.

J. Shen (1994), *Efficient spectral-Galerkin method I. Direct solvers of second- and fourth-order equations using Legendre polynomials*, SIAM J. Sci. Comp. **15** (6), 1489–1505.

J. Shen (1995), *Efficient spectral-Galerkin method II. Direct solvers of second- and fourth-order equations using Chebyshev polynomials*, SIAM J. Sci. Comp. **16** (1), 74–87.

J. Shen (1996), *Efficient Chebyshev-Legendre Galerkin methods for elliptic problems*, in Proc. 3rd ICOSAHOM, Houston J. Math. University of Houston, 233–239.

J. Shen and R. Temam (1995), *Nonlinear Galerkin method using Chebyshev and Legendre polynomials I. The one-dimensional case*, SIAM J. Numer. Anal. **32** (1), 215–234.

E.D. Siggia (1981), *Numerical study of small-scale intermittency in three-dimensional turbulence*, J. Fluid Mech. **107**, 375–406.

E.D. Siggia and G.S. Patterson (1978), *Intermittency effects in a numerical simulation of stationary three-dimensional turbulence*, J. Fluid Mech. **86** (3), 567–592.

J. Smagorinsky (1963), *General circulation experiments with the primitive equations. I. The basic experiment*, Monthly Weather Rev. **91**, 99–164.

L.M. Smith and W.C. Reynolds (1991), *The dissipation-range spectrum and the velocity-derivative skewness in turbulent flows*, Phys. Fluids A **3**, 992.

C.G. Speziale (1985), *Galilean invariance of subgrid-scale stress models in the large-eddy simulation of turbulence*, J. Fluid Mech. **156**, 52–62.

C.G. Speziale (1990), *Analytical methods for the development of Reynolds stress closures in turbulence*, ICASE Report **90-26**.

R. Temam (1984), *Navier–Stokes equations*, North-Holland, Amsterdam.

R. Temam (1986), *Remarks on Euler equations*, in Nonlinear functional analysis and its applications, F. Browder, ed., Proc. Symp. Pure Math. **45**, Amer. Math. Soc. 429–430.

R. Temam (1989), *Induced trajectories and approximate inertial manifolds*, Math. Model. and Numer. Anal. (M2AN) **23**, 541–561.

R. Temam (1990a), *Inertial manifolds*, The Mathematical Intelligencer **12** (4), 68–74.

R. Temam (1990b), *Inertial manifolds and multigrid methods*, SIAM J. Math. Anal. **21** (1), 154–178.

R. Temam (1991), *Stability analysis of the nonlinear Galerkin method*, Math. Comp. **57** (196), 477–505.

R. Temam (1993), *Méthodes multirésolutions en analyse numérique*, in Boundary value problems for partial differential equations and applications, C. Baiocchi and J.L. Lions, eds., Masson and Wiley.

R. Temam (1994), *Applications of inertial manifolds to scientific computing: a new insight in multilevel methods*, in Trends and perspectives in applied mathematics, volume in honor of Fritz John, J. Marsden and L. Sirovich, eds., Appl. Math. Ser. **100**, Springer-Verlag, 315–358.

R. Temam (1995), *Navier–Stokes equations and nonlinear functional analysis*, 2nd rev. ed., CBMS-NSF Regional Conf. Ser. Appl. Math., SIAM, Philadelphia.

R. Temam (1997), *Infinite dimensional dynamical systems in mechanics and physics*, 2nd ed., Appl. Math. Sci. Ser. **68**, Springer-Verlag, New York.

H. Tennekes and J.L. Lumley (1972), *A first course in turbulence*, MIT Press, Cambridge, Mass.

E.S. Titi (1990), *On approximate inertial manifolds to the Navier–Stokes equations*, J. Math. Anal. Appl. **149**, 540–557.

A. Vincent and M. Ménéguzzi (1991), *The spatial structure and statistical properties of homogeneous turbulence*, J. Fluid Mech. **225**, 1–20.

P.R. Voke and M.W. Collins (1983), *Large-eddy simulation: retrospect and prospect*, J. Phys. Chem. Hydrodynamics **4** (2), 119–161.

J. Werne (1995), *Incompressibility and no-slip boundaries in the Chebyshev-tau approximation: correction to Kleiser and Schumann's influence-matrix solution*, J. Comp. Phys. **120**, 260–265.

N. Wiener (1933), *The Fourier integral and certain of its applications*, Cambridge University Press, Cambridge, and Dover, New York.

P.R. Woodward, D.H. Porter, B.K. Edgar, S. Anderson, and G. Bassett (1995), *Parallel computation of turbulent fluid flow*, Comp. Appl. Math. **14** (1), 97–105.

A.M. Yaglom (1987), *Correlation theory of stationary and related random functions I: basic results*, Springer Series in Statistics, Springer-Verlag.

V. Yakhot and S.A. Orszag (1986), *Renormalization group analysis of turbulence. I. Basic theory*, J. Sci. Comp. **1** (1), 3–55.

Y. Zhou (1993a), *Degrees of locality of energy transfer in the inertial range*, Phys. Fluids A **5** (5), 1092–1094.

Y. Zhou (1993b), *Interacting scales and energy transfer in isotropic turbulence*, Phys. Fluids A **5** (10), 2511–2524.

Y. Zhou, M. Hossain, and G. Vahala (1989), *A critical look at the use of filters in large eddy simulation*, Phys. Lett. A **139** (7), 330–332.

Y. Zhou, G. Vahala, and M. Hossain (1988), *Renormalization-group theory for the eddy-viscosity in subgrid modeling*, Phys. Rev. A **37** (7), 2590–2598.

Y. Zhou, P.K. Yeung, and J.G. Brasseur (1994), *Scale disparity and spectral transfer in anisotropic numerical turbulence*, Phys. Rev. E. **53** (1), 1261–1264.

M. Ziane (1997), *Optimal bounds on the dimension of the attractor for the Navier–Stokes equations*, Physica D **105**, 1–19.

Index

V

W

Printed in the United States
By Bookmasters